Single-Mode Optical Fiber Measurement: Characterization and Sensing

The Artech House Optoelectronics Library

Brian Culshaw, Alan Rogers, and Henry Taylor, *Series Editors*

Acousto-Optic Signal Processing: Fundamentals and Applications, Pankaj Das

Amorphous and Microcrystalline Semiconductor Devices, Optoelectronic Devices, Jerzy Kanicki, ed.

Electro-Optical Systems Performance Modeling, Gary Waldman and John Wootton

Fiber-Optic Gyroscopes, Hervé Lefèvre

Field Theory of Acousto-Optic Signal Processing Devices, Craig Scott

Highly Coherent Semiconductor Lasers, Motoichi Ohtsu

Introduction to Electro-Optical Imaging and Tracking Systems, Khalil Seyrafi and S. A. Hovanessian

Introduction to Glass Integrated Optics, S. Iraj Najafi

Optical Control of Microwave Devices, Rainee N. Simons

Optical Fiber Sensors, Volume I: Principles and Components, John Dakin and Brian Culshaw, eds.

Optical Fiber Sensors, Volume II: Systems and Applicatons, Brian Culshaw and John Dakin, eds.

Optical Network Theory, Yitzhak Weissman

Principles of Modern Optical Systems, Volume I, I. Andonovic and D. Uttamchandani, eds.

Principles of Modern Optical Systems, Volume II, I. Andonovic and D. Uttamchandani, eds.

Reliability and Degradation of LEDs and Semiconductor Lasers, Mitsuo Fukuda

Single-Mode Optical Fiber Measurement: Characterization and Sensing, Giovanni Cancellieri

Single-Mode Optical Fiber Measurement: Characterization and Sensing

Giovanni Cancellieri
Editor

Artech House
Boston • London

Library of Congress Cataloging-in-Publication Data

Cancellieri, Giovanni.
Single-mode optical fiber measurement : characterization and sensing / Giovanni Cancellieri.
p. cm.
Includes bibliographical references and index.
ISBN 0-89006-602-7
1. Optical communications. 2. Fiber optics. 3. Optoelectronics. 4. Optical detectors. I. Title.

TA1800.C37 1993 92-32247
621.382'75—dc20 CIP

British Library Cataloguing in Publication Data

Cancellieri, G.
Single-mode Optical Fiber Measurement:
Characterization and Sensing
I. Title.
621.382

ISBN 0-89006-602-7

685 Canton Street
Norwood, MA 02062

International Standard Book Number: 0-89006-602-7
Library of Congress Catalog Card Number: 92-32247

10 9 8 7 6 5 4 3 2 1

Contents

Preface

This book is intended for graduate students and technicians involved in measurement operations in which fiber optics are employed. In recent years, the fields of optical fiber communication and optical fiber sensors, although separately developed, are converging, to the point where a unified treatment appears now possible. This is particularly true for single-mode fibers.

After the introduction of some basic concepts, a detailed description is given of experimental tests to be performed in laboratory or in manufacturing and field environments. Such tests are fundamentally subdivided into two classes: those performed in coherent light, and those not employing light with particular requirements about temporal coherence.

In strict connection with the measurement problems, an accurate characterization of many fiber-optic sensors is then presented, with the aim of showing the benefits resulting from the use of optical fibers in the wide field of sensing, control, and instrumentation. In these applications, it is possible to take advantage of some technical problems typically found in telecommunications, such as fiber sensitivity to environment. The expected unified treatment, therefore, appears the most natural one.

The authors wish to thank Pirelli Cavi - Divisione Italia and Teleco Cavi SpA for having provided some material on practical measurements.

F. Chiaraluce is grateful to Dr. E. Gambi, Università di Ancona (Italy), for his helpful assistance in the typing effort.

A. Galtarossa thanks Professor C. G. Someda, Università di Padova (Italy), for his continuous encouragement, Dr. F. Matera, Fondazione Ugo Bordoni - Rome (Italy), for helpful discussions, and F. Bergamin, SAIFO s.r.l. - Padova (Italy), for technical assistance.

E. Nava and G. Valentini wish to thank Dr. P. Gambini, CSELT - Torino (Italy), for his fruitful suggestions.

V. Annovazzi-Lodi and S. Merlo wish to express their sincere appreciation to Professor S. Donati, Università di Pavia (Italy), for his assistance and guidance throughout the course of their research work and in the preparation of Chapter 5.

S. Merlo is also thankful to Professor P. Yager and Dr. L. W. Burgess, University of Washington, Seattle (U.S.A.), for their guidance in the study of chemical sensors.

Chapter 1

Introduction

G. Cancellieri

In this introductory chapter, after a brief history of fiber optics, with particular attention to their applications in telecommunications and sensors, the present status of tests and characterization procedures, for either optical fibers or optoelectronic devices, is described. Single-mode fiber optics are considered, owing to their expected increasing importance in the future panorama of practical applications. Then the main recommendations from international bodies, in particular CCITT and IEC, are discussed. The importance of the definition of some standards, in both fiber optics manufacturing and testing, is stressed. Finally, the objectives of this book are summarized.

1.1 BRIEF HISTORY OF FIBER OPTICS IN TELECOMMUNICATION AND SENSOR APPLICATIONS

The first theoretical work on electromagnetic field propagation in a round dielectric cylinder dates from 1910, and was carried out by Hondros and Debye [1]. The idea of employing optical signals for information transfer by means of a light guide is however more recent, and is due to Buchholz [2]. About forty years ago fiber bundles were first employed for image reproduction in endoscopy [3]. Laser action in semiconductors, obtained in 1962 [4-6], finally gave the possibility of practical light generation and modulation, with acceptable launching efficiency.

After some theoretical works about the multi-mode nature of the propagation in optical fibers [7, 8], the first proposal of employing this transmission medium in long-distance communications was made in 1966 [9], under the assumption that silica-based glasses have an intrinsic loss, due to absorption and scattering, at some operating wavelengths, on the order of 20 dB/km. In 1974 the value of 2 dB/km was reached at 0.85 μm [10]; 0.47 dB/km was obtained at 1.3 μm in 1976 [11]; 0.2 dB/km was finally achieved at 1.55 μm in 1979 [12]. Various technological procedures for fiber manufacturing were proposed in those years [13–15].

Optimization of the index profile in graded-index multi-mode fibers [16] was necessary in order to reach transmission bandwidths on the order of 1 GHz over a reference distance of 1 km. Several models of the light propagation in multi-mode fibers were carried out, based either on a rigorous modal approach [17–19] or on ray optics [20–22]. Mode coupling was investigated [23, 24] with particular attention to the effects of microbending [25].

The first practical field trials of transmission systems based on multi-mode fibers were developed in the second half of the 1970s [26–28]. Optical cable structures, for outdoor and indoor telecommunication applications, were designed and optimized in the same years [29].

Nevertheless, the requirement of a much more extended available bandwidth, especially on long-haul high-capacity connections, made the use of single-mode fibers more and more preferable. Single-mode fibers had been investigated thoroughly for several years, and some aspects of material and waveguide dispersion were already known [30]. After some early problems in fiber-to-fiber practical joints [31] and in the effects of pure bend loss [32], such fibers became commonly used. Regeneration spacing as long as 200 km, with transmission speeds of the order of 1 Gbit/s, recently has been reached.

In parallel, the technology of the main optoelectronic devices, i.e., light sources and detectors, has been remarkably improved. In particular, continuous wave laser action at room temperature was possible from the early 1970s [33]. Laser devices with stable characteristics and long lifetimes became available some years after [34]. A compound semiconductor like GaInAsP was employed for having light emission at 1.3 μm and 1.55 μm. The laser linewidth was reduced progressively, with the purpose of making the fiber chromatic dispersion lower and lower. For this reason, the invention of the distributed feedback laser [35] was of fundamental importance.

Laser linewidths of the order of 1 MHz allowed the coherent modulation of light. Coherent systems based on heterodyne or homodyne light detection have been theoretically studied for about twenty years for applications with gas or solid-state lasers [36]. Nevertheless, the use of semiconductor lasers contributed to making such systems of practical interest [37], only more recently. Photodiode technology, both for PIN or for APD device fabrication, was continuously improved [38], first employing Si or Ge as the basic material, and then introducting compound semiconductors.

Passive optical directional couplers and branching devices, obtained either in all-fiber [39] or in micro-optic [40] solutions, were studied and practically fabricated with the advent of fiber-optic distributed access networks. *Wavelength division multiplexing*

(WDM), has been suggested in particular for broadcasting networks [41]. Active optical devices for direct signal processing, like acousto-optic, electro-optic, and magneto-optic devices, were developed, either in micro-optic [42] or in integrated-optic solution [43], for switch and external coherent modulation purposes. Integrated-optic components have the advantage of an intrinsic compatibility with single-mode operation. For example, geodesic lenses [44] are suitable for integrated spectrum analyzer functions.

More recently, all optical signal treatment has been investigated [45] in conjunction with a coherent modulation transmission and also with an optical *frequency division multiplexing* (FDM) [46].

In coherent modulation transmissions, it is necessary to have a known and stable polarization state of the received light. This can be obtained by means of suitable polarization control devices [47], placed at the fiber output, or by a proper polarization diversity technique [48]. Otherwise fibers with special polarization properties can be employed. They can be subdivided into polarization-maintaining [49] and single-polarization fibers [50].

Optical fiber sensors, developed in a completely uncorrelated way, have taken some advantage from the improved technology of fiber and device fabrication for telecommunication applications, in particular when a single-mode propagation is required and when a highly coherent light is to be employed. Practical optical fiber sensor application started in 1977 [51]. Since then, several kinds of sensors have been developed. They can be classified on the basis of their application as acoustic sensors [52]; magnetic field and electric current sensors [53]; gyroscopes [54]; temperature sensors [55]; chemical sensors [56]; acceleration, torque, displacement and pressure sensors [57]; and nuclear radiation dosimeters [58].

In all these applications of optical fiber sensors, the main limitations in their performance, such as those due to the laser source coherence time, phase noise and photodetector noise, are the same as in telecommunication applications of fiber optics. In this sense, it is convenient at the present time to develop a unitary treatment. Either in telecommunication or in sensor applications, it is necessary to separate the case in which light must have a remarkable degree of temporal coherence from that in which light incoherence is acceptable or sometimes preferable.

This separation has great importance in the field of tests and characterization procedures. Our attention will be devoted to single-mode fiber optics in particular, because of their increasing importance. The most peculiar difference of single-mode fiber and device measurements with respect to multi-mode ones [59] consists of the importance of an exact previous definition of the degree of temporal coherence characterizing the light to be employed for the actual operation and measurement.

1.2 PRESENT STATUS OF TESTS AND EXPERIMENTAL CHARACTERIZATION PROCEDURES

Measurement procedures characterized by high repeatability, like those performed on large numbers of devices or components, whose conformance with precise recommendations is to be tested, themselves must adhere to suitable recommendations. On the contrary, laboratory tests, to be performed on new devices or system parts with the objective of their adjustment or improvement, are to be prepared depending on the particular situation. In the latter case, we can only describe the principles on which the experimental tests are based.

In this section, we will outline some properties and the principles at the basis of their measurement, either of the one type or the other. The next section is devoted instead to discussing recommendations and standardization aspects.

It is necessary to be very precise about the techniques for light modulation and detection, since light is the vehicle through which the signal is transferred. In particular, as already stressed, light temporal coherence is of fundamental importance. For example, in single-mode optical fiber characterization, for use in transmission systems as well as in sensors, there are properties which are related to the propagation of a coherent light, such as time dispersion, modal noise, and, when the detection is polarization sensitive, polarization fading or depolarization. Other properties, such as attenuation, geometrical dimensions, and the cut-off wavelength, are instead independent of the light degree of temporal coherence.

Time dispersion, in a single-mode fiber, is mainly due to the wavelength dependence of the group delay per unit distance τ_g, which characterizes the fundamental mode. Such kind of time dispersion is called chromatic dispersion. In particular, we have material and waveguide effects [30]. The first ones are dominant far from cut-off, i.e., when the mode is well confined. The second ones can be appreciable sometimes near cut-off. There are second-order and third-order dispersion contributions, the former related to the first derivative of τ_g with respect to the wavelength λ, usually dominant, and the latter in which its second derivative is involved. Second-order material dispersion is naturally very small at $\lambda = 1.3$ μm. It can be compensated by an intentional increase of second-order waveguide dispersion at $\lambda = 1.55$ μm [60]. A very small contribution to time dispersion is usually due to polarization mode dispersion, which derives from the fact that both polarization states of the fundamental mode are usually in propagation with slightly different phase constants.

In multi-mode fibers, where time dispersion is several orders of magnitude greater, a direct measurement of this effect is possible, in the time or in the frequency domain [59]. In single-mode fibers, instead, we can only measure its causes separately. The experimental evaluation of chromatic dispersion, which is by far the dominant cause, is usually performed by selecting different single wavelengths and measuring their delays in traveling the fiber [61]. Polarization mode dispersion is usually evaluated by means of a selective excitation or detection of the two polarization states of the fundamental mode in the fiber [62].

Modal noise may occur when a multi-mode propagation is present and the light has a rather high degree of temporal coherence. The undesired multi-mode propagation can be due either to the pair of polarization states of the fundamental mode [63], or to a residual presence of the second-order mode [64], although severely attenuated, when the effective cut-off condition is exploited [65]. Modal noise affects only intensity-modulation direct detection transmissions, and is favored when a localized mode coupling is present, as occurs in connectors or splices. Its experimental evaluation can be performed only in the field, when the transmission system or the sensor has been placed in its definitive position.

Polarization fading and depolarization, on the other hand, may affect only coherent modulation transmissions, since they require a heterodyne or a homodyne detection scheme, which is typically polarization sensitive. Polarization fading is due to the coherent sum of the fields of the two polarization states, which may have opposite signs [37]. Depolarization may occur when the light degree of temporal coherence is not high enough to allow the definition of a precise polarization state [66].

Attenuation in single-mode fibers is due to intrinsic causes, like absorption and scattering [67], or to additional causes, such as bending and microbending. Scattering loss decreases as λ^{-4}, being a typical Rayleigh phenomenon, whereas absorption loss exhibits peaks, which at the present time are remarkably reduced. The spectral regions where absorption loss is minimum are called *windows*. The first window is at about $\lambda = 0.85$ μm, the second at $\lambda = 1.3$ μm and the third at $\lambda = 1.55$ μm. Pure bend loss is a consequence of the fact that a fraction of the light guided along a uniformly curved dielectric guide is radiated out from the side opposite to the curvature center [32]. Microbending is instead a statistical phenomenon, induced by random bends of very small radii of curvature and dependent on the roughness of fiber coating or cable support [25]. Bending and microbending losses are more and more negligible with increasing the field confinement of the fundamental mode.

The overall attenuation suffered by the signal traveling a single-mode fiber can be directly or indirectly measured. The first kind of measurement is based on the comparison between the optical power levels directly revealed at two different fiber sections. The second kind of measurement is performed through the reproduction of the power decay along the fiber length of a backscattered light. These tests, already well-defined and standardized, are practically equal to the analogous tests for multi-mode fibers [59]. They allow also an easy evaluation of the loss introduced by a fiber-to-fiber joint, either when it is a demountable connector or when it is a permanent fusion splice.

In order to introduce geometrical dimension measurements in a single-mode fiber, it is necessary to give a brief description of the fiber transverse section. It generally consists of two concentric regions: the inner one, called the *core*, has radius a, and the outer one, called the *cladding*, is characterized by a uniform value of the refractive index n_c. In the core, we can have a uniform refractive index (step-index fiber) or a nonuniform refractive index (graded-index fiber). Field confinement is due to the interface between core and cladding and sometimes to the form of the index distribution in the core. In this sense, the cladding diameter, which coincides with the fiber transverse size, is fixed only for mechanical purposes.

A fundamental parameter in the study of light propagation along a fiber of this type is that called normalized frequency and denoted with V. It results in

$$V = \frac{2\pi}{\lambda} a \sqrt{n_o^2 - n_c^2} \tag{1.1}$$

where n_o is the on-axis value of the refractive index. Any mode is guided above a certain value of V, called the cut-off V-value for this mode. The distribution of cut-off values depends on the index profile. In many index profiles, the fundamental mode, denoted LP_{01}, has null cut-off. In this case the upper boundary of single-mode operation is given by the value V_c, which represents the cut-off of the second-order mode, denoted LP_{11}. In a step-index fiber, the fundamental mode has null cut-off, whereas the LP_{11} mode exhibits $V_c = 2.405$.

The higher the V the better the field confinement. Since the field distribution Ψ in the radial coordinate r has a form which resembles a Gaussian curve, a parameter for the evaluation of its confinement, called the Gaussian mode spot-size W_o, can be introduced as follows

$$\Psi(r) = \frac{\Psi_o}{\sqrt{2\pi}\, W_o} \exp\left(- \frac{r^2}{2W_o^2} \right) \tag{1.2}$$

where Ψ_o is a suitable normalized field.

The optical power, which is experimentally measured at the fiber output section, is proportional to $|\Psi(r)|^2$; so we can write its distribution as

$$P(r) = P_o \exp\left(- \frac{r^2}{W_o^2} \right) \tag{1.3}$$

where P_o is a suitable constant. When the optical power distribution, in the theory or from experimental tests, departs significantly from the gaussian shape, we can introduce an r.m.s. definition of mode spot-size, usually different from W_o, as follows

$$W_{rms}^2 = 2 \frac{\int_0^\infty r^2 P(r) r \, dr}{\int_0^\infty P(r) \, dr} \tag{1.4}$$

where P(r) is here the true optical power distribution. This definition is of great importance for bending and microbending loss evaluation.

On the other hand, when joint loss is to be estimated, a more proper definition is the so-called differential spot-size (or Petermann #2 spot-size) [68], which can be obtained as

$$\bar{W}^2 = 2 \frac{\int_0^\infty \Psi^2(r) r \, dr}{\int_0^\infty \left(\frac{d\Psi}{dr}\right)^2 r \, dr} \tag{1.5}$$

If P(r) has a Gaussian shape, we have $W_{rms} = \bar{W} = W_o$.

For $V > V_c$ there is also an optical power distribution due to the LP_{11} mode, which appears in superimposition with that of the fundamental mode P(r). It is characterized by an annular maximum instead of a central maximum. We can obtain experimentally the behavior of the fundamental mode optical power distribution P(r) from a near-field measurement on the fiber output section [69]. Suitable treatment of these data enables us to evaluate mode spot-sizes according to any definition. There is also the possibility of calculating P(r) from proper far-field measurement [70]. Finally repeating this test at different wavelengths makes possible an experimental evaluation of the cut-off wavelength λ_c of the LP_{11} mode. Single-mode propagation occurs for $\lambda > \lambda_c$.

A direct measurement of the core diameter or of the index profile is difficult because of diffraction. A technique called *refracted near field scanning* [71], which is based on the excitation of radiation modes by a properly arranged optical source, allows the evaluation of the core diameter and of the index profile. In reality, there are several other possible techniques for such experimental evaluation. Some of them will be discussed in the following chapters.

1.3 RECOMMENDATIONS AND STANDARDIZATION

CCITT and IEC recommendations and standardization procedures interest essentially fiber optics for telecommunications applications. They regard either the transmission medium (optical cables) or some fiber-optic devices, but also deal with the techniques for their experimental characterization. Special applications and sensors, although they often make use of standard components, are usually not subjected to particular recommendations by themselves.

As an example we will report here some results from the CCITT recommendations regarding fibers for long-distance transmissions. The operation at $\lambda = 0.85\ \mu m$ is reserved to multi-mode fibers. At $\lambda = 1.3\ \mu m$ either multi-mode or single-mode fibers can operate. Finally the operation wavelength of $\lambda = 1.55\ \mu m$ is reserved for single-mode fibers only. At this last wavelength, dispersion optimization is frequently requested. We can have a dispersion-shifted fiber, for which the zero of second-order chromatic dispersion turns out to be in the neighborhood of 1.55 μm, or dispersion-flattened fibers, whose second-order chromatic dispersion curve exhibits values within a prefixed maximum tolerance between 1.3 μm and 1.55 μm.

Multi-mode fibers having 50-μm core diameter are subjected to recommendation G.651 of CCITT. It fixes the core diameter nominal value, the cladding diameter nominal value (125 μm), and their tolerances. Then a maximum concentricity error between core and cladding, a maximum core noncircularity, and cladding noncircularity are suggested. Among the transmission characteristics, some minimum values of attenuation and of total time dispersion are recommended at 0.85 μm and at 1.3 μm.

Recommendation G.652 of CCITT, which refers to single-mode fibers designed for operation at $\lambda = 1.3\ \mu m$, will be described here. Next an outline of other CCITT recommendations, still under development, will be presented.

We can list the following set of nominal values and maximum tolerances for geometrical parameters:

- mode field diameter: 10 μm ± 10%
- cladding diameter: 125 μm ± 2 μm
- concentricity error between mode field and cladding: < 1 μm
- mode field noncircularity: (not fixed)
- cladding noncircularity: < 2%
- cut-off wavelength: $1.1\ \mu m < \lambda_c < 1.28\ \mu m$

In Annex A to Recommendation G.652, in which several measurement techniques are proposed for the experimental evaluation of the mode field diameter, it is specified that such quantity coincides with the parameter $\bar{W}$, defined according to (1.5). The above range of permitted values for λ_c refers to the cut-off wavelength which can be measured on a fiber jacketed by the only primary coating. In a cabled fiber, we have an effective cut-off wavelength λ_{cc}, usually slightly smaller than λ_c. In this case $\lambda_{cc} < 1.27\ \mu m$ is recommended.

The refractive index profile does not generally need to be known, for this kind of single-mode fiber; however, it usually follows a step distribution. Considering the relationship

$$\lambda_c = \frac{2\pi}{V_c} a \sqrt{n_o^2 - n_c^2} \tag{1.6}$$

in which a = 5 μm, V_c = 2.405, n_c = 1.465 and λ_c = 1.28 μm are to be taken, we obtain a normalized index difference between core and cladding $\Delta \approx (n_0 - n_c)/n_0$ of about $2 \cdot 10^{-3}$.

The transmission characteristics of a single-mode fiber without any requirement of dispersion optimization are simply attenuation and chromatic dispersion. Attenuation is fixed at λ = 1.3 μm, and should be smaller than 1 dB/km, although values as low as 0.4 dB/km are well achievable.

The contribution of waveguide dispersion is assumed to be practically negligible with respect to that of material dispersion. Nevertheless it may be responsible for some shifts in the wavelength of zero total chromatic dispersion. Therefore chromatic dispersion data are organized as follows:

- the allowed range of variation for the zero-dispersion wavelength λ_o,
- the maximum permitted value of the slope S, which is expressed in ps/(nm^2·km), of the curve representing $\tau_g'(\lambda)$.

The zero-dispersion wavelength λ_o may vary from λ_{om} = 1300 nm to λ_{oM} = 1324 nm. The maximum slope of curve $\tau_g'(\lambda)$ is fixed in S_{max} = 0.093 ps/(nm^2·km).

Transmission characteristics of concatenated fiber pieces are still attenuation and chromatic dispersion. Chromatic dispersion increases linearly with the distance z along each fiber piece. In a cascade of several subsequent fiber pieces, especially those made from different technological procedures, a chromatic dispersion compensation may occur. In fact, when the operation wavelength is between the two zero-dispersion wavelengths λ_{o1} and λ_{o2} of two subsequent fiber pieces, negative and positive contributions may be suppressed in part. Nevertheless, because in most applications the various fiber pieces are taken from the same producer, overall chromatic dispersion is calculated as the extrapolation over the whole optical link length of the average value of chromatic dispersion measured on the single-fiber pieces.

Recommendations G.653 and G.654 of CCITT refer to dispersion-shifted single-mode fibers and loss-minimized single-mode fibers at λ = 1.55 μm, respectively. We can list the following main differences in some geometrical and transmission parameters, with respect to the values previously fixed for more traditional single-mode fibers.

First of all, the mode field diameter at λ = 1.55 μm, in dispersion-shifted fibers, can vary from 7.0 μm to 8.3 μm, with a maximum departure from its nominal value smaller than 10%. Furthermore, since at 1.55 μm this kind of fiber is expected to suffer some excess curvature loss (pure bend loss), the following performance has been suggested to be satisfied: the loss increase for 100 turns loosely wound with a 37.5-mm radius, measured at λ = 1.55 μm, should be less than 0.5 dB. As for chromatic dispersion values in the neighborhood of λ = 1.55 μm, the constraint

$$|\tau_g'(\lambda)| < 3.5 \text{ ps/(nm} \cdot \text{km)} \tag{1.7}$$

is imposed between $\lambda = 1525$ nm and $\lambda = 1575$ nm.

The only recommendation of interest to loss-minimized single-mode fibers at $\lambda = 1.55$ μm regards their attenuation, which should be lower than 0.25 dB/km, although, depending on the technological process employed for fiber fabrication, values smaller than 0.2 dB/km sometimes could be obtained. Chromatic dispersion is expected to be on the order of 20 ps/(nm·km), which means that no actions are made to reduce chromatic dispersion at such operation wavelength. The use of optical sources whose spectral width is extremely narrow, like DFB lasers, should guarantee against too large levels of signal distortion, even at very high transmission rates. There are no reasons for suggesting a reduced value for the mode field diameter in such fibers, which therefore should be of about 10.5 μm. The cut-off wavelength λ_c is finally expected to be of the order of 1530 nm, thus allowing operation in correspondence with the third window only.

No recommendations address index profile shape necessary for obtaining dispersion-shifted behavior or minimized-loss performance. In fact various solutions are acceptable in principle, depending also on the particular dopant employed for profile synthesis.

CCITT works are organized on 4-year study periods. In the study period from 1989 to 1992, Study Group VI deals with optical cables, and Study Group XV deals with tests and characterization procedures on fibers and related devices. In IEC the Technical Committee TC 86 is devoted to fiber optics. It is organized into six *working groups* (WG) whose activities are listed below:

86/WG1 – Fiber-optic terminology and symbology,
86/WG2 – Fiber-optic safety aspects,
86/WG3 – Now in SC 86 C,
86/WG4 – Fiber-optic test equipment calibration,
86/WG5 – Now in SC 86 C,
86/WG6 – Active fibers.

Furthermore there are two subcommittees, SC 86 A and SC 86 B, in turn organized into several WGs, according to the following list:

SC 86 A – Fibers and cables,
86 A/WG1 – Optical fibers and cables for short-distance links,
86 A/WG2 – Optical fibers for long-distance links,
86 A/WG3 – Optical fiber cables and installation methods;
SC 86 B – Fiber optic interconnection devices and passive components,
86 B/WG1 – Interconnecting devices for optical fibers and cables,
86 B/WG2 – Fiber-optic passive components,
86 B/WG3 – Quality assessment procedures for optical fiber interconnections.

Recently a third subcommittee has begun its activity, incorporating two WGs from TC 86. It is called SC 86 C and is organized as follows:

SC 86 C – Fiber-optic system specifications,
86 C/WG1 – (ex 86/WG3) Fiber-optic subsystem,
86 C/WG2 – (ex 86/WG5) Fiber-optic sensors.

This last WG confirms that, presently, sensors can be considered in the framework of more traditional applications of fiber optics.

1.4 OBJECTIVES OF THE BOOK

Light can be generated by optical sources with different degrees of temporal coherence. A pure monochromatic wave does not exist; however, a completely random electromagnetic field does not exist as well. Nevertheless we can have a light with high or low degree of coherence. On the other hand, a light detector able to reveal fields at optical frequencies (the equivalent of an antenna at radio frequencies) cannot be realized. We have, instead, only quadratic light detectors, although, by means of suitable arrangements, information carried by a coherently modulated light can be extracted from it.

A fundamental aspect to be carefully taken into account in the description of an optical component for transmission, signal processing, or sensor applications, is the definition of which type of light it has to treat, with particular attention to the property of temporal coherence. So we will separate the discussion of tests and characterization procedures, subdividing them into two main classes:

1. coherent light measurements,
2. measurements which do not require light with a particular degree of temporal coherence.

The first ones are devoted to evaluating the performances of fiber optic components (optical fibers, passive and active devices) whose operation is based on coherent light processing. The second ones, and at the moment their number is remarkably larger, regard fiber-optic components for which the light coherence properties are by far less important. However, the number of applications of the first type of measurement is expected to grow in future years.

In Chapter 2, after the introduction of some basic concepts, a description of interferometers, light generation, modulation and detection, the main properties of single-mode fibers, and those of the electro-optical devices is carried out. Chapter 3 is devoted to present measurements of fiber optics in incoherent light. Chapter 4 describes the measurements of fiber optics employing coherent light. Finally, Chapter 5 deals with fiber-optic sensors and their experimental characterization.

The present book describes single-mode fiber measurements, either in telecommunication or in sensor applications. In particular, Chapters 3 and 4 are mainly telecommunication oriented. Optical fiber communication represents a popular topic in recent scientific literature, and many books treat it extensively [72–78]. In this sense, it is not necessary to repeat here such treatment, except for some particular aspects which will be outlined during such two chapters. Chapter 5, in turn, faces the very wide panorama of fiber optic sensors, whose description, especially in comprehensive texts, is not so familiar. Furthermore in a book treating measurements, the problems of sensors appear relevant. In this sense, although some books have been published on fiber-optic sensors [79, 80], we have included a final chapter expressly devoted to their description and characterization, as a natural combination of the arguments developed in the previous chapters.

REFERENCES

[1] Hondros, D., and P. Debye, "Elektromagnetische Wellen an dielektrischen Drahten", *Ann. Phys.*, Vol. 32, 1910, pp. 465–476.

[2] Buchholz, H., "Die Quasioptik der Ultrakurzwellenweiter", *Elektrische Nachrichtentechnik*, Vol. 15, 1938, pp. 297–320.

[3] Kapany, N. S., *Fiber optics: Principles and applications*, Academic Press, New York, 1967.

[4] Hall, R. N., G. E. Fenner, J. D. Kingsley, T. J. Soltys, and R. O. Carlson, "Coherent light emission from GaAs junctions", *Phys. Rev. Letts.*, Vol. 9, 1962, pp. 366–368.

[5] Nathan, M. I., W. P. Dumke, G. Burns, F. H. Dill, and G. J. Lasher, "Stimulated emission of radiation from GaAs p-n junctions", *J. Appl. Phys.*, Vol. 1, 1962, pp. 62–64.

[6] Quist,T. M., R. H. Rediker, R. J. Keyes, W. E. Krag, B. Lax, A. L. McWhorter, and H. J. Zeiger, "Semiconductor maser of GaAs", *Appl. Phys. Letts.*, Vol. 9, 1962, pp. 91–92.

[7] Snitzer, E., "Cylindrical dielectric waveguide modes", *J. Opt. Soc. Am.*, Vol. 51, 1961, pp. 491–498.

[8] Kapany, N. S., and J. J. Burke, "Fiber optics. IX. Waveguide effects", *J. Opt. Soc. Am.*, Vol. 51, 1961, pp. 1067–1078.

[9] Kao, K. C., and G. A. Hockham, "Dielectric-fibre surface waveguides for optical frequencies", *Proc. IEE*, Vol. 113, 1966, pp. 1151–1158.

[10] French, W. G., J. P. MacChesney, P. B. O'Connor, and G. W. Tasker, "Optical waveguides with very low losses", *Bell Sys. Tech. J.*, Vol. 53, 1974, pp. 951–954.

[11] Horiguchi, M., and H. Osanai, "Spectral losses of low-OH-content optical fibres", *Electron. Lett.*, Vol. 12, 1976, pp. 310–312.

[12] Miya, T., Y. Terunuma, T. Hosaka, and T. Miyashita, "Ultimate low-loss single-mode fibre at 1.55 μm", *Electron. Lett.*, Vol. 15, 1979, pp. 106–108.

[13] Keck, D. B., and P. C. Shultz, "Method of producing optical waveguide fibers ", U. S. Patent 3 711 262.

[14] MacChesney, J. B., P. B. O'Connor, and H. M. Presbey, "A new technique for preparation of low-loss and graded-index optical fibers", *Proc. IEEE*, Vol. 62, 1974, pp. 1280–1281.

[15] Izawa, T., T. Miyashita, and F. Hanawa, "Continuous optical fiber preform fabrication", U. S. Patent 4 062 665.
[16] Gloge, D., and E. A. J. Marcatili, "Multimode theory of graded-core fibers", *Bell Sys. Tech. J.*, Vol. 52, 1973, pp. 1563–1578.
[17] Clarricoats, P. J. B., and K. B. Chan, "Electromagnetic-wave propagation along radially inhomogeneous dielectric cylinders", *Electron. Lett.*, Vol. 6, 1970, pp. 694–695.
[18] Olshansky, R., and D. B. Keck, "Pulse broadening in graded-index optical fibers", *Appl. Opt.*, Vol. 15, 1976, pp. 483–491.
[19] Bianciardi, E., and V. Rizzoli, "Propagation in graded-core fibres: a unified numerical description", *Opt. Quantum Electron.*, Vol. 9, 1977, pp. 121–133.
[20] Arnaud, J. A., "Pulse broadening in multimode optical fibres", *Bell Sys. Tech. J.*, Vol. 54, 1975, pp. 1179–1205.
[21] Ankiewicz, A., and C. Pask, "Geometric optics approach to light acceptance and propagation in graded index fibres", *Opt. Quantum Electron.,* Vol. 9, 1977, pp. 87–109.
[22] Cancellieri, G., "New ray-optics approach to the propagation in graded-index fibres", *Opt. Comm.*, Vol. 29, 1979, pp. 287–292.
[23] Marcuse, D., *Theory of dielectric optical waveguides*, Academic Press, New York, 1974.
[24] Olshansky, R., "Mode coupling effects in graded-index optical fibers", *Appl. Opt.*, Vol. 14, 1975, pp. 935–945.
[25] Gloge, D., "Optical-fiber packaging and its influence on fiber straightness and loss", *Bell Sys. Tech. J.*, Vol. 54, 1975, pp. 245–262.
[26] Epworth, R. E., "ITT 140 Mbit/s optical fiber system", *Proc. NTC 77*, Los Angeles, 1977, pp. 13–14.
[27] Miki, T., et al., "Experimental optical fiber transmission systems", *Trans. IECE of Japan*, Vol. E61, 1977, pp. 299–234.
[28] Brace, D. J., and I. A. Ravencroft, "Optical fibre transmission systems: the 8.448 Mb/s feasibility trial", *Post Office El. Eng. J.*, Vol. 70, 1977, pp. 146–153.
[29] Foord, S. G., W. E. Simpson, and A. Cook, "Some design principles for fibre optical cables", *Proc. 23-rd Int. Wire Cable Symp.*, Atlantic City, 1974, pp. 276–280.
[30] Gloge, D., "Dispersion in weakly-guiding fibers", *Appl. Opt.*, Vol. 10, 1971, pp. 2442–2445.
[31] Someda, C. G., "Simple low-loss joints between simple-mode optical fibers", *Bell Sys. Tech. J.*, Vol. 52, 1973, pp. 583–597.
[32] Gambling, W. A., H. Matsumura, and C. M. Ragdale, "Curvature and microbending losses in single-mode optical fibres", *Opt. Quantum Electron.*, Vol. 11, 1979, pp. 43–59.
[33] Hayashi, I., M. B. Panish, P. W. Foy, and S. Sumski, "Junction lasers which operate continuously at room temperature", *Appl. Phys. Lett.*, Vol. 17, 1970, pp. 109–111.

[34] Shen, C. C., J. J. Hsieh, and T. A. Lind, "1500-h continuous CW operation of double-heterostructure GaInAsP/InP lasers", *Appl. Phys. Lett.*, Vol. 30, 1977, pp. 353–356.

[35] Nakamura, M., K. Aiki, J. Umeda, and A. Yariv, "C. W. operation of distributed-feedback GaAs-GaAlAs diode lasers at temperatures up to 300 K", *Appl. Phys. Letts.*, Vol. 27, 1975, pp. 403–405.

[36] Pratt, W. K., *Laser communication systems*, Wiley, New York, 1969.

[37] Okoshi, T., and K. Kikuchi, "Heterodyne-type optical fiber communication", *J. Opt. Commun.*, Vol. 2, 1981, pp. 82–88.

[38] Pearsall, T. P., "Photodetectors for optical communication", *J. Opt. Commun.*, Vol. 2, 1981, pp. 42–48.

[39] Barnoski, M. K., "Data distribution using fiber optics", *Appl. Opt.*, Vol. 14, 1975, pp. 2571–2577.

[40] Rosemberg, D., "Micro-optic passive devices for multimode optical fiber communication systems", *Siemens Res. and Dev. Rept.*, Vol. 8, 1979, pp. 125–129.

[41] Tomlinson, W. J., and G. D. Aumiller, "Optical multiplexer for multimode fiber transmission systems", *Appl. Phys. Lett.*, Vol. 30, 1877, pp. 169–171.

[42] Kaminow, I. P., et al., "Thin film $LiNbO_3$ electro-optic directional coupler switch", *Appl. Phys. Lett.*, Vol. 22, 1973, pp. 540–542.

[43] Carenco, A., L. Menigaux, and P. Delpech, "GaAs electro-optical directional coupler switch", *Proc. 4-th ECOC*, Genova, 1978, pp. 404–409.

[44] Righini, G. C., V. Russo, and S. Sottini, "Geodesic optical systems for signal processing", *Proc. 4-th ECOC*, Genova, 1978, pp. 418–424.

[45] Fouckhardt, H., and K. J. Abeling, "All-optical switching and modulation in GaAlAs strip waveguides", *Proc. 12-th ECOC*, Barcellona, 1986, pp. 497–500.

[46] Bachus, E. J., and G. Heidt, "Application of coherent systems in the subscriber loop", *Proc. Inv. P. 12-th ECOC*, Barcellona, 1986, pp. 73–78.

[47] Okoshi, T., "Polarization-state control schemes for heterodyne or homodyne optical fiber communications", *IEEE J. Lightwave Technol.*, Vol. LT-3, 1985, pp. 1232–1237.

[48] Glance, B., "Polarization independent coherent optical receiver", *IEEE J. Lightwave Technol.*, Vol. LT-5, 1987, pp. 274–276.

[49] Kaminow, I. P., "Polarization in optical fibers", *IEEE J. Quantum Electron.*, Vol. QE-17, 1981, pp. 15–22.

[50] Varnham, M. P., et al., "Single-polarization operation of highly birefringent bow-tie optical fibres", *Electron. Lett.*, Vol. 19, 1983, pp. 246–247.

[51] Giallorenzi, T. G., et al., "Optical fiber sensor technology", *IEEE J. Quantum Electron.*, Vol. QE-18, 1982, pp. 626–664.

[52] Cole, J. H., R. L. Johnson, and P. B. Bhuta, "Fiber optic detection of sound", *J. Acoust. Soc. Am.*, Vol. 62, 1977, pp. 1136–1138.

[53] Rashleigh, S. C., "Magnetic field sensing with a single-mode fiber", *Opt. Lett.*, Vol. 6, 1981, pp. 19–21.

[54] Bergh, R. A., H. C. Lefevre, and H. J. Shaw, "An overview of fiber-optic gyroscope", *IEEE J. Lightwave Technol.*, Vol. LT-2, 1984, pp. 91–107.
[55] Kyuma, K., et al., "Fiber optical instrument for temperature measurements", *IEEE J. Quantum Electron.*, Vol. QE-18, 1982, pp. 676–679.
[56] Harmer, A., A. Scheggi, "Chemical, biochemical, and medical sensors", in *Optical fiber sensors*, ed. by B. Culshaw, J. Dakin, Artech House, Norwood, MA, 1989, pp. 599–652.
[57] Tveten, A. B., et al., "Fiber optic accelerometer", *Electron. Lett.*, Vol. 16, 1980, pp. 854–855.
[58] Evans, B. D., et al., "The fiber optic dosimeter on the navigational technology satellite 2", *IEEE Trans. Nucl. Sci.*, Vol. NS-25, 1978, pp. 1619–1624.
[59] Cancellieri, G., and U. Ravaioli, *Measurements of optical fibers and devices. Theory and experiments*, Artech House, Norwood, MA, 1984.
[60] Cohen, L. G., C. Lin, and W. G. French, "Tailoring zero chromatic dispersion into the 1.5–1.6 μm low loss spectral region of single-mode fibres", *Electron. Lett.*, Vol. 15, 1979, pp. 334–335.
[61] Cohen, L. G., and C. Lin, "Pulse delay measurements in the zero material dispersion wavelength region for optical fibers", *Appl. Opt.*, Vol. 16, 1977, pp. 3136–3139.
[62] Monerie, M., P. Lamouler, and L. Jeunhomme, "Polarization mode dispersion measurements in long single-mode fibres", *Electron. Lett.*, Vol. 16, 1980, pp. 907–908.
[63] Heckmann, S., "Modal noise in single-mode fibres operated slightly above cut-off", *Electron. Lett.*, Vol. 17, 1981, pp. 499–500.
[64] Petermann, K., "Nonlinear transmission behaviour of a single-mode fiber transmission line due to polarization coupling", *J. Opt. Commun.*, Vol. 2, 1981, pp. 59–64.
[65] Cancellieri, G., and A. Orfei, "Effective cut-off condition in single-mode optical fibres", *Proc. IOOC-ECOC 85*, Venice, 1985, pp. 659–662.
[66] Sakai, J., "Degree of polarization including the random mode conversion effect in anisotropic single-mode optical fibers", J. *Opt. Soc. Am.*, Vol. A-1, 1984, pp. 1007–1018.
[67] Marcuse, D., *Principles of optical fiber measurement*, Academic Press, New York, 1981.
[68] Petermann K., and R. Kuhne, "Upper and lower limits for microbending loss in single-mode fibers", *IEEE J. Lightwave Technol.*, Vol. LT-4, 1986, pp. 2–7.
[69] Coppa, G., P. Di Vita, and U. Rossi, "Characterization of single-mode fibres by near-field measurement", *Electron. Lett.*, Vol. 19, 1983, pp. 293–294.
[70] Artiglia, M., et al., "Mode field diameter in single-mode optical fibers", *IEEE J. Lightwave Technol.*, Vol. LT-7, 1989, pp. 1139–1152.
[71] Stewart, W.J., "A new technique for measuring the refractive index profile of graded index optical fibres", *Proc. IOOC 77*, Tokyo, 1977, pp. 395–398.
[72] Midwinter, J. E., *Optical fibers for transmissions*, Wiley, New York, 1979.

[73] Technical Staff of CSELT, *Optical fiber communication*, McGraw-Hill, New York, 1980.
[74] Kressel, H., editor, *Semiconductor devices for optical communication*, Springer, Berlin, 1980.
[75] Cherin, A. H., *An introduction to optical fibers*, McGraw-Hill, Tokyo, 1983.
[76] Keiser, G., *Optical fiber communications*, McGraw-Hill, Tokyo, 1983.
[77] Siemens, *Optical communications: a telecommunication review*, Wiley, New York, 1986.
[78] Edwards, T., *Fiber-optic systems network applications*, Wiley, New York, 1989.
[79] Dakin, J., and B. Culshaw, editors, *Optical fiber sensors, volume I: principles and components*, Artech House, Norwood, MA, 1988.
[80] Dakin, J., and B. Culshaw, editors, *Optical fiber sensors, volume II: systems and applications*, Artech House, Norwood, MA, 1989.

Chapter 2

Single-Mode Optical Fibers and Devices

F. Chiaraluce

2.1 BASIC CONCEPTS OF OPTICAL RADIATION

The problem of electromagnetic field representation involves scalar and vector quantities. When polarization is uniform, we can follow a completely scalar approach, simplifying greatly any field representation. A high degree of temporal coherence is necessary to define a uniform polarization state, except when this is linear. Linearly polarized light is, when possible, preferable for most types of light processing. Spatial coherence and diffraction phenomena may limit the quality of optical signal processing. Finally, interference is a fundamental property of light, related to its temporal coherence, which is exploited in several experimental test procedures.

In this section, after an outline about the main techniques for electromagnetic field representation, we deal with temporal coherence, polarization, spatial coherence, diffraction, and interference.

2.1.1 Field Representation

The vectorial nature of electromagnetic field propagation results directly from Maxwell's equation [1]. Nevertheless, when propagation occurs in isotropic media, and when weakly guiding structures are employed for the lateral confinement of the light, a uniform linear polarization characterizes the propagation field. In this case, we can use a completely scalar approach for field description. This assumption is made throughout the present subsection.

The scalar description of an electromagnetic field in motion along a direction z, in free space, can be performed by means of a function $\psi(t,z)$, where t is the time. Such function satisfies the wave equation

$$\frac{\partial^2\psi}{\partial z^2} - \frac{1}{c^2}\frac{\partial^2\psi}{\partial t^2} = 0 \tag{2.1}$$

in which c represents the free-space velocity of light. A general solution of (2.1), considering only the forward propagating wave, is

$$\psi(t,z) = \psi^{+}\left(t - \frac{z}{c}\right) \tag{2.2}$$

where function ψ^{+} is imposed by a proper initial condition.

A particular case is that called monochromatic field, for which

$$\psi(t,z) = A_o \cos(\omega_o t + \phi_o - k_o z) \tag{2.3}$$

where A_o, ϕ_o and ω_o are respectively its maximum amplitude, initial phase and circular frequency. Finally

$$k_o = \frac{\omega_o}{c} \tag{2.4}$$

represents the free-space wavenumber. Expression (2.3) suggests another representation, based on complex quantities. This can be set in the form

$$\psi(t,z) = \mathrm{Re}\left[\dot{\psi}(z)\, e^{i\omega_o t}\right] \tag{2.5}$$

where

$$\dot{\psi}(z) = A_o\, e^{i\phi_o} e^{-ik_o z} \tag{2.6}$$

A monochromatic field is a purely theoretical concept. Nevertheless it can be considered as a limit situation of an infinitely coherent field.

When the field is no longer monochromatic, (2.3) can be replaced by a suitable superimposition of contributions

$$\psi(t,z) = \frac{1}{2\pi} \int_{-\infty}^{+\infty} \Psi(\omega,z)\, e^{i\omega t}\, d\omega \tag{2.7}$$

where $\Psi(\omega,z)$ is a complex field spectral distribution. The inverse of (2.7) can be written as

$$\Psi(\omega,z) = \int_{-\infty}^{+\infty} \psi(t,z)\, e^{-i\omega t}\, dt \tag{2.8}$$

thus showing how $\psi(t,z)$ and $\Psi(\omega,z)$ are a Fourier transform pair. The property expressed by (2.2) leads to the following relationship

$$\psi^{+}\left(t - \frac{z}{c}\right) = \frac{1}{2\pi} \int_{-\infty}^{+\infty} \Psi^{+}(\omega,z)\, e^{i\omega(t - z/c)}\, d\omega \tag{2.9}$$

where Ψ^{+} is the Fourier transform of ψ^{+}.

Relationship (2.7) is useful when time dispersion effects are present. In this case, instead of (2.1) it is preferable to assume a Fourier-transformed wave equation, of the type

$$\frac{\partial^2 \Psi}{\partial z^2} + \beta^2 \Psi = 0 \tag{2.10}$$

in which β represents a general phase constant, where material and waveguide effects are taken into account. Since β is usually expressed as a function of ω, in this situation it is more convenient to follow the evolution of $\Psi(\omega,z)$ along z as

$$\Psi(\omega,z) = \Psi^{+}(\omega)\, e^{-i\beta(\omega)z} \tag{2.11}$$

Comparison with (2.9) shows how, in free space, $\beta(\omega)$ reduces to ω/c.

A real situation, common in practice, is that of a quasi-monochromatic field. Instead of (2.3), we can write here, for a wave propagating in free space,

$$\psi(t,z) = A(t)\cos[\omega_0 t + \phi(t) - k_0 z] \tag{2.12}$$

where A(t) represents amplitude fluctuations and $\phi(t)$ phase fluctuations, both slow with respect to the period $2\pi/\omega_0$ of the sinusoidal oscillation. Bandpass representation of (2.12), analogous to (2.5), becomes here

$$\psi(t,z) = \mathrm{Re}\left[\dot{\psi}(t,z)\, e^{i\omega_0 t}\right] \tag{2.13}$$

where

$$\dot{\psi}(t,z) = A(t)\, e^{i\phi(t)} e^{-ik_0 z} \tag{2.14}$$

A(t) and $\phi(t)$ can be induced by an intentional modulation, or can be the result of a natural random fluctuation, which affects any real optical source. In particular, for laser sources, this phenomenon is called *phase noise*. Nevertheless, A(t) and $\phi(t)$ can be assumed as stationary stochastic processes.

A Fourier transformation also can be applied to the complex signal $\dot{\psi}(t,z)\, e^{i\omega_0 t}$, leading to

$$\dot{\Psi}(\omega,z) = \int_{-\infty}^{+\infty} \dot{\psi}(t,z) e^{i\omega_0 t} e^{-i\omega t}\, dt \tag{2.15}$$

This enables us to modify the well-known time-differentiation property of Fourier transform, which consists of multiplying by $i\omega$ the transform of the signal when this is differentiated with respect to the time. Modification consists of replacing here $i\omega$ with $i(\omega - \omega_o)$.

When time dispersion is present, expression (2.14) is no longer valid. Nevertheless, we can take into account slow variation of $\dot{\psi}(t,z)$ along z, besides those due to $\beta(\omega_o)z$, in the following way

$$\dot{\psi}(t,z) = \dot{\psi}_o(t,z)\, e^{-i\beta(\omega_o)z} \tag{2.16}$$

where $\dot{\psi}_o(t,z)$ varies not appreciably along a distance of the order of $2\pi/\beta(\omega_o)$. Such a description takes into account the signal distortion suffered by A(t) and $\phi(t)$, when they are due to an intentional modulation. Comparison between (2.12) and (2.16), taking into account that $e^{-i\beta(\omega_o)z}$ replaces e^{-ik_oz}, enables us to write

$$\dot{\psi}_o(t,0) = A(t)\, e^{i\phi(t)} \tag{2.17}$$

Furthermore $\beta(\omega)$, for a quasi-monochromatic wave, can be expanded in power series, in the neighborhood of ω_o, leading to

$$\beta(\omega) \cong \beta(\omega_o)+\beta'(\omega_o)(\omega-\omega_o) + \frac{\beta''(\omega_o)}{2}(\omega-\omega_o)^2 + \frac{\beta'''(\omega_o)}{6}(\omega-\omega_o)^3+\ldots \tag{2.18}$$

Substitution into (2.10), and taking the inverse Fourier transformation, considering the above time-differentiation property, yields the following equation in $\dot{\psi}_o(t,z)$

$$\frac{\partial\dot{\psi}_o}{\partial z} + \beta'(\omega_o)\frac{\partial\dot{\psi}_o}{\partial t} - \frac{i}{2}\beta''(\omega_o)\frac{\partial^2\dot{\psi}_o}{\partial t^2} - \frac{1}{6}\beta'''(\omega_o)\frac{\partial^3\dot{\psi}_o}{\partial t^3} = 0 \tag{2.19}$$

in which terms of order higher than three have been neglected. This equation is approximated, because the second derivative of $\dot{\psi}_o(t,z)$ with respect to z has been ignored, owing to its slow spatial variation.

2.1.2 Field Polarization

The scalar representation of an electric field $\bar{E}(t,z)$, by means of a function $\psi(t,z)$, is possible only when field polarization is uniform. In order to describe polarization properties of a propagating field, it is necessary to consider the vectorial nature of this field. A monochromatic wave represented by the transverse electric field $\bar{E}(t,z)$ in propagation in free space along an axis z, can be decomposed according to a Cartesian reference set of coordinates x, y, using a symbolism analogous to that introduced in 2.1.1, which leads to

$$\bar{E}(t,z) = E_{xo}\cos(\omega_o t + \phi_{ox} - k_o z)\hat{x} + E_{yo}\cos(\omega_o t + \phi_{oy} - k_o z)\hat{y} \tag{2.20}$$

where $\hat{x}$ and $\hat{y}$ are suitable unit vectors.

For a given value of z, in the most general case, the free extremum of $\bar{E}(t,z)$ runs over an ellipse in the plane xy. This ellipse degenerates into a straight line (linear polarization) when $E_{xo} = 0$, or $E_{yo} = 0$, or $\phi_{ox} = \phi_{oy}$. It degenerates into a circle (circular polarization) when $E_{xo} = E_{yo}$ and $\phi_{ox} = \phi_{oy} \pm \pi/2$. Finally the sense of rotation on the ellipse or on the circle, looking toward higher values of z, is clockwise when $\phi_{ox} > \phi_{oy}$, counter-clockwise when $\phi_{ox} < \phi_{oy}$. The state of polarization, however, is in this case uniform, since it does not depend on the particular section z considered.

A different situation occurs when the two linear polarizations are subject to different phase constants β_{xo} and β_{yo}. This is a typical situation when the medium exhibits a property called birefringence. In this case (2.20) is replaced by

$$\bar{E}(t,z) = E_{xo}\cos(\omega_o t+\phi_{ox}-\beta_{xo}z)\hat{x}+E_{yo}\cos(\omega_o t+\phi_{oy} - \beta_{yo}z)\hat{y} \tag{2.21}$$

Here the state of polarization depends on the particular section z considered, and the above discussion can be repeated by substitution of ϕ_{ox} with $(\phi_{ox} - \beta_{xo}z)$, and of ϕ_{oy} with $(\phi_{oy} - \beta_{yo}z)$. Nevertheless, when $E_{xo} = 0$ or $E_{yo} = 0$, the polarization is maintained linear.

An example of intentional application of this property is that regarding the transformation of a linear polarization into a circular one. When an electric field of amplitude E_o incides at $\pi/4$ on the input end of a proper birefringent guide, it creates a propagating field which can be written as

$$\bar{E}(t,z) = \frac{E_o}{\sqrt{2}}\cos(\omega_o t - \beta_{xo}z)\hat{x} + \frac{E_o}{\sqrt{2}}\cos(\omega_o t - \beta_{yo}z)\hat{y} \tag{2.22}$$

At a distance

$$z = \frac{1}{\beta_{xo} - \beta_{yo}} \frac{\pi}{2} \tag{2.23}$$

the polarization state is transformed into a circular one.

An alternative description for the polarization state of a monochromatic field is that based on Stokes parameters. The ellipse on which $\underline{E}(t,z)$ varies exhibits a pair of intrinsic axes which are tilted with respect to the pair xy by an angle ϕ. Letting

$$\cos\delta = \frac{E_{xo}^2 - E_{yo}^2}{2E_{xo}E_{yo}} \tan(2\phi) \tag{2.24}$$

where δ is the phase difference between the two orthogonal components of the electric field, by supposing that $E_{xo} > E_{yo}$, the following four quantities are called Stokes parameters for polarization state description

$$\begin{aligned} s_0 &= E_{xo}^2 + E_{yo}^2 \\ s_1 &= E_{xo}^2 - E_{yo}^2 \\ s_2 &= 2E_{xo}E_{yo}\cos\delta \\ s_3 &= 2E_{xo}E_{yo}\sin\delta \end{aligned} \tag{2.25}$$

These parameters satisfy the relationship

$$s_0^2 = s_1^2 + s_2^2 + s_3^2 \tag{2.26}$$

So in a space s_1, s_2, s_3, a state of polarization is described by a point on a spherical surface, with center in the origin (0,0,0) and radius s_0. Such a locus is called Poincarè sphere.

If the field is not monochromatic, its polarization state cannot be defined so accurately, except when it is uniformly linearly polarized. It is then necessary to introduce a degree of polarization, strictly related with the light degree of temporal coherence. This will be done in the next subsection.

2.1.3 Temporal Coherence

For a field $\psi(t)$, independent of z for simplicity, we can introduce the intensity I according to

$$I = \lim_{\Delta t \to \infty} \frac{1}{\Delta t} \int_{-\Delta t/2}^{\Delta t/2} \psi^2(t)\, dt \tag{2.27}$$

This quantity is finite when $\psi(t)$ is purely sinusoidal, i.e., for a monochromatic field, since the overall integral coincides with that calculated over a single period. Even when, considering a quasi-monochromatic field, the amplitude fluctuations A(t) and phase fluctuations $\phi(t)$ are taken as stationary stochastic processes, I turns out to be finite. In fact substitution of statistical averaging to temporal averaging enables us to write

$$I = \langle \psi^2(t) \rangle = \langle A^2(t) \cos^2 [\omega_o t + \phi(t)] \rangle \tag{2.28}$$

Now, assuming as usual that $\phi(t)$ is uniformly distributed over an interval $[0,2\pi]$, the mean value of $\cos^2 [\omega_o t + \phi(t)]$ can be taken equal to 1/2; therefore, we obtain

$$I = \frac{1}{2} \langle A^2(t) \rangle = \frac{1}{2} \langle |\dot{\psi}(t)|^2 \rangle \tag{2.29}$$

in which any time dependence has been averaged out.

The intensity of a field so defined is very important, because it can be revealed by photodetectors. Unfortunately, at optical frequencies there are no detectors able to reveal fields; but even when the parameters of a field (amplitude, phase or frequency) are to be demodulated, we have to use intensity-sensitive detectors, although properly arranged.

Now, let us consider two fields $\psi_1(t)$ and $\psi_2(t)$. Their scalar sum occurs only when they have equal direction of propagation and state of polarization. Under such conditions, the field resulting from their superimposition turns out to be

$$\psi(t) = \psi_1(t) + \psi_2(t) \tag{2.30}$$

whose intensity is

$$I(t) = \langle \psi_1^2(t) \rangle + \langle \psi_2^2(t) \rangle + 2\langle \psi_1(t)\psi_2(t) \rangle \tag{2.31}$$

Owing to the presence of the last term, in general such intensity depends on the time t.

When $\psi_1(t)$ and $\psi_2(t)$ have the same circular frequency ω_o, in the product $\psi_1(t)\psi_2(t)$, besides a constant term, a double circular frequency $2\omega_o$ is present, which, however, is suppressed by averaging. So, taking $A_1(t)$, $A_2(t)$ as amplitudes and $\phi_1(t)$, $\phi_2(t)$ as phases, according to quasi-monochromatic field representation, we can write

$$I(t) = \frac{1}{2}\{\langle A_1^2(t)\rangle + \langle A_2^2(t)\rangle + 2\langle A_1(t)A_2(t)\cos[\phi_1(t) - \phi_2(t)] \rangle\} \tag{2.32}$$

From (2.29) we can also rewrite this expression as

$$I(t) = I_1 + I_2 + \mathrm{Re}[\langle \dot{\psi}_1(t)\dot{\psi}_2^*(t) \rangle] \tag{2.33}$$

It is possible to introduce a complex degree of coherence between such two fields, which results in

$$\Gamma_{12}(t) = \frac{\langle \dot{\psi}_1(t)\dot{\psi}_2^*(t) \rangle}{2\sqrt{I_1 I_2}} = \frac{\langle \dot{\psi}_1(t)\dot{\psi}_2^*(t) \rangle}{\sqrt{\langle |\dot{\psi}_1(t)|^2 \rangle \langle |\dot{\psi}_2(t)|^2 \rangle}} \tag{2.34}$$

leading to

$$I(t) = I_1 + I_2 + 2\sqrt{I_1 I_2}\ \mathrm{Re}[\Gamma_{12}(t)] \tag{2.35}$$

The real part of $\Gamma_{12}(t)$ can be also written in the form

$$\mathrm{Re}[\Gamma_{12}(t)] = \frac{\langle A_1(t)A_2(t)\cos[\phi_1(t) - \phi_2(t)] \rangle}{\sqrt{\langle A_1^2(t)\rangle \langle A_2^2(t)\rangle}} \tag{2.36}$$

which shows how the condition for having maximum variation in this parameter is reached when A_1, A_2, ϕ_1, ϕ_2 are fixed values and then the fields are purely monochromatic. In this situation, as a function of the difference $\phi_1 - \phi_2$, $\mathrm{Re}[\Gamma_{12}(t)]$ oscillates between – 1 and + 1, and the two fields exhibit maximum mutual coherence.

$\mathrm{Re}[\Gamma_{12}(t)]$ vanishes, instead, when the two fields exhibit complete mutual incoherence. In this situation the superimposition of fields entails the sum of the two optical intensities: $I = I_1 + I_2$, independently of t.

Another important parameter to describe the mutual coherence property between two fields is the contrast η defined as

$$\eta = \frac{I_{max} - I_{min}}{I_{max} + I_{min}} \tag{2.37}$$

where I_{max} and I_{min} represent respectively the maximum and minimum value of I(t), as given by (2.35). When the two fields have complete mutual incoherence, we obtain $I_{max} = I_{min} = I_1 + I_2$, and hence $\eta = 0$. On the opposite side, when $\mathrm{Re}[\Gamma_{12}(t)]$ varies between –1 and +1, we have $I_{max} = I_1 + I_2 + 2\sqrt{I_1 I_2}$, reached for $\mathrm{Re}[\Gamma_{12}(t)] = +1$, and $I_{min} = I_1 + I_2 - 2\sqrt{I_1 I_2}$, reached for $\mathrm{Re}[\Gamma_{12}(t)] = -1$, so that

$$\eta = \frac{2\sqrt{I_1 I_2}}{I_1 + I_2} \tag{2.38}$$

It is easy to demonstrate that the maximum value of η, in the latter situation, is achieved when $I_1 = I_2$, and results in $\eta = 1$. In conclusion, the most general pair of fields $\psi_1(t)$ and $\psi_2(t)$ are always characterized by mutual coherence properties which leads to a contrast η, whose possible value varies from 0 to 1, reaching such extrema respectively when the two fields have complete mutual incoherence and when they are perfectly coherent between them.

For a single field, subdivided into two beams of equal intensity, then properly recombined, we can write

$$\dot{\psi}(t) = \dot{\psi}_1(t) + \dot{\psi}_1(t + \tau)\, e^{-i\omega_0 \tau} \tag{2.39}$$

where τ is the delay between them. The overall intensity results in a function of τ

$$I(\tau) = \frac{1}{2}\langle |\dot{\psi}(t)|^2 \rangle = \frac{1}{2}\Big\{ \langle |\dot{\psi}_1(t)|^2 \rangle + \langle |\dot{\psi}_1(t+\tau)|^2 \rangle$$

$$+ 2\mathrm{Re}[\langle \dot{\psi}_1(t)\dot{\psi}_1^*(t+\tau)e^{i\omega_o\tau} \rangle] \Big\} \tag{2.40}$$

Since $\dot{\psi}(t)$ is stationary, we have

$$\langle |\dot{\psi}_1(t)|^2 \rangle = \langle |\dot{\psi}_1(t+\tau)|^2 \rangle = 2I_1 \tag{2.41}$$

and hence

$$I(\tau) = 2I_1 \Big\{ 1 + \mathrm{Re}[\Gamma_{12}(\tau)] \Big\} \tag{2.42}$$

where $\Gamma_{12}(\tau)$ is, in this case, the complex degree of coherence between $\dot{\psi}_1(t)$ and $\dot{\psi}_1(t + \tau)e^{-i\omega_o t}$.

In the limit $\tau \to 0$, for any optical source, we have

$$I(\tau) = 2I_1\,[1 + \cos(\omega_o\tau)] \tag{2.43}$$

For larger values of τ, the evolution of $I(\tau)$ depends on the temporal coherence of the light considered. For a pure monochromatic wave, whose temporal coherence is infinitely great, (2.43) holds for any value of τ, as depicted in Figure 2.1(a).

For the light emitted by a real optical source, we have to calculate the autocorrelation of the field $\psi_1(t)$, whose definition, owing to ergodicity, can be expressed as

$$\langle \dot{\psi}_1(t)\dot{\psi}_1^*(t+\tau) \rangle = \lim_{\Delta t\to\infty} \frac{1}{\Delta t} \int_{-\Delta t/2}^{\Delta t/2} \dot{\psi}_1(t)\dot{\psi}_1^*(t+\tau)\, dt \tag{2.44}$$

So we obtain a behavior of the type depicted in Figure 2.1(b). $I(\tau)$ is here still oscillating, but with amplitude of oscillation which decreases with increasing τ. The value of τ at which this amplitude turns out to be reduced by a factor 1/e is called coherence time t_c.

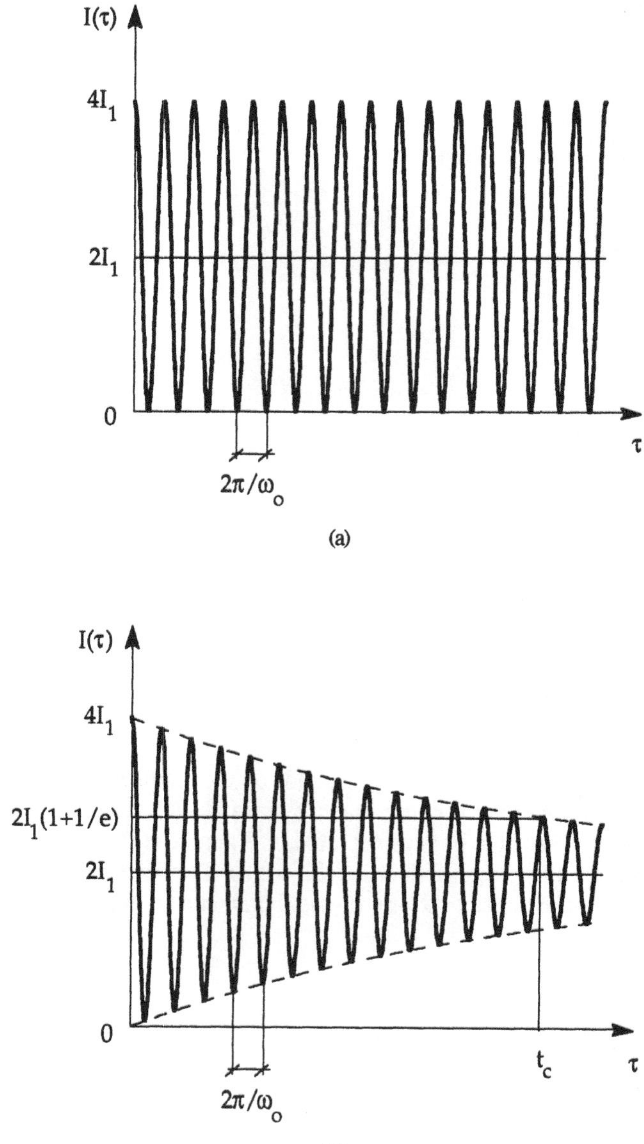

Figure 2.1 $I(\tau)$ for (a) an infinitely coherent source and for (b) a real source whose temporal coherence is t_c.

For a semiconductor laser, an analytical expression is available for $I(\tau)$. It is calculated by taking into account the process of spontaneous photon emission, which is always superimposed to stimulated photon emission [2]. This phenomenon is also called the *phase noise* of the laser source. According to this analysis we can write

$$\langle \dot{\psi}_1(t)\dot{\psi}_1{}^*(t+\tau) \rangle = 2I_1\, e^{-\alpha\,[(1+b^2)/4I_1]\,|\tau|} \tag{2.45}$$

where α represents the average rate of spontaneous photon emission and b the laser anti-guiding coefficient. From (2.45) we obtain

$$I(\tau) = 2I_1\left[1 + e^{-\alpha\,[(1+b^2)/4I_1]\,|\tau|}\cos(\omega_o\tau)\right] \tag{2.46}$$

According to the previous definition, the coherence time t_c turns out to be

$$t_c = \frac{4I_1}{\alpha(1+b^2)} \tag{2.47}$$

Expression (2.45) represents the autocorrelation of such field. The property of a light to be more or less monochromatic (that is, more or less coherent) can be easily evaluated through its power spectrum, which coincides with Fourier transform of the real part of the field autocorrelation, multiplied by $e^{i\omega_o\tau}$ [3]. So we obtain the power spectrum

$$p(\omega) = F\{\mathrm{Re}[\langle \dot{\psi}_1(t)\dot{\psi}_1{}^*(t+\tau)e^{i\omega_o\tau} \rangle]\} \tag{2.48}$$

which assumes consequently, the following Lorentzian shape

$$p(\omega) = \frac{2I_1t_c}{1 + (\omega - \omega_o)^2\, t_c{}^2} \tag{2.49}$$

The r.m.s. width of this power spectrum, measured in Hz, thus becomes

$$\Delta\nu = \frac{1}{\pi t_c} \tag{2.50}$$

As already stressed, the wider the power spectrum the higher the difficulty in the possibility of definition of a precise state of polarization for the correspondent field, except when the polarization is uniformly linear. With reference to a field decomposition according to a Cartesian set of transverse coordinates x, y,

$$\bar{E}(t) = E_x(t)\hat{x} + E_y(t)\hat{y} = E_{xo}\cos(\omega_o t+\phi_{ox})\hat{x} + E_{yo}\cos(\omega_o t+\phi_{oy})\hat{y} \tag{2.51}$$

in which E_{xo}, E_{yo}, ϕ_{ox}, ϕ_{oy} depend on t slowly, we can introduce a parameter Π, called degree of polarization, which represents the ratio between polarized light intensity and the overall light intensity. This parameter can be written as [4]

$$\Pi = \sqrt{1 - 4\,\langle E_{xo}{}^2 \rangle \langle E_{yo}{}^2 \rangle (1 - |\Gamma_{xy}|^2)} \tag{2.52}$$

where Γ_{xy} is the complex degree of coherence between the fields $E_x(t)$ and $E_y(t)$ while E_{xo} and E_{y0} have been normalized in such a way as to have $\langle E_{xo}{}^2\rangle + \langle E_{yo}{}^2\rangle = 1$.

2.1.4 Spatial Coherence and Diffraction

Let us consider a point optical source S, a screen with two pinholes O_1, O_2 separated by a distance d and a detector D, as shown in Figure 2.2. The field emitted by S is assumed to have a coherence time much longer than any time delay produced by the path difference between the trajectories SO_1, and SO_2, as well as between the trajectories O_1D and O_2D. This enables us to neglect any temporal coherence effect, in order to appreciate only spatial coherence effects.

After the screen, we have two equivalent point sources, according to Huygens' principle. Their lights, properly superimposed, give rise to an interference pattern. If the distance between O_1 or O_2 and S is much higher than the wavelength λ and the distance d, it is possible to perform a scalar summation of the two fields, in the form

$$\psi(t) = \psi_1(t) + \psi_2(t - \Delta\tau) \tag{2.53}$$

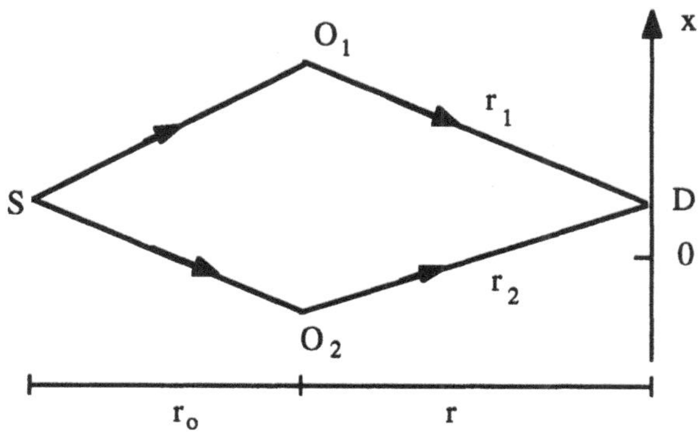

Figure 2.2 Experiment to introduce spatial coherence (Young interferometer).

In free space, characterized by the velocity of light c, we can write

$$\Delta\tau = \frac{r_2 - r_1}{c} \tag{2.54}$$

where r_1 and r_2 are the lengths of the segments O_1D and O_2D respectively. Assuming a linear displacement x, parallel to the line O_1O_2, with respect to a reference central point O, we can write

$$r_2 - r_1 \cong \frac{d}{r} x \tag{2.55}$$

where r » d is the distance between the screen and the plane over which the detector can move. Before reaching O_1 and O_2 the light emitted by S in general experiences unequal time delays. Denoting $\Delta\tau_0$ the difference between them, and ignoring possible amplitude differences between the two waves, we have

$$\psi_1(t) = \psi_2(t - \Delta\tau_0) \tag{2.56}$$

The interference pattern, which can be scanned by the movement of D, is imposed by the sum of the delay differences $\Delta\tau_0 + \Delta\tau$. The effect of $\Delta\tau$ is that of a classical interference between two point sources, as will be described in the next subsection with more details. $\Delta\tau_0$, instead, takes into account the source position with respect to O_1 and

O_2. By analogy with (2.55) $\Delta\tau_0$ will be proportional to d/r_0, where r_0 represents the distance between S and the screen.

If we have an extended source, superimposition principle enables us to repeat this calculation for a spatial distribution of points S, and then integrate the result obtained. This reduces the visibility of interference fringes in D, in a way similar to that originated by a finite temporal coherence. Since $\Delta\tau_0$ is proportional to d/r_0, such visibility reduction in the fringes is less and less significant with reduced distance d.

In order to characterize the light emitted by a certain optical source, from this point of view, it is possible to introduce the property of spatial coherence. Varying d, with O_1 fixed and O_2 variable, one obtains a circle in which the fringes contrast η measured by the detector is reduced by a factor higher than 1/e with respect to its maximum value. This circle, whose radius is d_c, represents the coherence area of the optical source considered on our screen. The latter depends not only on the spatial extension of the optical source in comparison with the distance r_0, but also on the phase relationship between the fields emitted from different points on the source area. A true point source has a coherence area of infinite extension. For example a star seen from the Earth exhibits a coherence radius d_c that is much larger than the size of the Earth, so that its light appears spatially coherent in any point of the world.

Another phenomenon, related to the spatial extension which characterizes the emitting area of an optical source or of a pupil in a screen, in comparison with the wavelength λ, is that called diffraction.

Assuming an extended source whose generic point is Q, and an observation point P, we can vary the relative position of such points taking two parallel planes, distant z between them, as shown in Figure 2.3. A Cartesian pair of coordinates x',y' describes the position of Q, whereas another Cartesian pair x,y describes the position of P.

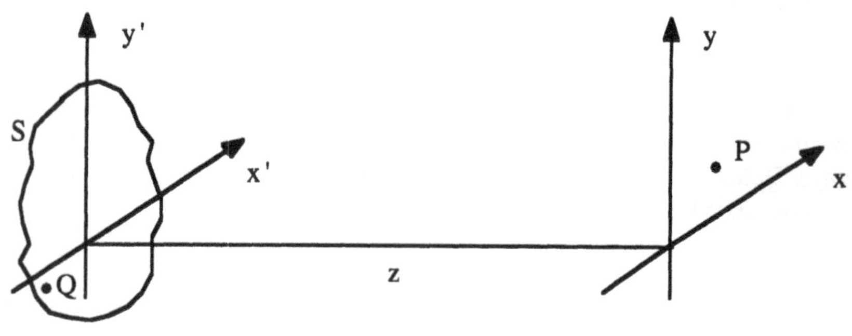

Figure 2.3 Coordinate systems for the description of diffraction properties.

When $z \gg x, y, x', y'$ and also $z \gg \lambda$, the field in P can be calculated by means of a suitable double Fourier transform applied to an equivalent field $\Psi(Q)$. Precisely we have [5]

$$\Psi(P) = \frac{i}{\lambda} e^{-ik_0 z} e^{-ik_0 (x^2 + y^2)/2z} \int_S \int \Psi(Q) \, e^{ik_0 (xx' + yy')/z} \, dx'dy' \qquad (2.57)$$

where S is the surface over which Q can vary. Equation (2.57) is obtained under the first-order approximation

$$\sqrt{z^2 + (x - x')^2 + (y - y')^2} \cong z \left[1 + \frac{1}{2} \left(\frac{x-x'}{z} \right)^2 + \frac{1}{2} \left(\frac{y-y'}{z} \right)^2 \right] \qquad (2.58)$$

In function $\Psi(Q)$ a phase factor $e^{-ik_0 (x'^2 + y'^2)/2z}$ is present. Approximation (2.58), and hence expansion (2.57) are valid in a region called first Fresnel zone. There are other Fresnel zones, whose extension is wider and wider, according to higher order expansions of the left-hand side of (2.58).

There is also a so called Fraunhofer zone, where (2.57) holds with the additional approximation

$$z \gg \frac{k_0}{2} (x'^2 + y'^2) \qquad (2.59)$$

In this situation, the phase factor $e^{-ik_0 (x'^2 + y'^2)/2z}$ has negligible effects on function $\Psi(Q)$.

2.1.5 Interference

When two fields having the same central circular frequency ω_0 are properly superimposed, as discussed in 2.1.3, we obtain an interference pattern characterized by fringes which consist of alternatively bright and dark regions. They can be collected on a screen or scanned by a suitable moving detector. The contrast η, introduced according to (2.37) and (2.38) represents a measure of the visibility of such fringes. As already discussed, η is bounded between 0 and 1, and increases with increasing the mutual coherence between the two fields.

Let us consider now a unique field, properly subdivided, and then recombined. In this case we obtain an interference pattern whose fringe contrast increases as the temporal coherence of such field increases. Here we have to also consider, for more generality, the case of waves having two different intensities I_1 and I_2 related to the intensity I_0 of the original field according to

$$I_1 = a^2 I_0 \quad , \quad I_2 = (1 - a^2) I_0 \tag{2.60}$$

where a is an amplitude ratio.

Extension of the (2.42) for the resulting intensity is

$$I(\tau) = I_0 + 2\, I_0\, a \sqrt{1 - a^2}\, \mathrm{Re}\, [\Gamma_{12}(\tau)] \tag{2.61}$$

In this situation η becomes

$$\eta = \frac{a \sqrt{1 - a^2}\, \{\mathrm{Re}\, [\Gamma_{12}(\tau)]_{max} - \mathrm{Re}\, [\Gamma_{12}(\tau)]_{min}\}}{1 + a \sqrt{1 - a^2}\, \{\mathrm{Re}\, [\Gamma_{12}(\tau)]_{max} + \mathrm{Re}\, [\Gamma_{12}(\tau)]_{min}\}} \tag{2.62}$$

which depends on a, besides on $\Gamma_{12}(\tau)$. It is easy to demonstrate that the best condition to increase η is $a^2 = 1/2$, which leads to

$$\eta = \frac{\frac{1}{2}\, \{\mathrm{Re}\, [\Gamma_{12}(\tau)]_{max} - \mathrm{Re}\, [\Gamma_{12}(\tau)]_{min}\}}{1 + \frac{1}{2}\, \{\mathrm{Re}\, [\Gamma_{12}(\tau)]_{max} + \mathrm{Re}\, [\Gamma_{12}(\tau)]_{min}\}} \tag{2.63}$$

An instrument which exploits interference effects to perform a measurement of various parameters is called an *interferometer*. These parameters can be related to optical, mechanical and chemical properties. Among the optical properties, we can consider the degree of temporal coherence of the optical source itself. In such condition, our objective is to transform $I(\tau)$ into another function $I(q)$, where q is a parameter which can be scanned by properly arranged equipment.

Keeping in mind that, in the introduction of such concepts, τ has been assumed as a relative delay between such two parts of the same field, with reference to a free-space propagation, we can write

$$\tau = \frac{\Delta z}{c} \qquad (2.64)$$

where c is the free-space velocity of light, and Δz is a path difference between the two fields. Varying Δz, one can obtain a picture like those reported in Figure 2.1 and then experimentally determine the coherence time t_c.

The simplest apparatus for performing this measurement is that shown in Figure 2.4. It is called a *Michelson interferometer*, and exploits a semitransparent mirror and two mirrors. Letting l_1 and l_2 the lengths of the two arms of such interferometer, it is straightforward to obtain $\Delta z = 2(l_2 - l_1)$. We can, for instance, move one mirror, maintaining the other one fixed, thereby varying Δz properly.

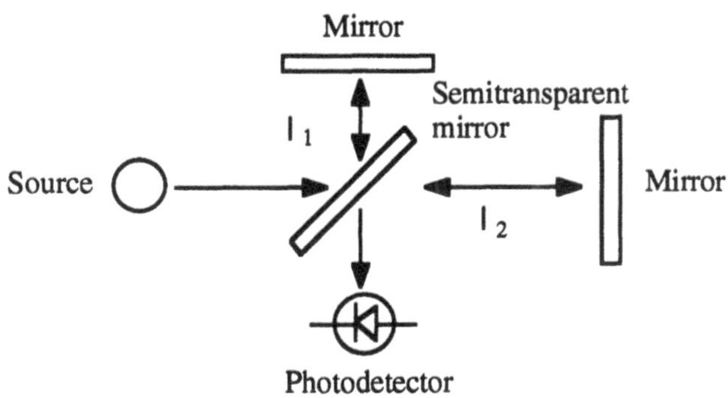

Figure 2.4 Michelson interferometer.

Even extremely incoherent optical sources have t_c orders of magnitude greater than $2\pi/\omega_0$. Therefore, since $2\pi/\omega_0$ corresponds to a variation in Δz equal to the free-space wavelength λ, we can consider that a path difference Δz covering many wavelengths λ is usually necessary for appreciating t_c. Furthermore, independently of t_c, we can assume that $\lambda/2$ represents the path difference separation between a maximum and a minimum in $I(\tau)$; that is, between a bright and a dark fringe adjacent to each other.

As another example, we can consider the situation shown in Figure 2.5 where we have an optical source and a screen set at very large distance from a plane parallel plate of refractive index n_2, and thickness d, whereas the upper medium has refractive index n_1. The incidence angle of the incoming plane wave ϑ_1 and the refraction angle ϑ_2 are related between them by means of Snell's law [1]

$$n_1 \sin\vartheta_1 = n_2 \sin\vartheta_2 \tag{2.65}$$

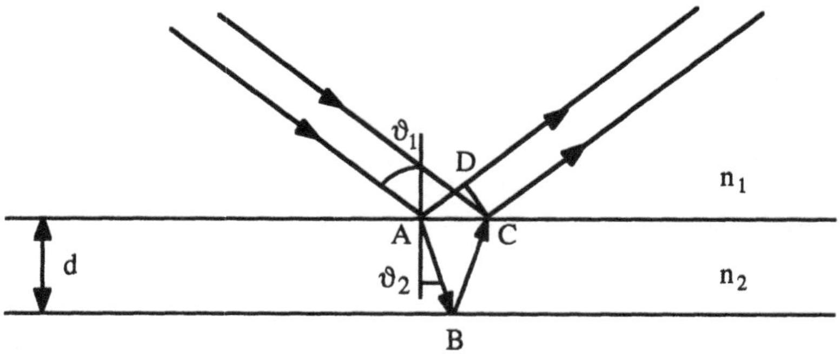

Figure 2.5 Selective reflection with source and observation planes at infinity.

The difference in the optical paths between such two waves can be written as

$$\Delta p = n_2 (AB + BC) - n_1 AD = n_2 \frac{2d}{\cos\vartheta_2} - n_1 2d \frac{\sin\vartheta_2}{\cos\vartheta_2} \sin\vartheta_1 \tag{2.66}$$

which from (2.65), after simple passages, becomes

$$\Delta p = n_2 2d\cos\vartheta_2 \tag{2.67}$$

So we have

$$\tau = \frac{1}{c} \Delta p = \frac{1}{c} n_2 2d\cos\vartheta_2 \tag{2.68}$$

It is also possible to consider multiple reflections and, taking the limit of an infinite series of contributions, we easily obtain [6], for an optical source whose coherence time t_c tends to infinity, i.e., an ideal monochromatic field,

$$I(\Theta) = \frac{T^2}{1 + R^2 - 2R\cos\Theta} I_0 \tag{2.69}$$

where I_0 is the overall intensity, and variable Θ is defined as follows

$$\Theta = \frac{4\pi}{\lambda} n_2 d\cos\vartheta_2 + \arg[r] \qquad R = |r|^2 \tag{2.70}$$

in which R and T are the power reflection and transmission coefficients. Finally, r is the field reflection coefficient. Expression (2.69) can also be rewritten as

$$I(\Theta) = a(\Theta)I_0 \qquad a(\Theta) = \frac{T^2}{1 + R^2 - 2R\cos\Theta} \tag{2.71}$$

in which $a(\Theta)$ represents the spectral response of this optical system.

When no loss is present, we can write $T + R = 1$. In this situation the behavior of $a(\Theta)$ against Θ with parameter R is of the type depicted in Figure 2.6. This is a periodic function whose period is 2π, represented between $\Theta \cong 2m\pi$ and $\Theta = 2(m+1)\pi$, m being an integer.

As for the contrast η we can write

$$I_{max} = \frac{T^2}{(1 - R)^2} I_0 \qquad I_{min} = \frac{T^2}{(1 + R)^2} I_0 \tag{2.72}$$

so that

$$\eta = \frac{2R}{1 + R^2} \tag{2.73}$$

which is independent of T, and confirms how the higher the R the sharper the peaks in $a(\Theta)$.

Since the optical frequency $\nu = c/\lambda$ exhibits a linear relationship with respect to Θ, for a new optical frequency ν', corresponding to Θ', we have

$$\nu - \nu' = \frac{c}{4\pi n_2 d \cos\vartheta_2}(\Theta' - \Theta) \tag{2.74}$$

in which r has been taken the same for ν and ν'. Because of superimposition principle, which certainly holds because the medium is linear, we can consider the curves of Figure 2.6 as a sort of impulse response for the light intensity in the ν domain. In fact, the response previously calculated corresponds to an ideal intensity distribution having the form of a Dirac delta function, in the optical frequency ν. In this sense, the response to a real intensity distribution $I(\nu)$ will be the convolution of $I(\nu)$ itself with $a(\nu)$.

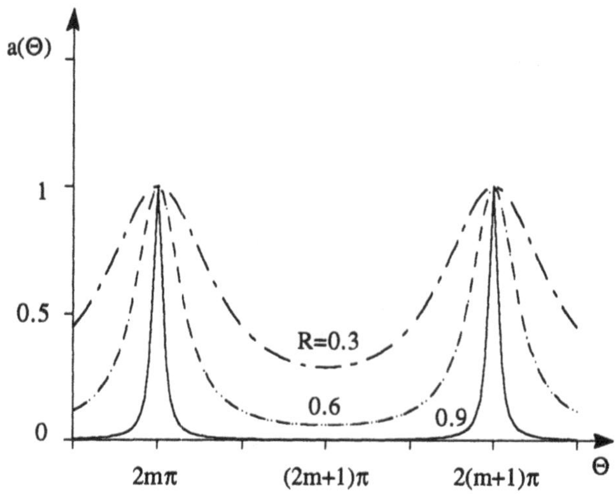

Figure 2.6 Typical behavior of $a(\Theta)$.

We can assume, for the sake of simplicity, $\vartheta_2 \cong 0$. So, the distance between two adjacent maxima of $a(\nu)$ is $c/(2n_2d)$. Such a frequency difference is called the free spectral range of this device. Instead, the distance $\Delta\nu_0$ between two points on the two sides of the same maximum, at which $a(\nu)$ is 0.5, is called the spectral resolution.

The ratio between the free spectral range and $\Delta\nu_0$ is finally defined as the finesse F, i.e.,

$$F = \frac{c}{2n_2 d \Delta\nu_0} \quad (2.75)$$

It turns out to be a function only of the reflectivity R, which precisely results in [6]

$$F = \pi \frac{\sqrt{R}}{1 - R} \quad (2.76)$$

The higher the F, the better the performance of such an instrument in the experimental evaluation of optical source spectral widths. The plane parallel plate so employed is called a Fabry-Perot etalon.

2.2 INTERFEROMETERS

In single-mode fiber and devices characterization, as well as in optical fiber sensors, interferometers are widely used. An interferometer is an apparatus in which two or more interfering fields propagate along different paths before being recombined. Such fields are usually obtained from a unique optical source by means of a suitable amplitude division. Recently, single-mode fibers have been used to increase the length of the optical paths in an interferometer. Polarization-maintaining or single-polarization fibers are to be employed in order to assure field superimposition with proper polarization alignment.

Interferometers are used in various fields of metrology. Among them: surface characterization, spectrometry, measurements of refractive indices, measurements of rotations.

2.2.1 Newton's Rings

In order to understand the mechanism of formation of fringes in an interferometer, let us consider the classical experiment shown in Figure 2.7, which produces a system of circular concentric fringes called *Newton's rings*. They are observed when a point source S emits a light which is transformed into an approximately plane wave. R represents the curvature radius of the convex lower surface of a lens which performs such transformation in contact with the upper plane surface S_1 of a glass cylinder. C is the point of contact between these two surfaces. Fringes will appear on the bottom plane surface S_2 of the cylinder as a consequence of the variable thickness of the air gap between the two glasses. Letting r be the distance between C and a generic point on the plane surface S_1, the air gap over such point has a thickness of

$$h = R - \sqrt{R^2 - r^2} \cong \frac{r^2}{2R} \tag{2.77}$$

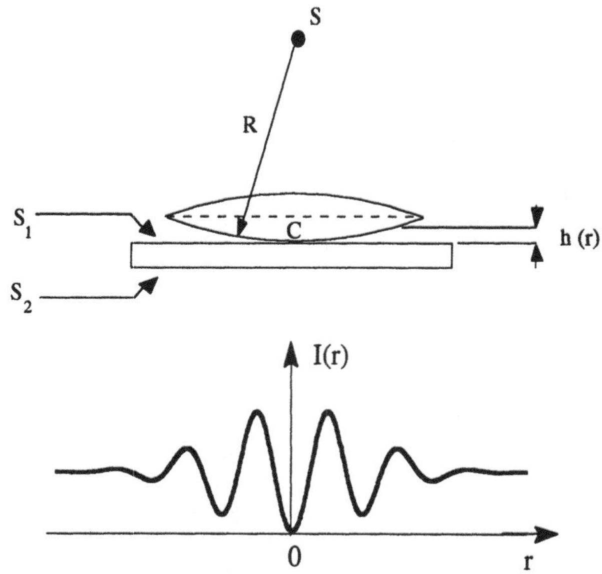

Figure 2.7 Experimental device for obtaining Newton's rings.

The air gap introduces a resonance, which, considering only one reflection, gives rise to dark fringes in correspondence with h integer multiples of $\lambda/2$. This leads to

$$r = \sqrt{mR\lambda} \tag{2.78}$$

So quasi-monocromatic fields can be spectrally analyzed, although only as a first approximation, by measuring the radius which characterizes this system of dark fringes.

White light exhibits a dark central spot and subsequent first-order annular maxima which appear colored in a characteristic sequence. This sequence of maxima was the object of an early experiment attributed to Newton.

Probably the simplest interferometer that can be conceived, in which the principle of Newton's ring formation is exploited, is the so-called Fizeau interferometer. It is based on a film of variable thickness. The light from a collimated beam is reflected by the two surfaces of the film and forms fringes which can be collected on a screen set in front of the film. Such fringes have a form which depends on the film thickness evolution. The Fizeau interferometer is then employed for surface characterization.

2.2.2 Michelson Interferometer

The classical scheme of a Michelson interferometer has been already described (Figure 2.4, subsection 2.1.5). In order to understand how fringes can be formed, let us consider Figure 2.8.

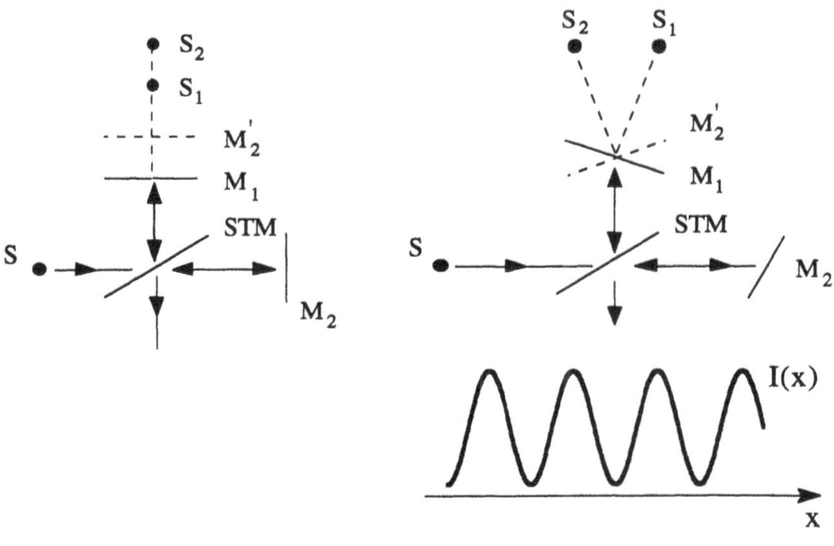

Figure 2.8 Michelson interferometer with perfectly parallel mirrors and with slightly tilted mirrors.

On the left, we observe that an image M_2' of mirror M_2 is formed behind mirror M_1, at a distance $(l_2 - l_1)$. This entails the presence of two images, S_1 and S_2, of the source S. S_1 and S_2 are aligned and distant $(l_2 - l_1)$ one another. On the right, a rather different situation is depicted. Here the two mirrors are slightly tilted in such a way as to form two images, S_1 and S_2, whose positions are laterally displaced from one another. In this situation a series of straight fringes is formed. In the small angle approximation, letting l_0 the distance between S and the semitransparent mirror, the two images can be considered distant $l_0 + l_1 = l_0 + l_2$ from a reference plane, placed across M_1 and M_2'.

The lateral shift between them, instead, results in

$$d \cong 2(l_0 + l_1)\delta \tag{2.79}$$

where δ is the tilt angle, assumed equal for the two mirrors.

Assuming an axis x on the reference plane, parallel to the displacement between S_1 and S_2, as shown in Figure 2.9, the path difference between the two fields reaching a point Q, identified by the generic coordinate x, can be calculated as

$$\Delta p = \sqrt{(l_0+l_1)^2 + \left(\frac{d}{2} + x\right)^2} - \sqrt{(l_0+l_1)^2 + \left(\frac{d}{2} - x\right)^2} \cong \frac{dx}{l_0 + l_1} \tag{2.80}$$

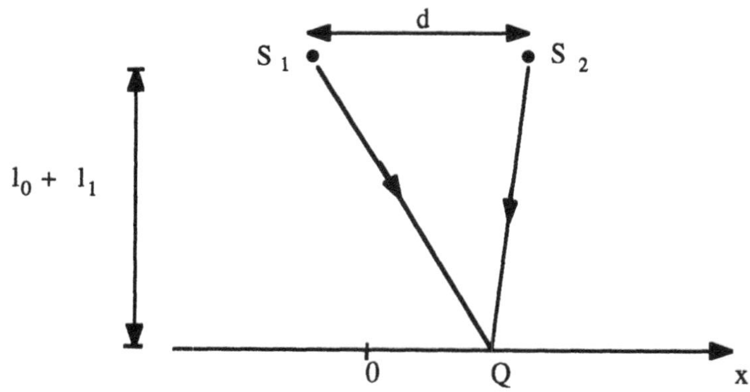

Figure 2.9 Equivalent situation in the small angle approximation.

The spatial period of the fringes so obtained, which are straight and oriented perpendicularly to the plane containing S_1, S_2 and axis x, can be determined by equating Δp to λ. This leads to

$$x_P = \lambda \frac{l_0 + l_1}{d} \cong \frac{\lambda}{2} \delta \quad (2.81)$$

in which (2.79) has been also used. Since δ is a rather small angle, x_P turns out to be very small.

2.2.3 Mach-Zehnder Interferometer

The scheme of a Mach-Zehnder interferometer is shown in Figure 2.10. It employes two semitransparent mirrors, STM_1 and STM_2, and two mirrors, M_1 and M_2, disposed at the vertices of a rectangle. If M_1 and M_2, instead of exactly $\pi/4$, are slightly tilted with respect to this position, we can consider to have two wavefronts, tilted from each other by the same angle. These two wavefronts produce an interference pattern similar to that described in Figure 2.10, which can be collected on a screen, either in O_1 or in O_2. The fringes will be straight and parallel to the line of intersection of the two wavefronts.

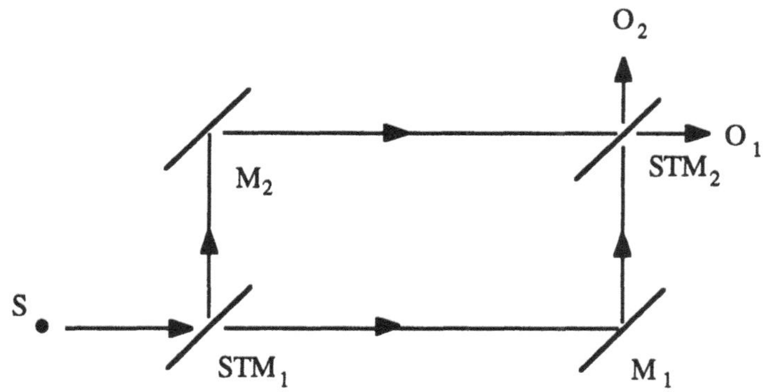

Figure 2.10 Scheme of a Mach-Zehnder interferometer.

The Mach-Zehnder interferometer is employed for the measurement of refractive indices in gases, liquids, or solids. Such materials are introduced along one of the two arms, inducing a variation in the optical path difference, from which one can determine the refractive index with high accuracy. This interferometer is more practical than the Michelson interferometer because the two paths are well separated, and traversed by the optical beam only one time. Nevertheless, as will be shown in detail in Chapter 5, the use of a Michelson interferometer may be preferable in some applications, such as remote measurements.

2.2.4 Spectrometers

The problem of spectrometry consists of evaluating experimentally the spectrum of an optical source. This can be done by means of gratings or prisms up to a spectral resolution $\lambda_0/\Delta\lambda$ on the order of 10^6. Assuming a central emission wavelength λ_0 on the order of 1 µm, $\Delta\lambda$ turns out to be not smaller than 10^{-3} nm, which corresponds to 300 MHz. In order to exceed this limit, it is necessary to use interferometric techniques, such as those based on a Fabry-Perot etalon, or those called wavelength meters.

A Fabry-Perot etalon, as discussed in 2.1.5, is by itself an interferometer. The plane parallel plate can be employed either with light reflection or with light transmission, the one operation being complementary to the other. With reference to one of the maxima in the impulse response $a(\nu)$ defined by (2.71), if the spectral resolution $\Delta\nu_0$ introduced in 2.1.5 is much smaller than the spectral linewidth $\Delta\nu$ of the optical source, its power spectrum can be accurately evaluated.

Let us consider now the situation shown in Figure 2.11 where a point source S emits a light that is collimated by a suitable lens in such a way as to allow its collection on a Fabry-Perot etalon perpendicularly to its surface. This arrangement, which is characterized by circular symmetry, produces a system of circular fringes.

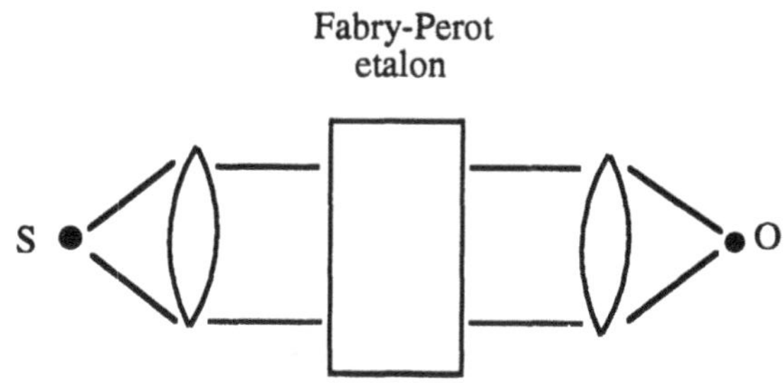

Figure 2.11 Formation of annular fringes in a Fabry-Perot interferometer.

Any spectral component of the light emitted by the optical source has a proper system of fringes. Therefore a monochromator must analyze the output in O, whose entrance slit takes only a small part of any annular fringe. In this way the spectral resolution is imposed by the value of $\Delta\nu_0$, which characterizes the Fabry-Perot etalon, and not by the monochromator spectral resolution.

This technique has the drawback that only a small part of the optical power available from the Fabry-Perot etalon is employed for the spectral analysis. There is a parameter, called the *etendue* of a spectrometer, which consists of the product between the area of the emitting surface and the solid angle in which the spectrometer can accept light to be analyzed. This parameter is a figure of merit for spectrometer performance, especially when $\Delta\nu_0$ is very fine and the spectral width $\Delta\nu$ of the optical source is large.

A Fabry-Perot spectrometer, because of its circular symmetry, has the advantage of a very high intrinsic etendue, which is however strongly reduced owing to the use of the monochromator at the output of the etalon. It is possible to consider another arrangement, called a scanning Fabry-Perot interferometer, in which the gap between the two plates of the etalon is filled by a gas with variable pressure (thereby introducing a change in refractive index n_2) or has a variable width d. This allows one to select different spectral components without the necessity of a monochromator and hence to exploit the entire etendue of the Fabry-Perot etalon.

The absolute evaluation of the central emission wavelength λ_0 of a given optical source can be obtained by an instrument called wavelength-meter. It is available in two versions:

- the dynamic wavelength-meter,
- the static wavelength-meter.

The scheme of a dynamic wavelength meter is shown in Figure 2.12. It is based on a movable cube-corner, two semitransparent mirrors, STM_1 and STM_2, two detectors, D_1 and D_2, and a reference laser whose emission spectrum is very narrow around λ_1. When the movable cube-corner exhibits a change in its position by a distance d, the number of maxima in the light intensity seen by detector D_1 is

$$N_1 = 2\frac{d}{\lambda_0} \tag{2.82}$$

whereas the one seen by detector D_2 is

$$N_2 = 2\frac{d}{\lambda_1} \tag{2.83}$$

so that

$$\lambda_0 = \frac{N_2}{N_1}\lambda_1 \tag{2.84}$$

This formula is truly valid only when this instrument operates in vacuo. When there is air along the optical paths a correction is necessary for taking into account the wavelength dependence of the air refractive index [6].

Several types of static wavelength meters have been proposed. They can employ one of the following ineterferometers:

1. a Fizeau interferometer,
2. a Michelson interferometer,
3. a Fabry-Perot interferometer.

In all these cases, a reference laser and a photodiode array are requested. In such a solution, instead of a movable reflection and a point detector for the measurement of a series of maxima related to reflection motion, the direct evaluation of a spatial fringes by means of a photodiode array is exploited.

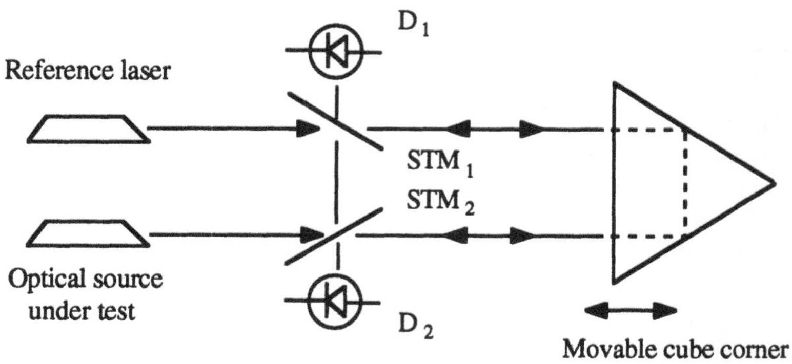

Figure 2.12 Scheme of a dynamic wavelength meter.

In both dynamic and static wavelength meters, the precision in the evaluation of λ_0 is imposed by the spectral linewidth of the reference laser. A static wavelength meter is also able to operate with pulsed test sources.

2.2.5 Sagnac Interferometer

A Sagnac interferometer is based on two optical beams which travel in opposite direction along a closed path. It can be realized employing mirrors or optical fibers. This interferometer is particularly suited for the experimental evaluation of rotation rate.

Michelson and Gale, by the use of an interferometer of this type with a perimeter of about one mile, measured the Earth's rotation rate [7]. The main problem of a traditional Sagnac interferometer is to keep the optical beam well collimated over very long distances. On the other hand, this problem does not exist using an optical fiber. Furthermore an optical fiber can be arranged to form a N-turn coil, as shown in Figure 2.13, which gives rise to an improved accuracy.

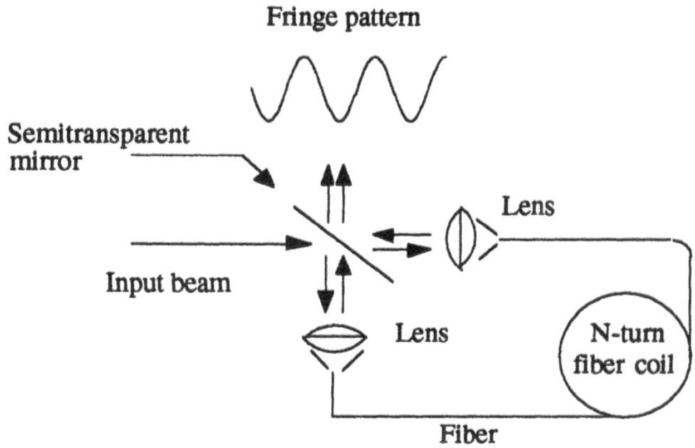

Figure 2.13 Scheme of a Sagnac interferometer realized with a single-mode optical fiber.

The time delay between the two counterpropagating waves responsible for fringe formation can be calculated as

$$\Delta\tau = \frac{n}{c}(2\pi + \Omega\, t_0)\, NR - \frac{n}{c}(2\pi - \Omega\, t_0)\, NR = 2\frac{n}{c}\Omega\, t_0\, NR \qquad (2.85)$$

where R is the coil radius, Ω the rotation rate to be measured, t_0 the propagation time necessary to cover a turn of the coil, and n is the effective refractive index of the propagation. In order to have a well-defined value of n, a single-mode fiber is to be employed.

From its definition, t_0 can be written as

$$t_0 = \frac{n}{c} 2\pi R \qquad (2.86)$$

which, substituted into (2.85), leads to

$$\Delta\tau = 4\pi\left(\frac{n}{c}\right)^2 NR^2\,\Omega \tag{2.87}$$

So the larger the product NR^2, the higher the precision in the evaluation of small rotation rates Ω.

2.3 LIGHT MODULATION AND DETECTION

The light of an optical source can be modulated either coherently or incoherently. In the former case the required extension of its spectral width is to be remarkably smaller than the modulation signal bandwidth. Light detection is performed depending on the technique employed for its modulation. Typically a coherent modulation is accompanied by a coherent detection, and an incoherent modulation, or simply intensity-modulation, by an incoherent detection, usually called *direct detection.*

Both types of light modulation can be performed either externally or internally with respect to the optical source. In semiconductor devices, which will be treated with particular attention here, an internal modulation is preferred because of the intrinsic compactness which can be achieved by the resulting component. On the contrary, for lamps, or lasers not based on semiconductor materials (e.g., gas or solid-state lasers) which are frequently used for optical characterization procedures, external modulation is necessary.

Direct detection is performed by a device whose response is proportional to the light intensity directly collected on its active area. Coherent detection is accomplished, instead, by a heterodyne or a homodyne arrangement, in which an optical wave mixing is required. Such a scheme induces a translation of the light spectrum to microwave frequencies (heterodyne scheme) or directly to baseband frequencies (homodyne scheme). In the latter case, possible only with a linear modulation format, the detection process includes signal demodulation. In both types of light detection various kinds of devices can be employed: thermal detectors, photoelectric detectors, photoconductive detectors, and photodiodes. Our attention will be focused on the latter type of photodetectors because of their better properties in sensitivity, speed of response, and reliability.

2.3.1 Intensity and Coherent Modulation

The most common method for transmitting information by means of an optical carrier is to associate the information with the intensity of the carrier. This technique is called baseband or intensity modulation and is widely used in optical fiber systems because it is both inexpensive and easy to implement.

Light intensity modulation is usually performed by directly varying the drive current of the optical source, by means of a waveform proportional to the signal to be transmitted. In this way the transmitter circuit turns out to be very compact, and, in the near future, we can expect to reach integration of electronic and optoelectronic devices on the same semiconductor chip. No particular degree of light coherence is required in this kind of systems, so that basically no attention is paid to the phase of the carrier. This implies that the original spectral spread of the optical carrier is usually much wider than the spread due to modulation. At the same time, the light beam is not particularly well collimated; this is not a serious problem for telecommunication systems using semiconductor sources, as the reduced size of the devices allows a direct injection of the optical power into the fiber link. Nevertheless, some applications in classical optical metrology may require a very collimated light beam exiting the source. Only in this case, in fact, it is possible to retain a satisfactory degree of collimation when the light beam is used in the measurement set, combined with other beams, and treated by mirrors and lenses. Also greater powers could be necessary for such purposes and, to this goal, the use of gas or solid-state lasers, for which external modulation is preferable, could be advisable.

Whereas a wide range of modulators fabricated from different materials (such as semiconductor, polymer, ferroelectric, etc.) are being actively researched, most high-speed modulators are currently based on $LiNbO_3$ waveguide technology. The operation of such modulators is based on the so called electro-optic, magneto-optic, or acousto-optic effects. These effects consist in the variation of the effective refractive index of the crystal, for example $LiNbO_3$, which constitutes the modulator, attained by means of the application of an electric field, a magnetic field, or a mechanical strain, respectively. By effective refractive index we mean the refractive index of an equivalent homogeneous unlimited medium in which the electromagnetic wave propagates with the same phase constant of the actual structure. In particular, the most extensively developed intensity modulator is based on a Mach-Zehnder-type structure. In this device the phase of two combining optical signals are controlled in such a way to have constructive or destructive interference and, therefore, an intensity modulated signal. As an example, bandwidth as large as 4 GHz can be obtained [8] for devices designed for operation at 1.5 μm with an active length of 20 mm and a drive power of ~ 5 V for a full modulation. The extinction ratio is typically greater than 20 dB while the insertion loss introduced in the light beam varies between 4 and 7 dB.

When a coherent modulation scheme is considered, the light degrees of freedom on which modulation can act are its state of polarization, phase, or frequency; an amplitude modulated signal, instead, can be indirectly obtained from the previous ones or can be derived by means of simple passive devices. Fixing our attention on digital signals, this leads to classical modulation formats identified as *polarization shift keying* (POLSK), *phase shift keying* (PSK), *differential phase shift keying* (DPSK), *frequency shift keying* (FSK), *amplitude shift keying* (ASK) or *on-off keying* (OOK), respectively.

As an example, we can consider the case of an electro-optic modulator conceived with the goal of realizing a PSK scheme. The refractive index variation Δn, induced by the electro-optic effect, gives rise to a phase modulation $\Delta\phi$ of the optical carrier, according to the relationship

$$\Delta\phi = \frac{2\pi L \Delta n}{\lambda_o} \tag{2.88}$$

where L is the device length along which the electro-optic effect takes place and λ_o is the wavelength in free space. Δn, in turn, depends on the light polarization state, on the intensity of the modulation field and, obviously, on the electro-optic coefficients, which are typical of the considered crystal. A device designed for operating at 1.5 μm with a modulation bandwidth of up to 4 GHz has typically an active length of 20 mm [8], an insertion loss variable between 3 and 5 dB, and requires ~ 6 V to produce the π phase change necessary for a bipolar scheme with antipodal signals.

The electro-optic effect can be effectively adopted also for obtaining polarization modulated signal. More precisely, when the modulation signal is absent, a linearly polarized plane wave entering the modulator travels along it, maintaining its polarization characteristics unchanged. On the contrary, when an external modulation signal is applied, one obtains, at the modulator output, a wave which is, in general, elliptically polarized, with an angle between the major axis of the ellipse and one of the coordinate axes, assumed as a reference, depending on the phase difference between the components along x and y of the electric output field vector. In magneto-optic modulators, instead, the wave coming out from the device is still linearly polarized but, because of the applied magnetic field, the polarization orientation is changed.

Phase modulation, frequency modulation, polarization modulation, and amplitude modulation can be coherently demodulated, provided that the frequency and phase of their waveforms are sufficiently stable. As will be shown in Section 2.3.4, the coherent demodulation arrangement acts on the received electromagnetic field, which is characterized by a temporal mean value equal to zero. The modulated signal, in fact, can assume either positive or negative values. On the contrary, when intensity modulation is used, the direct detection process acts on the received power which, obviously, can assume only positive values.

2.3.2 Photodiodes

An ideal photodiode should convert a photon into an electron-hole pair by means of a one-to-one process. In this way, the photogenerated current i_s should be expressed as a function of the optical power P incident on the photodiode active area as

$$i_s = \frac{q}{h\nu_o} P \tag{2.89}$$

where $h\nu_o$ is the energy of a photon, and q is the electron charge.

In practical photodiodes the process of current photogeneration is statistical. A relationship like (2.89) therefore holds only on average. Furthermore, there are photons that, although absorbed by the material, do not originate an electron-hole pair. Finally, the optical power P also will fluctuate, because of the statistical variations of the light intensity, due in part to classical phase fluctuations and in part to quantum-mechanical properties. In this sense, with reference to suitably averaged quantities $\langle i_s \rangle$ and $\langle P \rangle$, we can write

$$\langle i_s \rangle = R\langle P \rangle \qquad (2.90)$$

in which the proportionality factor *R*, called the *photodiode responsivity* and measured in A/W, is expressed as

$$R = \eta_o \frac{q}{h\nu_o} \qquad (2.91)$$

where η_o takes into account the nonideal behavior of the photodiode. It is called quantum efficiency, and is always smaller than 1 (typically on the order of 0.7).

η_o depends on the wavelength as a consequence of the bandgap energy of the device semiconductor, which is responsible of light absorption. This mechanism is exactly opposite to that of light generation in the neighborhood of a semiconductor *p-n* junction, but is usually less wavelength selective.

The wavelength of the radiation to which the photodiode is sensitive is determined by the bandgap energy of the material employed. A good spectral match between GaAlAs sources operating in the wavelength range of (0.8 μm – 0.9 μm) is assured by using photodiodes made of Si. Other materials, like Ge, GaAs, InGaAs and InGaAsP can also be adopted in this wavelength region, but Si is preferable because of its highly developed technology and noise characteristics, which permit high receiver sensitivity. In the longer wavelength region (1.3 μm – 1.6 μm), germanium photodiodes are used, instead, because the response of Si rapidly decreases when the wavelength increases beyond 1 μm. However, a variety of III-V semiconductor alloys such as InGaAsP, GaAlSb, InGaAs, GaSb and GaAsSb have also been investigated for long-wave applications. Some of them are characterized by very high quantum efficiency, fast response speeds and low additional noise.

The average values $\langle P \rangle$ and $\langle i_s \rangle$ in equation (2.90) are obtained because the device exhibits an intrinsic low-pass response, imposed by the presence of a junction capacity C_d. After such low-pass response, whose bandwidth may range from few Hz in metrological applications to some GHz in telecommunication system receiver applications, the fundamental optical oscillation at frequency ν_o and some random fluctuations due to the statistical nature of the light are averaged out. Nevertheless, if the optical power P(t) is intensity modulated, provided that the receiver circuit is properly arranged, its variations can be followed by $i_s(t)$, thus obtaining a direct detection of the modulation signal.

On the other hand, when the field of a certain light is coherently modulated, in the sense discussed in the previous section, in order to detect the modulation signal present in it, we have to arrange a superimposition of two fields, one of them usually locally generated, thus leading to a coherent detection for such signal. This procedure, which gives rise to heterodyne or homodyne detection schemes, will be analyzed more in detail in Section 2.3.4. Such detection arrangements, however, employ one or more than one photodiodes, which operate according to equation (2.90). The only difference is that here P(t) is the optical power associated to the superimposition of the two fields.

In a photodiode, the *p-n* junction is usually disposed on a plane parallel to the device active area on which the light is to be collected. The layer between the junction and the active area, generally p-doped, has a thickness that is not too large, in order to enhance photon absorption able to generate electron-hole pairs. In a simple *p-n* device, commonly called a *solar cell,* a p-layer that is too thin leads to a very low sensitivity of the whole device, and thus to a high reverse current which becomes a dark current when a reverse bias voltage is applied in order to obtain a linear voltage response, as will be discussed in the next subsection. A better optimization of design is possible in a *p-i-n* structure. Here the resistivity of the insulating layer, placed between the p- and the n-layer, contributes to increase the overall device resistivity, originating very low reverse current even under a rather high reverse voltage.

PIN photodiodes are commonly used either in metrological applications or as photodetectors in the receiver circuits of telecommunication systems. Typical values of the responsivity R for such devices are 0.5 A/W, when Si is employed for their fabrication, and 1 A/W, in the case of Ge or gallium-arsenide materials, suitable for use at $\lambda > 1$ μm. The junction capacitance C_d has values which increase when the area of the *p-n* junction plane, and hence the device active area, is increased. With very small active areas, like those on the order of 10^{-2} mm^2 that characterize typical single-mode fiber optics applications, C_d can reach values as small as 1 pF.

Avalanche photodiodes (APDs) are sometimes used to obtain an internal current amplification for the photogenerated signal. They are characterized by a region of strong electric field in which an avalanche multiplication of the photogenerated carriers takes place under the action of a high reverse-bias voltage.

In this situation, there is an additional random process whose effects are superimposed on those previously considered: the avalanche multiplication M experienced by any photogenerated electron-hole pair. In this sense, equation (2.90), for an APD photodetector, must be replaced by

$$\langle i_s \rangle = R \langle M \rangle \langle P \rangle \tag{2.92}$$

where $\langle M \rangle$ denotes the mean value of M. A statistical independence between such process and the random fluctuations of P can be assumed, leading to an overall mean as the product of the two means.

The mean square $\langle M^2 \rangle$ of such process will influence the noise superimposed on the signal. Such mean square is usually related to $\langle M \rangle$ through a quadratic relationship of the type

$$\langle M^2 \rangle = F(\langle M \rangle)\langle M \rangle^2 \tag{2.93}$$

in which factor $F(\langle M \rangle)$, that is defined as the noise figure, can be written as [10]

$$F(\langle M \rangle) = \langle M \rangle \left[1 - (1 - k) \left(\frac{\langle M \rangle - 1}{\langle M \rangle} \right)^2 \right] \tag{2.94}$$

where k is a constant expressing the ratio between the smallest and the largest ionization coefficients, for the two types of carriers. For Si photodiodes, k is of the order of 0.02; for other materials, it ranges from 0.1 to 1.

2.3.3 Direct Detection

The photogenerated signal obtained from (2.90) or from (2.92), respectively with a PIN or with an APD photodetector, is usually processed as an electric voltage. Therefore a load resistance R_L must be set in front of the photodiode. This resistance must be much smaller than the nonlinear internal resistance of the photodiode, in order to have a voltage v_s linearly related to $\langle i_s \rangle$. Considering also the possibility of the application of a reverse bias voltage V_b, we have a circuit of the type shown in Figure 2.14, on the right.

In the same figure a voltage-current diagram is depicted. The first quadrant, where v_s and i_o are both positive, is that of photovoltaic operation, where V_b is null and the photodiode acts as a power generator. In the quadrant where v_s is negative and i_o is positive, we have $i_o \approx \langle i_s \rangle$, and then a linear relationship between v_s and $\langle i_s \rangle$, imposed by a straight line, which can be calculated as

$$v_s = -V_b + R_L \langle i_s \rangle \tag{2.95}$$

In order to have a linear response, it is necessary that v_s is kept negative. This implies the existence of an upper bound for R_L, which however becomes higher as V_b is increased. The value of such reverse voltage fixes the maximum dynamical range of the photodetector so realized.

Note that the uniform spacing of the characteristics in such a quadrant is a consequence of the assumption that optoelectric transduction obeys an equation of the type (2.90) or (2.92). Such equations are no longer true, when operating in the first quadrant, for the current i_o flowing through the load resistance because a larger and larger fraction of $\langle i_s \rangle$ passes through the nonlinear internal resistance of the photodiode as R_L increases.

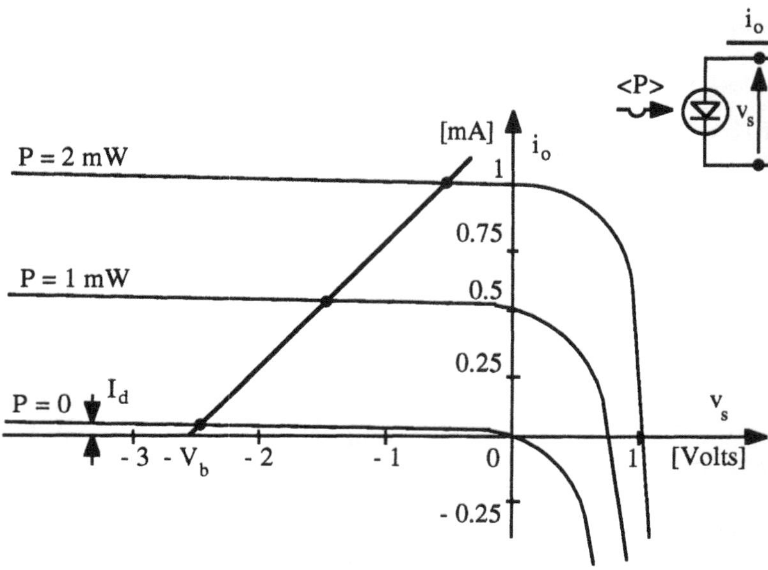

Figure 2.14 Voltage-current characteristics of direct detection.

In conclusion, assuming a sufficiently high value of V_b, the larger the R_L the stronger the voltage response to optical power variation. A limit in this sense can be due to the reduction in the width of the baseband response of the photodetector, which has a cut-off frequency inversely proportional to $R_L C_d$. By contrast, the sensitivity of the response becomes better and better with increasing R_L, as will be shown in the following. A trade-off, in this sense, can be found. A certain relaxation in the constraints to be imposed for chosing the optimum value of R_L is obtained by the use of a transimpedance preamplifier.

From Figure 2.14, the role played by the reverse photodiode voltage I_d, which has the meaning of a *dark current* since it is present even when we have no incident optical power

$\langle P \rangle$, becomes evident. Measuring the useful signal through the current i_o which, in linear regime, flows through the load resistance, we can then write

$$i_o = I_d + \langle i_s \rangle \tag{2.96}$$

The noise affecting such signal can be subdivided into two contributions: the one is a shot noise, due to the presence of the useful current itself, the other is a thermal noise. The power of such noise, expressed in A^2, can be written as

$$\langle i_n^2 \rangle = \left(2qI + 4\,\frac{KT}{R_L} \right) B_N \tag{2.97}$$

where K is Boltzmann's constant, T the absolute temperature, B_N the noise equivalent bandwidth, and I is the shot noise current. For a PIN photodiode I coincides with i_o, so that

$$I = I_d + \mathcal{R}\langle P \rangle \tag{2.98}$$

For an avalanche photodiode, instead, we have to write

$$I = I_{dn} + \langle M \rangle^2 (I_{dm} + \mathcal{R}\langle P \rangle) = I_{dn} + F(\langle M \rangle)\langle M \rangle^2 (I_{dm} + \mathcal{R}\langle P \rangle) \tag{2.99}$$

where I_{dn} is the dark current not subjected to avalanche multiplication, whereas I_{dm} is the part subjected to such multiplication.

On the other hand, the power of the useful signal results in

$$\langle i_s^2 \rangle = \mathcal{R}^2 \langle M \rangle^2 \langle P \rangle^2 \tag{2.100}$$

which holds for an APD, and reduces to the expression valid for a PIN by taking $\langle M \rangle = 1$.

So the signal-to-noise ratio S/N, in the two cases, can be written as

$$\left(\frac{S}{N}\right)_{PIN} = \frac{R^2\langle P\rangle^2}{\left[2q\,(I_d + R\langle P\rangle) + 4\,\frac{KT}{R_L}\right] B_N}$$

$$\left(\frac{S}{N}\right)_{APD} = \frac{R^2\langle M\rangle^2\langle P\rangle^2}{\left\{2q\,[I_{dn} + F(\langle M\rangle)\langle M\rangle^2(I_{dm} + R\langle P\rangle)] + 4\,\frac{KT}{R_L}\right\} B_N} \tag{2.101}$$

From both of these expressions we can conclude how a high load resistance R_L contributes to improve the S/N, up to a limit called the *shot noise limit*, beyond which thermal noise becomes negligible. In such a situation (which, however, can be reached only for rather strong useful signals, neglecting also any dark current effect) we obtain the limit values

$$\left(\frac{S}{N}\right)_{PIN} = \frac{R\langle P\rangle}{2qB_N} = \eta_o \frac{\langle P\rangle}{2h\nu_o B_N}$$

$$\left(\frac{S}{N}\right)_{APD} = \frac{R\langle P\rangle}{2qF(\langle M\rangle)B_N} = \eta_o \frac{\langle P\rangle}{2h\nu_o F(\langle M\rangle)B_N} \tag{2.102}$$

This behavior is equivalent to a situation in which the noise is characterized by a power spectral density $2h\nu_o/\eta_o$, multiplied by $F(\langle M\rangle)$ for an APD, thus justifying the definition of noise figure for such parameter.

Although those expressed by (2.102) represent the best conditions of operation respectively for a PIN and for an APD photodetector, the most critical situation in which they have to operate is when the useful signal is low. In such a situation, characterized by detectable thermal noise and dark current shot noise, the sensitivity of photodetection is called the minimum level of optical power $\langle P\rangle$ that is possible to detect with a prefixed quality of reproduction. For the reception of analog signals, a typical required S/N is on the order of 40 dB. When we have a digital signal, we can fix a value of S/N as low as 10 dB.

For metrological purposes, the reference value of S/N = 0 dB is often assumed, which means an equivalent optical power which equals the noise power. Such a figure, called *noise equivalent power*, or NEP, is defined for a reference noise bandwidth of 1 Hz so that, considering expression (2.101), it is measured in $W/\sqrt{Hz}$. It gives an idea of the minimum detectable optical power that can be obtained using a very selective amplifier, whose bandwidth is of the order of 1 Hz. An amplifier of this type is effectively available for metrological purposes and corresponds to the well-known characteristics of a lock-in amplifier [11].

In such an evaluation, besides the sources of noise already included in expression (2.101), the intrinsic noise of the lock-in amplifier or any preamplifier, in whichever circuital arrangement (for instance, one based on a transimpedance preamplifier), must be considered too. In most cases, however, the dominant noise term in this application, especially for PIN photodiodes, turns out to be the thermal noise, thus stressing the importance of the best choice for the load resistance R_L.

As clearly shown in (2.97), the power of shot and thermal noise is higher and higher for increasing values of the noise equivalent bandwith. In the case of digital transmissions (the most interesting in practice) B_N can be assumed to be equal to half the transmission capacity C. Thus typical curves of the sensitivity of photodetection can be plotted as a function of C for different configurations of the receiver device. They show a monotonically increasing behavior with different slopes, depending on which noise contribution is dominant. The adoption of an APD detector is usually preferable, since it allows one to improve the receiver sensitivity. In this sense, however, the optimization of $\langle M \rangle$ also could be, at least in some cases, advisable.

2.3.4 Coherent Detection

When the optical wave is coherently modulated, a heterodyne or homodyne detection scheme must be employed. The most general structure of a coherent system is shown in Figure 2.15, which shows also the transmitter side and the optical link. The system includes two lasers: laser T, used as source of the optical radiation at the transmitter, and laser LO, employed as local oscillator at the receiver. The transmitted optical signal propagates along the optical fiber link. Once it arrives at the receiver section, it combines with the signal produced by the local oscillator and the resulting signal is collected by the photodetector active area. Polarization alignment is necessary between such two electromagnetic waves. At the output of the photodetector, there is an electrical signal, centered on the intermediate frequency f_{IF}, which is expressed as

$$f_{IF} = |\nu_T - \nu_{LO}| \qquad (2.103)$$

where ν_T and ν_{LO} are the central emission frequencies of lasers T and LO, respectively. More specifically, in the heterodyne case, f_{IF} is not zero, usually assuming a value between $3F_s$ and $5F_s$, where F_s is the symbol rate. In the homodyne case, $f_{IF} = 0$ and laser LO is usually phase locked to laser T by means of a circuit for the carrier recovery (such as a *phase-locked loop* (PLL)).

The performance obtainable by means of homodyne systems is nominally superior to that of the corresponding heterodyne schemes. Nevertheless it is much more critical and sensitive to possible deviations from the ideal behavior due to the characteristics of real components. Therefore, at present, only heterodyne systems appear feasible. Moreover, among the latter a further subdivision in two groups is necessary, owing to the possibility that the signal processing at the intermediate frequency is done in a synchronous way or not. In heterodyne receivers based on a synchronous processing of the signal at the intermediate frequency, the carrier at f_{IF} is recovered itself by means of a suitable electrical circuit (for example, of the PLL type). The carrier thus regenerated is then used for the subsequent postdetection processing, and the overall system can be defined as completely coherent. Much more simple is the solution based on a nonsynchronous processing of the signal at intermediate frequency. In this kind of receiver, in fact, the carrier at f_{IF} is not recovered at all. For this reason, such systems can be defined as only weakly coherent and their performance is necessarily worse (but in some cases only slightly) than that of the corresponding completely coherent systems.

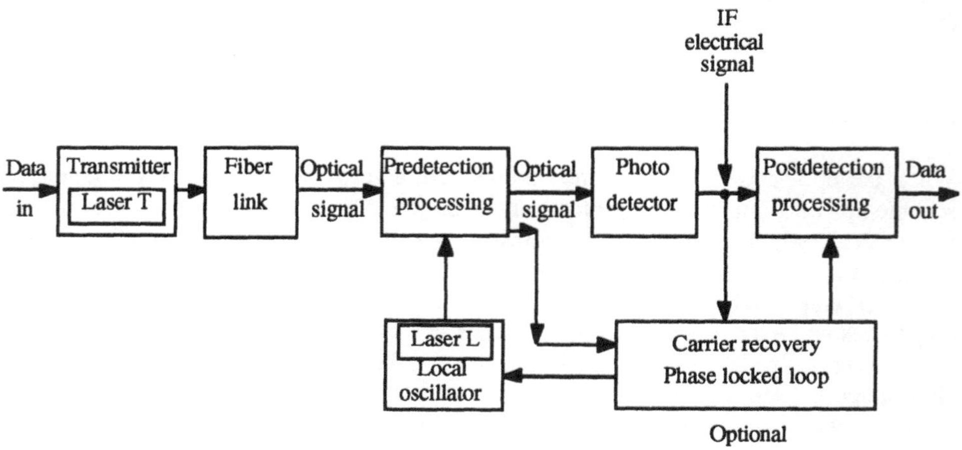

Figure 2.15 Schematic diagram of the most general coherent system.

With reference, for the sake of generality, to the heterodyne case, the total electric field incident onto the photodiode active area is given by

$$E_t = E_r\cos(2\pi\nu_T t + \phi_r) + E_{LO}\cos(2\pi\nu_{LO} t + \phi_{LO}) \qquad (2.104)$$

where ϕ_r and ϕ_{LO} represent the initial phase of the received and of the locally generated signal, while E_r and E_{LO} are the amplitudes of such waves. Under the hypothesis that field superimposition expressed by (2.104) is made in air, they are related to the powers of the received and of the locally generated wave by the following relationships

$$P_r = \sqrt{\frac{\varepsilon_o}{\mu_o}} \frac{E_r^2}{2} \qquad P_{LO} = \sqrt{\frac{\varepsilon_o}{\mu_o}} \frac{E_{LO}^2}{2} \qquad (2.105)$$

where ε_o, and μ_o are, respectively, the permittivity and permeability values of the free space. In ideal conditions, these waves have the same direction of propagation and polarization plane.

Following a procedure quite similar to that described in the previous section for a direct detection scheme, considering an APD photodetector, we can derive the power of the useful signal, which, considering also the case of a homodyne detection, results in

$$\langle i_s^2 \rangle = 2R^2 \langle M \rangle^2 P_r P_{LO} \qquad \text{(heterodyne detection)}$$

$$(2.106)$$

$$\langle i_s^2 \rangle = 4R^2 \langle M \rangle^2 P_r P_{LO} \qquad \text{(homodyne detection)}$$

The case of a PIN photodiode can be easily obtained by setting $\langle M \rangle = 1$ in these expressions.

(2.106) clearly shows the main advantage offered by the adoption of a coherent system, with respect to an intensity-modulation direct detection scheme; that is, the possibility of having, by suppressing the constant component and by a suitable high-pass filtering, a signal at intermediate frequency whose power is proportional to the product between the power of the received signal and that of the local oscillator. Therefore, it can be increased enough, acting on the power emitted by laser LO, to make thermal noise almost negligible. In other terms, the shot noise limit can be here achieved also for rather weak useful signals and the following expression for the signal-to-noise ratio results

$$\left(\frac{S}{N}\right) = \eta_o \frac{P_r}{h\nu_o F(\langle M \rangle) B_N} \qquad \text{(heterodyne detection)}$$

$$(2.107)$$

$$\left(\frac{S}{N}\right) = 2\eta_o \frac{P_r}{h\nu_o F(\langle M \rangle) B_N} \qquad \text{(homodyne detection)}$$

in which also (2.91) has been taken into account. As usual, when a PIN photodiode is adopted, $F(\langle M\rangle) = 1$ is to be assumed.

The sensitivity of a coherent receiver, also here defined as the minimum received optical power necessary for achieving a prefixed quality of reproduction, depends, on the other hand, on the particular modulation format and demodulation technique adopted. Only digital schemes are considered as their importance in practical systems is much greater than the analog ones. In this sense a measure of the transmission quality is given by the bit error rate, which cannot exceed a prefixed limit. In the following, we will consider, for simplicity, only binary modulation-demodulation schemes. The extension to M-ary transmissions can be made without conceptual difficulties and leads to similar conclusions as regards the performance comparative analysis described below.

The most efficient binary systems characterized by the minimum request of received power are those based on PSK or ASK, using a completely coherent demodulation scheme. In the considered binary case, the transmitted waveforms are the same for both the schemes and given by

$$\begin{aligned} s_1(t) &= s_o\cos(2\pi\nu_o t) \\ s_2(t) &= -\, s_o\cos(2\pi\nu_o t) \end{aligned} \tag{2.108}$$

The bit error rate, in turn, can be easily evaluated, leading to

$$P_E = \frac{1}{2}\operatorname{erfc}\left(\sqrt{\frac{S}{N}}\right) \tag{2.109}$$

The DPSK technique, which can be considered as only weakly coherent in that the phase reference for demodulation is derived from the phase of the carrier during the preceding signaling interval (and the receiver decodes the digital information based on the differential phase), offers performance only slightly worse, especially if the signal-to-noise ratio is not too small. More precisely, the bit error rate is in this case

$$P_E = \frac{1}{2}\exp\left(-\frac{S}{N}\right) \tag{2.110}$$

Finally, the FSK solution with completely coherent demodulation, under the hypothesis of orthogonality in the elementary transmitted waveforms

$$s_1(t) = s_o \cos(2\pi \nu_1 t)$$
$$s_2(t) = s_o \cos(2\pi \nu_2 t) \tag{2.111}$$

is characterized by a penalty of 3 dB, which becomes a little greater when a weakly coherent demodulation is adopted. In the first case, in fact, we have

$$P_E = \frac{1}{2}\,\mathrm{erfc}\left(\sqrt{\frac{S}{2N}}\right) \tag{2.112}$$

while in the weakly coherent situation the following expression holds

$$P_E = \frac{1}{2}\exp\left(-\frac{S}{2N}\right) \tag{2.113}$$

Some of these conclusions have to be modified when the effects of the phase noise introduced by the laser sources are taken into account. The consequence of this noise mechanism, which is due to the phase fluctuations produced by spontaneous emission and by relaxation oscillation in the device semiconductor material, is to impress a random phase modulation on the otherwise monochromatic output of the laser, thereby impairing its performance, especially when an angle-modulation setup is considered. Laser phase noise is usually characterized in terms of the linewidth Δn of the laser emission spectrum. This must be suitably limited, with respect to the symbol rate F_s, in such a way that the penalty introduced by phase noise is not too high. More precisely, when it is desired to use the phase information on the signal, as in PSK or in homodyne detection, a ratio $F_s/\Delta\nu$ typically greater than 6×10^3 has to be ensured, in order to have a penalty less than 0.5 dB at a bit error rate of 10^{-9}. Much broader linewidths may be used in heterodyne weakly coherent OOK (whose waveforms are obtained from (2.108) by letting $s_2(t) = 0$) or FSK systems, in which phase information is not used. In this case a ratio $F_s/\Delta\nu$ greater than 30 seems sufficient for having the same penalty discussed above [12].

The existence of the above constraints indicates the need for suitably narrow-linewidth lasers, which can be now designed, although some technological problems still remain. In this sense, it could be useful to resort in some cases to different solutions for reducing phase noise effects, like those based on the employment of suitable error-correcting codes or those facing such undesired signal degradation in unitary fashion together with polarization fluctuations, which will be discussed in Section 2.5.3.

2.3.5 Statistical Properties of Light Detection

The limit situation described by (2.102) in direct detection schemes is reached only when a high value of S/N is prefixed. Typical applications for which such expressions hold include analog transmissions. For digital transmissions and even metrological applications, where the signal-to-noise ratio can be remarkably smaller and there is the possibility of extremely weak light pulses impinging on the photodetector active area, circuit noise and dark-current shot noise become dominant.

In a hypothetical very highly sensitive receiver, we can consider the possibility of making thermal noise negligible by properly cooling the electric preamplifier of the receiver. Furthermore, dark-current shot noise can be almost eliminated by a suitable biasing of the APD, synchronized with the optical pulse rate [13, 14]. In such a situation the receiver can be considered as a sort of photon counter [15]. For the analysis of a theoretical receiver thus conceived, it is necessary to take carefully into account the statistics of photon arrivals.

The measurement of the statistical properties of various photon counting distributions can be performed. This consists of estimating the probability p(n) of having n photons in a light pulse of time duration T_r, characterized by N photons on average. In classical light statistics, we can have two extreme situations:

- a Bose-Einstein distribution, characterized by

$$p(n) = \frac{1}{1 + N}\left(\frac{N}{1 + N}\right)^n \tag{2.114}$$

- a Poisson distribution, characterized by

$$p(n) = \frac{N^n}{n!} e^{-N} \tag{2.115}$$

Bose-Einstein distribution (2.114) is pertinent to (spatial and temporal) single-mode black-body radiation. Poisson distribution (2.115) corresponds to coherent light that is both fully coherent (an idealized source represented by a constant phasor $Ae^{i\phi}$) and of constant amplitude (a source represented by a phasor whose modulus A(t) is completely determined and whose phase ϕ is completely random; such a case is closely approximated by a laser source stabilized in amplitude). Poisson distribution is also the statistics of black-body radiation as measured by any practical detector (i.e., a detector of finite area and response time which gives the average on many spatial and temporal modes). Let us note that the intrinsic noise limit in detection of Poissonian light corresponds to the well-known shot noise level.

From a quantum mechanical point of view, optical radiation can exist in a number of possible states other than classical light. The uncertainty principle fixes the minimum value of the product of the fluctuations affecting a pair of conjugate variables of any physical object. For optical radiation, the number of photons N in a given observation time and the correspondent optical phase are conjugate parameters; therefore, their uncertainties $\langle \Delta N^2 \rangle$, $\langle \Delta \phi^2 \rangle$ satisfy the relationship [16]

$$\langle \Delta N^2 \rangle \langle \Delta \phi^2 \rangle \geq \frac{1}{4} \tag{2.116}$$

In terms of the two quadrature components a_p, a_q of the electrical field, that are conjugate variables as well, the uncertainty principle reads

$$\langle \Delta a_p{}^2 \rangle \langle \Delta a_q{}^2 \rangle \geq \frac{1}{16} \tag{2.117}$$

A coherent state of radiation, which corresponds to a classical coherent wave, is a minimum uncertainty state in the sense that Eq. (2.117) holds with the equality sign; in addition, the uncertainties are symmetrically distributed on the two conjugate variables, so that

$$\langle \Delta a_p{}^2 \rangle = \langle \Delta a_q{}^2 \rangle = \frac{1}{4} \tag{2.118}$$

Furthermore, $\langle \Delta N^2 \rangle = \langle N \rangle$, as classically expected for a Poisson field.

Nevertheless it is possible to satisfy Eq. (2.117) also with values of $\langle \Delta a_p{}^2 \rangle$, $\langle \Delta a_q{}^2 \rangle$ that are different between them. A state of this type is called *phase-quadrature squeezed state*. In this case the uncertainty on one component is reduced at the expense of increasing that on the other component. Such situation can be exploited, e.g., in interferometrical measurements or in homodyne detection where one field component can be selected to carry information so that performance better than the semiclassical shot noise limit can be accomplished.

Another class of squeezed states are the so-called *number-phase squeezed states*. In this case, the fluctuations $\langle \Delta N^2 \rangle$ of the photon number are reduced below the Poissonian value (that is, $\langle \Delta N^2 \rangle < \langle N \rangle$) at the expense of the phase fluctuation, which is increased. In the limit $\Delta N \rightarrow 0$, a *number state* is obtained in which the photon number is completely determined while the phase is completely random. This kind of light would be useful for experiments based on photon counting. In practice, an uncertainty on N smaller than 0.1 would be sufficient for realizing an effective photon counter.

Though the potential applications of squeezed light appear attractive, it shall be noted that the experimental capability of making squeezed sources, both phase-quadrature and number-phase, is presently very limited.

The performance of a photon counting system, on the other hand, depends on the particular modulation format adopted. For example, using a conventional on-off transmission, based on two equiprobable levels, one with zero photons (low level) and the other with N photons on average (high level), almost 9 photons per bit are necessary, with a coherent field, in order to obtain an error probability on the received symbol smaller than 10^{-8}. The inverse of this quantity, which has the same meaning of the sensitivity parameter introduced in the previous subsection with reference to direct and coherent detection, called *transmission efficiency*, is usually noted by ρ and expressed in bits per photon. It is possible to verify that in the present case its value is much greater than that obtainable for traditional intensity-modulation direct detection transmission systems and comparable with that of a 2-PSK, completely coherent, homodyne detection system.

Nevertheless, in order to maximize ρ, more involved modulation formats can be considered. In particular, employing finite energy symbols, subdivision of symbol time duration into n equal adjacent time slots enables us to obtain an M-ary orthogonal set of symbols (with $M = n$), each characterized by the high transmission level in only one time slot of the subdivision. This basic modulation format, proposed in [15] as optical *pulse position modulation* (PPM), has been object of several evolution attempts. In [17], for example, the usage of k high transmission levels per symbol time duration ($1 < k \leq n$), instead of one, has been proposed in the so called *multi-pulse position modulation* (M-PPM), while in [18] an extended version of the PPM set of symbols, including the symbol with zero photons at all time slots, has been introduced as PPAM channel. In [19], on the other hand, it has been shown that, under the hypotheses of the Shannon theorem [20], higher transmission efficiencies can be achieved by means of a more general set, named *complete multi-pulse position modulation* (CM-PPM). It consists of all the sequences having at most (and not exactly, as in the M-PPM case) k high transmission levels per symbol time duration.

All these transmission schemes exhibit an improvement in the transmision efficiency at the expense of the required bandwith, which is greatly expanded. This problem can be partly solved by applying suitable error-correcting codes chosen among those particularly efficient in eliminating erasures [21].

An alternative approach consists in assuming different input probabilities for the two symbols of a binary transmission [22]. Making the transmission of the high level less probable than that of the low level, transmission efficency can be remarkably increased, although a certain bandwidth expansion has to be tolerated.

Any measurement action introduces a perturbation in the optical radiation to be received, according to the uncertainty principle. Nevertheless, in a particular state of such radiation, we can perform a measurement on one parameter of a conjugate pair without perturbing it but introducing an increase in the uncertainty of the other one. For example, it is possible to measure N without absorbing the light pulse by simply increasing the $\Delta\phi$. Such principle of operation, already experimentally proved [23], is called quantum nondemolition detection. In the first proposed, physically feasible configuration, it is shown in Figure 2.16.

The probe wave is sensitive to the refractive index change, which, in turn, is proportional to the signal intensity in the Kerr medium. The latter is a typical nonlinear effect, which can be efficiently adopted in the most general context of integrated signal processing, particularly for realizing optically bistable devices, switches, upper and lower threshold devices, optical limiters, and so on. Some of these structures will be described in Sections 2.6 and 2.7. The reflectivity of the mirrors M1 and M2 is zero at the signal frequency and unity at the probe frequency. In this way the device acts as an interferometer only for the probe wave, which is phase modulated without affecting the signal wave.

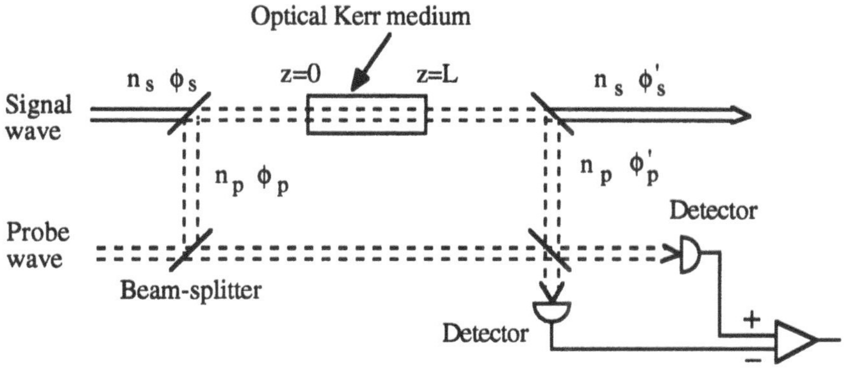

Figure 2.16 Configuration of the quantum nondemolition measurement for the evaluation of the number of photons N without absorbing the light pulse.

2.4 PROPERTIES OF SINGLE-MODE FIBERS INDEPENDENT OF THE LIGHT DEGREE OF TEMPORAL COHERENCE

As already stressed in Chapter 1, taking measurement requirements into consideration, the properties of single-mode fibers, themselves the object of the measurement, can be subdivided into two classes: those independent of the light degree of temporal coherence and those dependent on it.

For the first class, we can perform the measurement without particular constraints on the emission spectrum of the test optical source. For the second class, such constraints must be instead carefully considered. Note that, in some cases, the measurements on the fiber should be preferably performed with the same optical source employed for its subsequent use. Nevertheless, most properties and their measurements are defined by recommendations or standardizations suggested in CCITT or IEC documents, independently of the subsequent fiber use.

For instance, the mechanical and geometrical properties of standard single-mode fibers, as well as their attenuation characteristics, conform to precise recommendations, as discussed in 1.3. The single-mode operation wavelength range is not so precisely fixed, but also can be considered as a property substantially independent of the degree of temporal coherence characterizing the optical source. This is not true for chromatic dispersion properties, although a wide-spectrum optical source, properly filtered, can be assumed for their evaluation. Finally, some polarization-preserving properties of certain special single-mode fibers to be employed with coherent transmission systems or in sensor applications are strongly dependent on the degree of temporal coherence of the light source, so that their experimental evaluation is to be made devoting particular cares to the choice of a suitable optical source also for the measurement setup.

After a brief outline of the propagation in a weakly guiding round-core fiber, we will focus our attention on attenuation and chromatic dispersion properties in general. Some aspects regarding birefringence properties or polarization preserving features will be discussed instead in the following section, together with some chromatic dispersion effects acting on coherent transmissions.

2.4.1 Outline of the Propagation in a Weakly Guiding Round-Core Fiber

The assumption of weak guidance, i.e., of a small index difference between core and cladding, enables us to neglect any term on the right-hand side of the vector-wave equation [24] which describes the propagation of an electromagnetic field in an optical fiber. This entails the possibility of existence of linearly polarized (LP) quasi-modes [25]. They are expressed by a scalar function of the transverse coordinates, which represents the transverse electric or magnetic field.

The scalar wave equation so obtained can be written as

$$\nabla_t^2 \Psi + (k^2 - \beta^2)\Psi = 0 \tag{2.119}$$

where ∇_t^2 denotes the transverse Laplacian and

$$k^2 = k_o^2 n^2 = \left(\frac{2\pi}{\lambda}\right)^2 n^2 \tag{2.120}$$

is the square wavenumber, dependent on the refractive index distribution n in the fiber transverse section, β is the phase constant, finally Ψ is the scalar field function.

Solution of (2.119) depends on the particular shape of n. For a step distribution having round symmetry (round-core step-index fibers), it can be expressed in terms of Bessel functions. In this case, single-mode operation occurs for

$$0 < V \leq 2.405 \tag{2.121}$$

From (1.1), in which n_o and n_c can be assumed as the refractive index value in the core and in the cladding respectively, the left bound corresponds to $\lambda \rightarrow \infty$, whereas the right bound is associated with

$$\lambda_c = \frac{2\pi}{2.405} a \sqrt{n_o^2 - n_c^2} \tag{2.122}$$

Round-core fibers having refractive index distribution not step shaped exhibit different values of the right bound in (2.121). The same thing sometimes occurs also in its left hand, which means that the wavelength interval of single-mode operation includes only finite wavelengths. For practical purposes, however, only the right extremum for V is of interest. It enables us to define the cut-off wavelength λ_c of the fundamental mode, called LP_{01}, in propagation along such structure.

For $\lambda < \lambda_c$, the propagation becomes multimodal. Given λ_o the wavelength of operation, first of all, we have to satisfy the constraint

$$\lambda_o < \lambda_c \tag{2.123}$$

Furthermore, the higher the difference $(\lambda_o - \lambda_c)$ the larger the mode spot-size, as introduced in (1.2).

As long as the fiber can be considered to have a round core, the field of the fundamental LP_{01} mode is a function of the only radial coordinate r, that is, $\Psi = \Psi(r)$. Furthermore, the two linear polarizations of such mode, the one having electric field oriented as x and the other having electric field oriented as y, x and y being Cartesian coordinates in the transverse plane, are degenerate. This means that they exhibit the same phase constant β.

The second mode, called LP_{11}, exhibits Ψ function of an azimuthal coordinate ϕ besides the usual r. The phase constant β associated to such mode is always smaller than that of the fundamental mode.

When the refractive index distribution is not circularly symmetric, or some strains are present in the fiber transverse section (thus originating anisotropic behavior of the glass), the two linear polarizations of the LP_{01} mode are not degenerate. This phenomenon is called *modal birefringence*, and it can have a geometrical or a material nature, depending on its cause. When geometrical birefringence is present, the fiber core can be considered to be elliptically deformed. Modal birefringence, which is sometimes intentionally induced, in order to reach particular polarization properties, will be discussed in Section 2.5.3 in more detail.

2.4.2 Attenuation

Attenuation in a single-mode fiber is due to intrinsic causes, like absorption or scattering, or to external effects, among which bending and microbending are the most important.

Light absorption arises as a consequence of two distinct resonance mechanisms occurring, respectively, in the ultraviolet region and in the medium infrared region: electronic transitions between quantum energy levels and vibrational motions of the silica molecule. More precisely, absorption occurs when a photon interacts with an electron in the valence band and excites it to a higher energy level. Owing to the band gap organization of the structure of the amorphous glass materials, a resonance wavelength can be defined as

$$\lambda_o = \frac{hc}{E_2 - E_1} \tag{2.124}$$

where h is Planck's constant, c the free-space velocity of light, E_2 and E_1 the initial and final energy state of the electron or of the molecular vibrational motion. Obviously, the presence of viscous forces also has to be taken into account, so that the resonances are not, in general, very selective.

The spectral attenuation curve, externally to the resonance peaks, exhibits in the visible and near infrared regions an exponential decay with respect to the optical frequency ν. It can be described by the so called Urbach's law [26], which gives curves whose slopes are rather steep, so that in the region from about 0.6 μm to 1.7 μm the attenuation is less than 1 dB/km and becomes smaller than 0.1 dB/km in the region from 0.9 μm to 1.6 μm.

More important in determining the absorption loss is the presence of impurities in fiber material. Impurity absorption is fundamentally due to the presence of transition metal ions such as iron, chromium, cobalt and copper, and of OH^- (water) ions, during the glass melting process which precedes the fiber preform fabrication. The rather broad absorption peaks of the various transition metal impurities are often overlapped, which leads to extended absorption bands in the visible and in the near infrared region.

The presence of OH^- ion impurities in fiber preforms results mainly from the air contained in the environment which surrounds the glass during its fabrication or growth.

Significant progress has been made recently in order to reduce the residual OH^- content of fibers to less than 0.1 ppm. Of course, this result can be obtained by a refinement of the preform production process.

In any case, the absorption loss due to OH^- is rather selective. The fundamental absorption peak lies near 2.7 μm while the first, second, and third overtone occur at 1400, 950, and 725 nm, respectively. Combinations with the infrared resonance frequencies of pure silica are also possible, leading to an intense absorption peak at 1.23 μm. As a result of such phenomenons, the spectral attenuation curve exhibits, in the region between λ = 1.3 μm and λ = 1.55 μm, a maximum value which is not smaller than 1 dB/km, even when the percentage of residual OH^- content is reduced to the level said above.

Also the presence of dopants, included in the silica glass to vary the refractive index distribution, may induce some further absorption loss, which depends on the type and on the concentration of the dopant itself. In this sense, the worst overall effects are attained when GeO_2 or P_2O_5 are used for doping the core. It is in fact quite obvious from a qualitative point of view that when the mode propagating along the fiber is confined, the loss is almost completely determined by the core region characteristics. Better performance is then obtainable by realizing the desired index distribution, whenever possible (for example in the step-index case), through the doping of the cladding instead of the core. This implies the usage of B_2O_3 or F as dopants.

F doping seems particularly attractive because it offers the possibility of reducing another cause of attenuation in the fiber: the excess absorption loss due to the presence of stresses at the core-cladding interface. In fact, the F molar percentage required for achieving a prefixed refractive index variation is the smallest among the dopants commonly used. This implies the possibility of minimizing the dopant concentration and, consequently, all the losses related to high dopant concentration gradient at the core-cladding interface. In addition, the fundamental Si-F vibrational resonance occurs near 13.8 μm, that is, in a region which is not of interest for actual transmissions. The same consideration about the peak resonance allocation is valid for GeO_2, while the adoption of P_2O_5 or B_2O_3 may take place to some problems in this sense, especially in the neighborhood of λ = 1.55 μm.

A further important contribution to the fiber intrinsic loss is given by scattering, which is originated by microscopic variations in the material density, compositional fluctuations, and structural inhomogeneities or defects consequent to the fiber manufacturing process. In particular, the random distribution of the molecules of the glass, which is responsible for its characteristic amorphous structure, together with the fluctuations in composition due to the presence of several oxides, give rise to refractive index variations which occur over distances that are small compared with the wavelengths of interest. These index variations are in turn responsible for Rayleigh-type scattering of light. The fraction of scattered power exceeding a certain limit angle becomes radiation power and is lost for propagation.

A detailed analysis of Rayleigh scattering is possible considering local microscopic variations of refractive index. This allows us to determine the variation $\Delta\bar{s}$ of Poynting vector $\bar{s} = 1/2\ \bar{E}\times\bar{H}^*$, for a plane wave incident on the microscopic region ΔV where the variation of the refractive index is not null. In order to evaluate $\Delta\bar{s}$ in a deterministic way, it is necessary to know the index fluctuations in detail. In reality, taking into account the statistical nature of such fluctuations, what is possible to evaluate is the mean value, $\langle\Delta\bar{s}\rangle$, of the scattered wave Poynting vector. By integrating $\langle\Delta\bar{s}\rangle$ over a sphere of radius r, one obtains the overall scattered power and, from this, the field attenuation constant, which is given by

$$\alpha_s = \frac{16\pi^3 n_o^2}{3\lambda^4}\langle\Delta n^2\rangle D^3 \tag{2.125}$$

In this expression, Δn is the refractive index variation, which can be seen as a stochastic process; n_o is the nominal, unperturbed, value of the refractive index; and D represents the correlation length of Δn. (2.125) has to be slightly modified for taking into account the effects of dopant concentration fluctuations.

Expression (2.125) shows the main characteristic of this kind of scattering: the proportionality of its effect to λ^{-4}. Though unavoidable, this cause of loss can be reduced by increasing the transmission wavelength. This suggests the adoption of medium infrared (MIR) fibers, which are, at present, widely investigated in the perspective of obtaining ultra-low-loss fibers. In addition, in order to evaluate α_s, one has to determine the product $\langle\Delta n^2\rangle D^3$. This depends on the entity of the refractive index fluctuations and on their statistics and can be derived by the knowledge of the glass solidification temperature, photoelastic coefficient, and isothermal compressibility. In conclusion, at the wavelengths of interest for transmission, the following practical formula, independent of the dopant additional effects, holds

$$A_s \approx \frac{0.76}{\lambda^4} \tag{2.126}$$

where A_s is measured in dB/km and λ in μm. It becomes dominant from about 0.6 to about 1.6 μm.

Rayleigh scattering is a typical linear phenomenon. Nevertheless, structural inhomogeneities can also cause an anelastic, and then nonlinear, interaction between the incident light and the glass structure, in which also the vibrational energy of the silica molecule is involved. As a function of the power level we can have Brillouin or Raman scattering [27]. Both produce a wavelength shift between incident and scattered light (much more pronounced in the Raman case) which can be intentionally exploited for light amplification. However they need high levels of optical energy traveling the fiber, so that they lead to a true limitation in the design of a transmission system only when a laser light is pulsed and exhibits very high peak power in the fiber.

Another type of scattering may exist due to defects created during the fiber fabrication. In general, these defects are in the form of trapped gas bubbles, unreacted starting material, and crystallized regions in the glass. The recent evolution of the preform manufacturing methods has, however, greatly reduced the incidence of such mechanisms to the point where they are now almost completely negligible.

When the fiber is cabled, a further excess absorption loss may take place. In fact, as a consequence of electrolytic corrosion of metal parts in the cable or of the degradation in coating and filler materials, hydrogen permeation may occur. When H_2 molecules remain uncombined with the silica matrix, in interstitial positions, we have a pure absorption loss, whose fundamental vibrational resonance has its first overtone at $\lambda = 1.24$ μm. It causes an absorption attenuation of the order of 1.7 dB/km with a H_2 concentration of only 1 ppm. However, it can be arbitrarily reduced by lowering the H_2 pressure outside the fiber.

On the contrary, when hydrogen combines with oxygen, it increases the OH^- ions concentration, making the corresponding absorption loss mechanism more effective. It is important to stress that a fundamental role in the production of combined hydrogen is played by the dopants, in that the energy of activation of the bond SiOH is greater than that of the bonds GeOH or POH. The adoption of F-doped fibers seems preferable also because F is included in the silica matrix at pure atomic level, without the necessity of oxygen. In any case, the adoption of fibers particularly resistant to hydrogen diffusion as well as the use of cable structure in which possible sources of free hydrogen has been removed seems advisable. In order to ensure complete protection of the cable against the water, single metal sleeves, combined with the accurate stuffing of all the interstitial places with jelly, are usually adopted, while synthetic material such as silicone or nylon-silicone are avoided as a rule because they produce free hydrogen during their degradation. Finally, hydrogen chemical absorbers in the cable structure are also sometimes employed in order to increase the protection.

Turning our attention to bending and microbending, we can observe how these causes of additional losses are closely related to the field confinement which characterizes the operation condition. In both cases, the higher the field confinement the smaller the entity of such undesired effects. Pure bend losses are usually very strong and occur in correspondence with uniform curvatures to which the fiber is subjected. For dispersion-shifted single-mode fibers, operating at $\lambda_o = 1.55$ μm, which are the most critical from this point of view because of the relatively high value of the difference $(\lambda_o - \lambda_c)$ characterizing them, a curvature radius, above which any effect of pure bend is practically negligible, is of the order of 10 cm. This value can be carefully considered in designing the enclosures inside which fiber-to-fiber joints are housed during installation procedures.

Microbending losses are much smaller, and become appreciable only since they are distributed along the whole fiber length. They are caused by the strain to which a fiber can be subjected because of temperature-induced length variations with respect to the remaining cable components. Random bends, having very small curvature radii thus take place. Their qualities and size depend on the roughness of the fiber coating and cable support.

The process of power exchange between the fundamental mode and a packet of strongly attenuated radiation modes can be described as a stochastic process whose correlation length is to be compared with the difference between the phase constant β of the LP_{01} mode and that of a packet of radiation modes [28].

Since the LP_{11} mode, when guided, is much more strongly affected by microbending than the LP_{01} mode, in some situations the former could be in propagation, although severely attenuated. This condition, which should be avoided in practice because it may induce intermodal time dispersion or modal noise [29], gives rise to a phenomenon called *effective cut-off wavelength.* The cut-off of the LP_{11} mode, defined as the wavelength above which this mode is practically absent, increases with the distance from the optical source because microbending loss severely penalizes such mode close to the true cut-off wavelength λ_c.

2.4.3 Chromatic Dispersion Effects in Intensity-Modulation Direct Detection Schemes

Intensity-modulation direct detection schemes for either telecommunication applications or sensors applications are characterized by a propagation of an optical power. Because the fiber pieces, the terminal connectors, and the fiber-to-fiber joints are linear in the electromagnetic fields and not in the optical powers, some undesired signal fluctuations could arise when the propagation is not rigorously single-mode. This phenomenon is called *modal noise* and is strictly related to the temporal coherence properties of the optical source. It will be discussed in the next section.

Assuming that the propagation is rigorously single-mode, letting β the phase constant of the unique mode in propagation, the group delay per unit distance τ_g can be written as

$$\tau_g = \frac{d\beta}{d\omega} \tag{2.127}$$

where ω is the circular frequency. Alternatively

$$\tau_g = -\frac{\lambda^2}{2\pi c}\frac{d\beta}{d\lambda} \tag{2.128}$$

Expression (2.127), in which τ_g and β are functions of ω, is more suited for coherent transmissions, where the spectral width of the optical source is expressed in frequency. Expression (2.128) is appropriate, instead, for intensity-modulation transmissions, where such spectral width is expressed in wavelength.

Function $\beta(\lambda)$ is imposed mainly by two effects:

1. material dispersion related to the wavelength dependence of the refractive index n,
2. waveguide dispersion due to the field confinement in the transverse section.

The derivative

$$\tau_g' = \frac{d\tau_g}{d\lambda} \tag{2.129}$$

calculated in correspondence with the central emission wavelength of the emission spectrum λ_o, is responsible for second-order chromatic dispersion effects. As a consequence of points 1 and 2, the value of τ_g' also depends mainly on material and waveguide contribution. An example is reported in Figure 2.17, where a typical behavior of the dispersion curve for a single-mode step index fiber has been plotted.

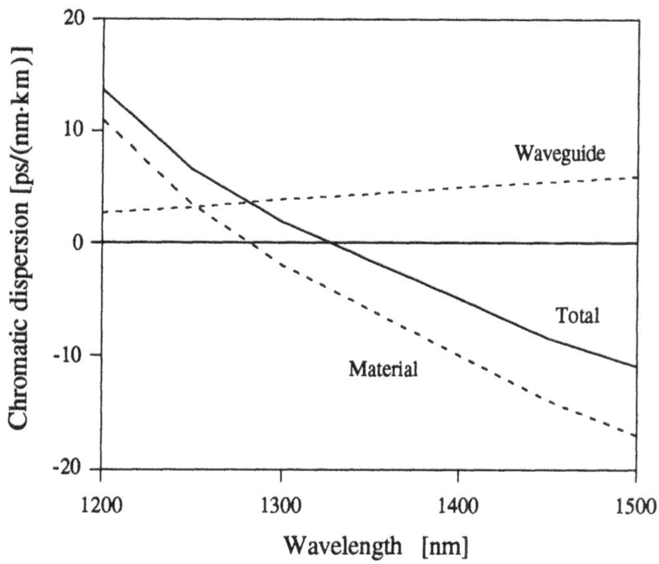

Figure 2.17 Typical behavior of the second-order chromatic dispersion for a step-index fiber.

Both material and waveguide dispersion have been evidenced and then combined in the figure. They can have different signs and their effects can be cancelled. In conventional germanium-doped fibers this occurs at nearly 1.3 μm. Nevertheless, the adoption of suitable refractive index profile shapes, more involved than the step-index one allows the zero-dispersion wavelength to be shifted to 1.55 μm, thus obtaining dispersion-shifted or dispersion-flattened fibers, already introduced in Section 1.3.

Allowing $p(\lambda)$ to be the power spectrum of the optical source, we can assume that it is independent of the modulation signal, expressed through a function of time $s(t)$. Thus we obtain a received optical power [30] which can be written as

$$P_r(t) = \frac{1}{2\pi} \int_{-\infty}^{+\infty} S(\omega_m) e^{i\omega_m(t - \tau_g z)} \left[\int_{-\infty}^{+\infty} p(\lambda) e^{-i\omega_m \tau_g'(\lambda - \lambda_o) z} \, d\lambda \right] d\omega_m \qquad (2.130)$$

where $S(\omega_m)$ is the Fourier transform of $s(t)$ and ω_m the modulation circular frequency. In such an expression both τ_g and τ_g' are calculated at $\lambda = \lambda_o$.

Expression (2.130) can be qualitatively justified by considering that, because of the direct modulation of the optical power, its transit time along the fiber is dependent on the group delay of the mode in propagation. Furthermore, an intramodal second-order time dispersion effect due to the bandwith limitation imposed by the optical source has to be taken into account. This is responsible for the λ-integration in the above expression.

The wider the extension of both $p(\lambda)$ and $S(\omega_m)$, the higher the chromatic dispersion effect. Such a property is summarized by the dimension of τ_g' itself, which is ps/(nm·km). We can assume, in fact, from (2.130), a linear increase with the spectral width $\Delta\lambda$ of the optical source and with the distance. The resulting delay distortion $\tau_g'\Delta\lambda z$ is to be compared with the typical times over which signal $s(t)$ varies appreciably.

$\tau_g'(\lambda_o)$ vanishes at approximately $\lambda_o = 1.27$ μm. In this situation third-order chromatic dispersion related to $\tau_g''(\lambda_o)$ comes out which, however, has always extremely small distortion effects. Typical values of $\tau_g'(\lambda_o)$ at $\lambda_o = 1.55$ μm are on the order of – 20 ps/(nm·km). Such an effect is dominated by material dispersion. An intentional increase of waveguide dispersion, obtained in dispersion-shifted fibers, can reduce $|\tau_g'(\lambda_o)|$ remarkably. Typical values of $\tau_g'(\lambda_o)$ at $\lambda_o = 1.55$ μm for this kind of fiber fall within the interval between – 1 and + 1 ps/(nm·km).

2.4.4 Active Fibers

In recent years an increasing interest has been devoted to optical amplification performed by means of fibers doped with rare earths. This is due to several good features that make them attractive with respect to traditional optical amplifiers (Fabry-Perot or traveling wave), such as

1. insensitivity to light polarization;
2. low cross-talk;
3. high-saturation power;
4. low noise figure;
5. wide spectral response;
6. low insertion loss.

The last property in particular follows from the good geometrical matching with the fibers employed in the optical link.

The reason of such a trend, which has rapidly led to commercially available devices, can be found in the possibility, made actual by the technology improvements, of realizing single-mode fibers characterized by very low attenuation, which include well-defined and reproducible concentrations of neodymium and erbium. The first relevant result, in this sense, was obtained in 1985 with the demonstration of a low-threshold, high-slope-efficiency, neodymium fiber laser [31].

The successful application of the same techniques to erbium-doped fibers, thus demonstrating the feasibility of amplifiers and lasers operating in the range 1520–1580 nm, has confirmed the entrance of active fiber components in the area of optical communications, where such devices could become of fundamental importance for transmissions over very long distances and/or characterized by very high bit rate. Though also applicable in principle to coherent transmission systems, the active fiber amplifiers are at present particularly interesting for their application in intensity-modulation direct detection systems, where they allow a remarkable enhancement of the distance between subsequent regeneration sections.

In its basic configuration (see Figure 2.18), the optical amplifier consists of a pump source, a dichroic coupler, and a piece of erbium-doped fiber whose length is properly chosen. The pump light, propagating through the coupler, is transferred to the same fiber which supports the propagation of the useful signal at 1.5 μm, arriving with it at the active fiber input. Inside the active fiber, an almost complete population inversion occurs: photons at 1.5 μm interacting with the inverted medium cause the stimulated emission of other photons perfectly in phase with the incident ones. As a consequence, an amplification of the useful signal occurs.

The gain values achievable by such a device can vary with the input signal level because of some saturation phenomena, which are always present. For signals of the order of –40 dBm, gains greater than 35 dB, in correspondence of pump powers variable between 10 and 15 mW, at 980 nm, have been experimentally observed. When the input signal level is very high (on the order of –3 dBm or more) the amplifier operates in saturation conditions. In this case, it is preferable to characterize its performance by means of a conversion efficiency parameter ε, defined as the ratio between the incoming pump power and the outgoing amplified signal power. Values of ε greater than 0.5 has been obtained in industrial laboratories [32].

As regards the noise introduced by the amplifier, it is noticeable that the almost complete active ion population inversion, together with the very long fluorescence time typical of erbium, greatly limits the spontaneous emission and any other noise contibution, contrary to what happens in similar structures based on semiconductors. The noise figure of an erbium-doped optical amplifier [33] is very close to the theoretical limit of 3 dB derived from the quantum analysis of the electromagnetic field inside the active medium.

Finally, such devices have been proved to be particularly suited for the amplification of modulated signals at very high bit rates and for very long distances, as well as for enhancing the levels of femtosecond impulses. For all these reasons they appear as very promising components also for optical transmission future developments.

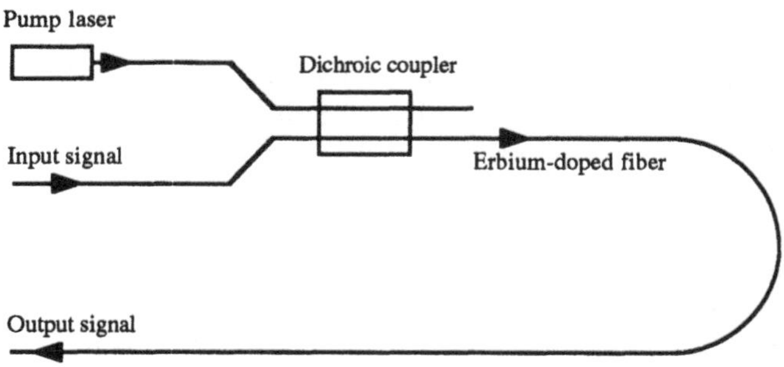

Figure 2.18 Schematic configuration of an optical amplifier based on an erbium-doped fiber.

2.5 PROPERTIES OF SINGLE-MODE FIBERS DEPENDENT ON THE LIGHT DEGREE OF TEMPORAL COHERENCE

In this section, after a brief discussion about modal noise, which affects intensity-modulation direct detection schemes, we will consider typical aspects regarding coherent transmissions, such as those related to geometrical and material birefringence, and the phenomenon of polarization instabilities. Chromatic dispersion will be analyzed again, this time with particular attention to its effects on coherent transmission.

The peculiar characteristic of a coherent transmission is in the detection of the signal which is performed by a heterodyne or a homodyne scheme. It consists of a local oscillator whose circular frequency ω_L is slightly different from that ω_0 of the carrier associated with the received field. A stable polarization is required to such a field in order to be coherently superimposed on that locally generated. This requirement can be satisfied by means of special single-mode fibers (polarization maintaining or single polarization) or, alternatively, by the addition of polarization control devices at output or by the use of particular modulation-demodulation formats.

2.5.1 Modal Noise Effects

When some effects of modal birefringence are present, the two polarizations of the fundamental mode travel the fibers with different phase constants β_x and β_y. In any fiber-to-fiber joint along the optical link, and also at the link demountable connector, such two polarizations are coupled.

Assuming that the optical source excites both polarizations by means of the same intensity-modulated signal, at the link output we will have several field contributions whose phase delays will be different from each other. The photodiode performs a direct detection of the overall received power, which is proportional to the total field squared. Thus a linear response and an undesired nonlinear distortion term appears.

The linear response can be denoted as $P_l(t)$ and is independent of the optical circular frequency ω, except for chromatic dispersion effects. The nonlinear distortion term $P_{nl}(t,\omega)$ instead depends strongly on ω, owing to the presence of factors of the type $\cos[(\beta_x - \beta_y)L_j]$, where L_j represents the length of the j-th fiber piece. The difference $(\beta_x - \beta_y)$ varies rapidly with ω, since both phase constants are roughly proportional to ω itself.

So, in order to obtain a sort of ω-averaged estimate of such noise, we can write

$$P_{nl}(t) = \langle P_{nl}(t,\omega) \rangle = \int_{-\infty}^{+\infty} P_{nl}(t,\omega) p_o(\omega) d\omega \tag{2.131}$$

where $p_o(\omega)$ is the power spectrum of the optical source, in terms of circular frequencies.

Now, $p_o(\omega)$ cannot be monochromatic because modulation itself induces an increase in its spectral width $\Delta\omega$. On the opposite side, when $\Delta\omega$ is very large the extremely high number of contributions due to the different spectral components tends to cancel $\langle P_{nl}(t,\omega)\rangle$. Therefore we have an intermediate range of values for $\Delta\omega$, correspondent to an optical source whose temporal coherence is high but not too high, that is particularly unfavorable from this point of view.

The same problem occurs when, instead of the two polarizations of a single mode, we have the LP_{11} mode, although weakly attenuated, in addition to the fundamental LP_{01} mode. In both cases here described, the phenomenon is originated by a multi-mode propagation. Thus the noise which affects the received signal, and whose power is estimated by (2.131), is usually called *modal noise*.

2.5.2 Geometrical and Material Birefringence

The difference between the phase constants β_x and β_y of the two polarizations of the fundamental mode LP_{01} gives rise to a phenomenon usually called *birefringence*. This can be originated by geometrical causes, such as an elliptical deformation of the fiber core, or by material causes. In the latter situation a distribution of strains in the fiber cross section induces an anisotropic behavior in the glass.

The parameter

$$B_n = \frac{\beta_x - \beta_y}{k_0} \tag{2.132}$$

being $k_0 = 2\pi/\lambda$ the free-space wavenumber, is called *modal birefringence* and gives a measure of this phenomenon. Although the axis on which a possible elliptical deformation of the core and that along which the distribution of strains takes place rarely coincide, we can assume an additive law for the effects of the two causes, of the type

$$B_n = B_{ng} + B_{nm} \tag{2.133}$$

where B_{ng} accounts for geometrical effects, B_{nm} for material.

Sometimes modal birefringence is intentionally induced. The purpose, in this case, is to maintain a polarization; in fact, when random perturbations affect the fiber, polarization coupling occurs only when the power spectrum of such perturbation, seen as a stationary stochastic process, is not negligible in correspondence with a spatial frequency equal to $(\beta_x - \beta_y)$. Since such a power spectrum exhibits a low-pass behavior, the larger the difference $(\beta_x - \beta_y)$, the higher the probability of avoiding polarization coupling.

Otherwise, modal birefringence is considered a defect of the fiber. Even when the core exhibits a very good circularity and extremely low residual strains affect the fiber cross section, B_n is not null. A very small value for this parameter is of the order of 10^{-10}.

Modal birefringence induces polarization mode dispersion, and, when the light is coherently modulated, it is also responsible for the undesired fluctuations which may affect the received field.

In intensity-modulation direct detection schemes, polarization mode dispersion can be evaluated through the difference between the two delays per unit distance τ_{gx} and τ_{gy} associated with the two polarizations. Such difference [34] can be expressed as

$$\Delta\tau_g = \frac{1}{c}\left(B_n + k_0 \frac{dB_n}{dk_0}\right) \tag{2.134}$$

Typical values of $\Delta\tau_g$ are of the order of 1 ps/km.

The possible fluctuations in the received field, when the light temporal coherence is strong, as in coherent modulation transmissions, are originated by a variation in the state of polarization of the light. This phenomenon is due to the fact that in a system of this type the heterodyne scheme employed for light detection can operate correctly only if the state of polarization of the received field is known and stable. Some possible tools for circumventing such problem will be briefly discussed in the next section.

2.5.3 Polarization Fluctuations

Considering the two perpendicular plane polarizations of the fundamental mode represented by the fields E_x and E_y, we can expect that, in the absence of any polarization coupling, if the optical source is able to excite only one of them, the same remains alone in propagation along the fiber. Unfortunately, polarization coupling occurs because of undesired longitudinal nonuniformities of the fiber itself; in particular the continuous twists and localized mutual rotations at fiber-to-fiber joints are involved in this action. Furthermore, temperature fluctuations induce a continuous variation of the state of polarization at the output, since it is elliptical in general but can degenerate into a linear one at distances that are multiples of $\pi/(\beta_x - \beta_y)$, where β_x and β_y are the phase constants of such two polarizations.

With reference to a polarization matching with the locally generated field, assumed parallel to the xz-plane, we would have an interest in maximizing E_x with respect to E_y. Nevertheless E_x fluctuates, although its fluctuations are complementary with respect to the E_x fluctuations. This suggests the possibility of employing a pair of detectors, arranged as a polarization diversity scheme [35].

Alternatively, since the polarization state is in any case elliptical, we can employ a polarization control device, at the fiber output, which consists of an optical arrangement able to transform the polarization state of the field at the fiber output into the desired one. We can have three types of such a device [36]:

1. a random elliptical polarization can be transformed into a prefixed polarization and then reduced to a linear one along an assigned plane by employing lateral pressure or the electro-optic effect;
2. a random elliptical polarization can be transformed into a known linear one and then properly rotated by exploiting bend-induced intentional birefringence or by using suitable phase-plates;
3. the desired field polarization at the fiber output can be obtained by employing Faraday rotators.

Some of these devices do not ensure the endless polarization control capability which is indispensable, for example, in order to obtain endless polarization state matching between the signal light and the local oscillator light of a coherent system. Particular solutions have to be conceived for this purpose, including one based on three or four phase shifters connected in series with fast axes rotated by 45 deg with reference to each other [37, 38] and controlled by special algorithms, as well as one consisting of two devices which operate as a quarter-wave plate and a half-wave plate, respectively [39, 40]. By optimizing the angle of the two wave plates, endless polarization control can be attained. In any case, when electromechanical components are used for realizing the control, some polarization tracking problems could arise. The time response of such devices, in fact, could be not suitable for compensating very fast polarization changes without loss of information.This problem, however, can be solved by resorting to a completely electronic feedback.

There are also other transmission techniques which exploit directly the double channel offered by the two polarization states by means of a suitable polarization modulation [41]. Such transmission systems have also some advantages over traditional systems for dealing with the effects of phase noise.

Either the use of polarization diversity or that of a polarization control device, as well as the introduction of a true polarization modulation, fail when the degree of temporal coherence of the optical source is not sufficiently high. In this situation, a phenomenon commonly called depolarization of the transmitted light occurs. Nevertheless, this fact becomes appreciable well after the condition in which phase noise is strong enough to make any type of coherent modulation and demodulation practically impossible.

On the other hand, a technique that is capable of avoiding the undesired polarization being excited at all, could be preferable in principle. It is based on the use of polarization-maintaining or single-polarization fibers and is applied frequently for sensor applications. A first solution consists of devising an axially nonsymmetrical structure in which one of the two orthogonal LP_{01} modes is cut off. This result can be achieved, at least in principle, by exploiting a geometrical as well as a material rather strong birefringence. We refer to these structures as absolutely single-mode single-polarization fibers.

The most simple geometry which can be conceived for this purpose is the elliptical-core fiber but more involved structures, like the side-pit [42] or the side-tunnel fibers [43] have also been proposed and fabricated. In all such cases the birefringence is essentially geometrical, though a residual material contribution always exists due to the nonidealities of the fabrication process. In addition, the two deep index valleys present in the side-tunnel fiber induce a distribution of stress over the fiber transverse section which leads to a significant increase in the relative interval of single-polarization operation $(V_{cx} - V_{cy})/V_{cx}$. This parameter, which can be assumed as a measure of the fiber capability of ensuring the propagation of only one polarization state, is on the order of some percent either for elliptical-core fibers (even when very high eccentricities are adopted) or for side-pit fibers, and also when the core-cladding refractive index difference is very large. On the contrary, values of $(V_{cx} - V_{cy})/V_{cx}$ of the order of ten percent appear feasible by using side-tunnel fibers with a normalized index difference Δ of about 0.01. Nevertheless, it is worthwhile to stress that, at the current state of technology, single-polarization fibers introduce an additional attenuation also on the desired polarization. This appears as a rather strong obstacle to a wide diffusion of this kind of fibers for telecommunication systems.

When the single-polarization regime is not possible, a polarization-maintaining scheme can be adopted that consists in making the difference between the phase constants of the two orthogonally polarized LP_{01} modes as large as possible. Again, this can be achieved by an axially nonsymmetrical refractive index distribution or by a nonsymmetrical stress distribution. Such fibers are called linearly birefringent or polarization-maintaining fibers, in the sense that a launched LP_{01} mode can be propagated in the given polarization over a long distance when the phase constant difference is large. More precisely, if the length $2\pi/(\beta_x - \beta_y)$ is much shorter than the typical spatial period of geometrical perturbations (for example, bends) in the fiber, the energy transfer between the two states of polarization of the fundamental mode is practically suppressed.

On the other hand, circularly birefringent fibers (also called circularly single-polarization) can be attained [44]. In this case, a suitable twist is given to an axially symmetrical fiber in order to decouple the clockwise and counterclockwise circular polarizations, taking advantage of twist-induced optical activity. In particular, the phase difference is proportional to the twist angle per unit length.

2.5.4 Chromatic Dispersion Effects in Coherent Transmissions

Function $\beta(\omega)$ can be expanded in power series in the neighborhood of the central emission circular frequency ω_o. This leads to

$$\beta(\omega) \cong \beta(\omega_o) + \beta'(\omega_o)(\omega-\omega_o) + \frac{\beta''(\omega_o)}{2}(\omega-\omega_o)^2 + \frac{\beta'''(\omega_o)}{6}(\omega-\omega_o)^3 + \ldots \quad (2.135)$$

where $\beta'(\omega_o)$, $\beta''(\omega_o)$, $\beta'''(\omega_o)$ represent subsequent derivatives of such a function, all taken for $\omega = \omega_o$. The phase delay per unit distance is related to $\beta(\omega_o)$ according to

$$\tau_f(\omega_o) = \frac{\beta(\omega_o)}{\omega_o} \quad (2.136)$$

whereas the group delay $\tau_g(\omega_o)$, following (2.127), becomes

$$\tau_g(\omega_o) = \beta'(\omega_o) \quad (2.137)$$

$\beta''(\omega_o)$ and $\beta'''(\omega_o)$ are responsible of second-order and third order chromatic dispersion, respectively. $\beta''(\omega_o)$ is proportional to $\frac{d\tau_g}{d\lambda}$, therefore it vanishes at approximately $\lambda_o = 1.27\ \mu m$.

When the spectral width of the optical source is negligible with respect to the modulation bandwidth of the signal, centered on ω_o, the received field [30] can be written as

$$\Psi_r(t) = \frac{1}{2\pi} e^{-i\omega_o \tau_f z} \int_{-\infty}^{+\infty} \Psi(\omega)\, e^{-i\tau_g(\omega - \omega_o)z}\, e^{i\omega t}\, d\omega \tag{2.138}$$

where ω is the optical frequency and $\Psi(\omega)$ the Fourier transform of the modulated field. In (2.138) both $\beta''(\omega_o)$ and $\beta'''(\omega_o)$ have been considered negligible. In this situation, the bandpass signal exhibits no distortion, while τ_f and τ_g, both calculated at $\omega = \omega_o$, represent the delay per unit distance suffered by the carrier and by the modulation envelope respectively. $\beta''(\omega_o)$, produced by $(\omega - \omega_o)^2 z$, gives rise to a distortion which increases as the square root of the distance z. Analogously $\beta'''(\omega_o)$, produced by $(\omega - \omega_o)^3 z$, originates a distortion which increases as the cubic root of z.

2.6 DEVICES FOR INCOHERENT LIGHT GENERATION AND PROCESSING

Incoherent light sources, in conventional terms, are those whose spectrum is wider than few nm. So lamps, employed sometimes in experimental bench arrangements as wide-spectrum sources, and semiconductor *light emitting diodes* (LEDs), similarly used but frequently employed also in telecommunication and field test applications, only populate this class of optical sources. Nevertheless, with the advent of highly coherent semiconductor lasers, also conventional heterojunction lasers are usually classified as sources of incoherent light. In this sense, the threshold for the separation between incoherent and coherent light generators can be set now at a value for the radiation spectral width of the order of 0.1 nm. Solid-state and gas lasers, whose use is typical in traditional optical benches, remain classified as coherent light source since their selectivity, imposed by the overall length of their cavity and necessary for laser action, is intrinsically remarkable.

A necessary condition for a light source to generate a highly coherent light consists in operating in a single-mode region. This means that resonance of the device is to be favored on a single transverse mode as well as a single longitudinal mode. The second requirement can be satisfied by properly selecting the optical path length within the open resonator. The first one can be controlled by acting on the thickness and refractive indices of the various layers used to fabricate a semiconductor laser or inserting into the laser resonator a circular aperture. The size of this aperture has to be chosen in such a way as to allow the fundamental mode to experience little diffraction loss while forcing the higher order modes to suffer appreciable attenuation.

Single-mode operation can be facilitated also by replacing one of the partially reflecting mirrors (usually a plane mirror) constituting the Fabry-Perot cavity with a spherical mirror. The remaining plane mirror is arranged approximately in the curvature center of the sphere and the overall scheme is called hemispherical [45]. Single-mode regime can be easily attained by a suitable choice of the distance between the two mirrors but, as a counterpart, the output optical power is only about half the maximum obtainable.

A particular case is that of tunable optical sources. Tunability can be reached, for example, by means of a monochromator inserted after a wide-spectrum lamp or an LED. Nevertheless, such a solution usually gives rise to very low optical power levels, especially when a rather thin wavelength spectrum is to be selected by the monochromator. A different solution may consist in adopting dye lasers. Laser action in these devices takes place between the broad energy bands characteristic of complex strongly absorbing and fluorescing molecules, commonly called *dyes*, in condensed liquid solvents. More than 200 substances are actively used as laser dyes, each fluorescing in a specific wavelength region. The tunability of the dye laser, due to the broadened electronic states, is then realized by inserting different dyes into the laser cavity [46].

Tunability over a frequency range of the order of 100 THz can be accomplished by using fiber-Raman lasers based on the stimulated Raman scattering nonlinear process in optical fibers. This consists in the conversion effect of a small fraction of the optical power incident on a nonlinear medium to another optical beam at a frequency downshift by an amount determined by the vibrational modes of the medium itself. Thus the incident light acts as a pump to generate the frequency-shifted light, which is called the *Stokes wave*. In its classical configuration, a tunable fiber-Raman laser is formed by a single-mode fiber placed inside a Fabry-Perot cavity including a prism. Tuning is made possible by the intracavity prism, which spatially disperses the various Stokes wavelengths. These are finally selected by rotating the mirror that terminates the Fabry-Perot cavity. Alternatively, the laser wavelength can be tuned by simply changing the cavity length. This technique is referred to as *time-dispersion tuning*; the previous method is called *spatial* [27].

As light processing we can define any operation acting on an optical radiation with the purpose of varying intentionally its parameters. A fundamental aspect of light processing (not only in telecommunication applications) is, for example, its intensity or coherent modulation. Such operation, and the main devices able to ensure it, have been described in Section 2.3.1.

Light processing, in general, can be accomplished by a two-port, a three-port, or an N-port device, with $N > 3$. A typical two-port device is a semiconductor optical source, in which the drive current crossing the device generates and modulates the light intensity of the emitted radiation. An example of three-port device is an external modulator, in which an electric signal induces variations in the light crossing the device itself.

Light processing, in extended terms, can also be considered to include the operation, performed by means of suitable passive devices, through which two beams are combined into a unique beam (like, for example, that required by a heterodyne detection scheme), or one beam is subdivided into some beams. In operations of this type, all-fiber devices appear very practical. They are intrinsically not wavelength-selective components, therefore are mainly employed with not particularly coherent-light beams. Also *wavelength division multplexing* (WDM) devices have a similar characteristic.

A completely different situation is that of *frequency division multplexing* (FDM) devices, where highly-coherent-light beams are combined or extracted, within spectral distances of the order of one GHz, or even hundreds of MHz.

Among the remaining functions that a light processing component can carry out, it is possible to list:

1. optical switching,
2. optical filtering,
3. optical isolation,
4. polarization control.

All such functions are typically made by active components. The control signal can be either electrical or optical.

Except for all-fiber devices, whose use is essentially confined to passive optical branching (depending on the size of the devices themselves and especially on their fabrication technology), we can consider the following two classes of components for light processing: micro-optic devices and integrated-optic devices. Both of them can be employed with either incoherent light or coherent light.

2.6.1 LED Sources

An LED consists of a broad bandgap semiconductor diode realized by bonding a p-type semiconductor material to an n-type semiconductor. When the device is forward-biased, current flows through the junction as a consequence of the flow of holes from the hole-rich p material to the n material and, conversely, of the flow of electrons from the n- to the p-type semiconductor. Light emission results from the radiative recombination processes between holes and electrons which take place in the two regions. However, radiative recombination in the p region is more efficient, so that LEDs are physically built in such a way that most emitted radiation escapes from the p material. In any case, the recombinations are substantially single, so that such a phenomenon, which is usually called spontaneous emission, leads to a rather broad spectrum of emitted wavelengths and to an almost isotropic radiation pattern inside the material.

The bandgap energy, which is typical of the material itself, imposes the emission wavelength in the sense that the emitted photons have an energy approximately equal to this bandgap. GaAlAs devices are commonly used in the neighborhood of $\lambda = 0.85$ μm, while InGaAsP devices, with their emission wavelengths in the region between 1.0 and 1.7 μm, are commercially available for applications at $\lambda = 1.3$ μm and $\lambda = 1.55$ μm.

Two fundamental LED configurations are commonly used in fiber-optic applications: surface emitters (or Burrus or front emitters) and edge emitters. Surface emitters have the plane of the active light-emitting region perpendicular to the axis of the fiber; the emission pattern is almost isotropic with a half-power beam width equal to 120 deg. Edge emitters, on the other hand, are characterized by the presence of an active junction region which is sandwiched between two guiding layers. The guiding effect is obtained by adopting, for such layers, a refractive index lower than that of the active region but higher than the index of the surrounding material. In this way, the incoherent light emitted by the active junction region is guided towards the fiber core. As a result, the emission pattern of the edge emitter is more directional than that of the surface emitter, with a half-power beam width in the plane perpendicular to the junction not greater than 30 deg. In this way, the coupling with the fiber is obviously made more efficient.

LED sources exhibit a linear characteristic for the emitted power P as a function of the drive current I and the wavelength shift due to current variations is usually small with respect to the spectral extension of their linewidth, which often exceeds 10 nm. In addition, it must be noted that the equivalent electric circuit of an optical source always contains a capacitance which, placed in parallel to its dynamic resistance, leads to a low-pass baseband response. This capacitance is related to the finite lifetime of carriers injected into the recombination region and fixes the maximum value of the modulation bandwidth. In case of an LED source the latter is of the order of 500 MHz. Also important is the value of the power that an LED source can launch into the single-mode fiber realizing the transmission medium: it rarely exceeds 0.1 mW, even in case of edge emitters.

2.6.2 Light Sources with Moderate Temporal Coherence

If a higher temporal and spatial coherence than that provided by an LED source is required, it is necessary to use a laser source. This can occur also in intensity-modulation systems in the presence of very stringent requirements of chromatic dispersion lowering.

A semiconductor laser source is characterized by several similarities with an LED but also by some important differences that should be considered when selecting the best source for a specific fiber-optic communication system or sensor.

Similarly to an LED, in a laser made of semiconductor materials light generation is due to the mechanism of electron-hole pair recombination which takes place in correspondence of a *p-n* junction. Nevertheless, in contrast to the LED, carrier injection produces population inversion and the emitted photons are in phase with the photons associated to an incident radiation. In this way, the latter is amplified and the resulting emission, highly monochromatic and able to support a very directional light beam, is known as *stimulated emission*. Thus, temporal and spatial coherence can be remarkable.

Laser sources are characterized by a threshold in their function P(I), whose value may range between I = 20 mA to I = 100 mA. Only above such a threshold can stimulated emission of radiation take place. The modulation bandwidth is greater than that offered by an LED. This is related to the phenomenon of stimulated emission, which causes a remarkable reduction of the carriers lifetime.

In its classical configuration, a laser can be seen as a resonator structure. A mode of a resonator can be defined as a self-consistent field configuration. This means that the optical field distribution reproduces itself after one round trip in the resonator. We can identify a set of longitudinal (or axial) modes whose spatial energy distribution in a transverse plane is the same and is imposed by the transverse geometry of the waveguide but have different axial distributions corresponding to different numbers of half-wavelengths along the axis of the resonator. These longitudinal modes are spaced in frequency by c/2L, where c is the velocity of light and L is the optical path length within the resonator.

The laser linewidth depends on the number of longitudinal modes excited. Each individual longitudinal mode, in fact, is characterized by a spectral linewidth of the order of 0.1 Å. So the presence of three or four lines, as it is typical for many devices operation conditions, leads to an overall spectral width of 0.5 to 1 nm within the half-intensity points. For rather low drive currents, that are greater than the threshold value, the number of longitudinal modes may be greatly increased and overall spectra as large as 4 nm may result in some cases.

To obtain laser operation at a single longitudinal mode, which is typically required in order to have a laser emission linewidth smaller than 0.1 nm, it usually necessary to design a resonator sufficiently short, so that c/2L is greater than the bandwidth of the gain. Alternatively, one can design a resonator characterized by high loss for all modes within the oscillation bandwidth except for the desired one.

It is possible to show that for each longitudinal mode there is a set of self-consistent solutions for the light energy inside the resonator corresponding to different distributions in the plane perpendicular to the resonator axis. They are called the *transverse modes* of the laser. In order to obtain oscillation in a single transverse mode as well, the property experienced by higher order modes of being spread further from the resonator axis can be effectively used. It is then sufficient to insert a circular aperture into the laser whose size is such that the fundamental mode is subjected to little diffraction loss while higher order modes suffer appreciable attenuation.

As regards the spectral linewidth of semiconductor single-mode lasers, when a standard Fabry-Perot solution is adopted, it is typically on the order of 100 MHz. For coherent optical communication systems this broad linewidth is unsatisfactory. The main difficulty in obtaining narrower linewidths in semiconductor lasers is that there is a modulation of the refractive index due to fluctuations of the excited population caused by spontaneous transitions. This has an effect of broadening the spectral linewidth of the lasing transition by a factor $(1 + b^2)$, where b represents the antiguiding coefficient of the device material. Typical values of b range between 4 and 8. The laser spectral linewidth can be easily derived by the curve of the laser power spectrum versus frequency. Some examples are qualitatively reported in Figure 2.19, where the effect of increasing b has been evidenced.

For the sake of clarity, the laser power spectrum has been normalized with respect to the maximum value which falls in correspondence of the central emission frequency f_0. It must be noted that besides the antiguiding coefficient, the values of the average spontaneous emmission rate and of the optical intensity also play an important role in determining the laser linewidth broadening. More precisely, the smaller the average spontaneous emission rate and the higher the optical intensity, the smaller the laser linewidth. From a different point of view, it is possible to verify that the spectral linewidth of a semiconductor laser is proportional to the ratio $[\log_e(1/R)]^2/L^2$, where L is the cavity length and R is the reflectivity of one terminal facet. When inherently narrow spectral linewidths are necessary, as in coherent communication systems, particular laser structures and techniques have to be conceived. Some solutions to this problem will be discussed in Section 2.7.

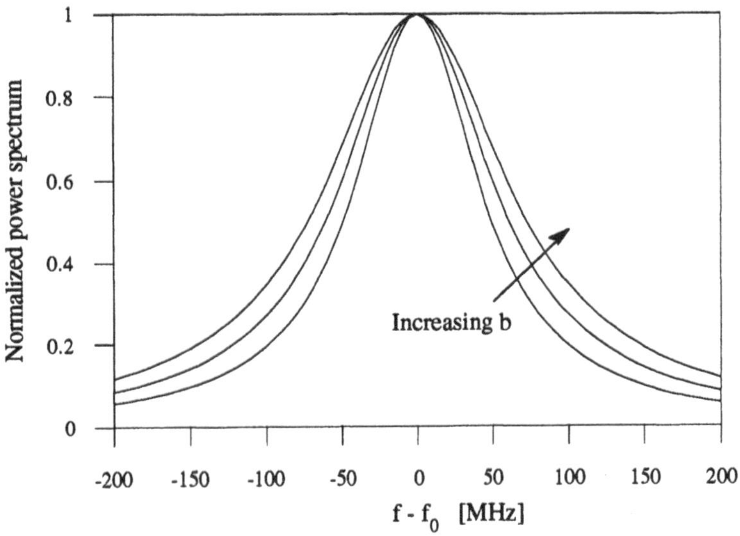

Figure 2.19 Typical behavior of the normalized power spectrum for a semiconductor laser.

2.6.3 All-Fiber Devices for Incoherent Light Processing

All-fiber devices are usually passive components in which the optical signals can be combined processing their powers. They act as branching elements, with three ports, four ports, or N ports.

An example of a three-port device is a beam-splitter [47]. It is represented in Figure 2.20(a). The power entering port 1 is subdivided into two equal fractions, except for possible losses, which reach port 2 and port 3, respectively. In the opposite direction, the powers entering port 2 and port 3 are superimposed and collected on port 1.

One type of four-port device is called a *directional coupler* [48]. It is represented in Figure 2.20(b) and acts as a similar component at microwave frequencies. The power entering one port on the one side is subdivided into two fractions, according to a ratio R which characterizes the directional coupler itself. These fractions outgo exit two ports on the other side.

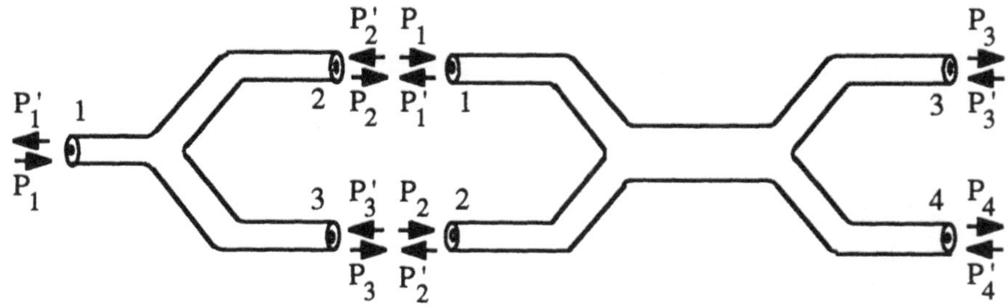

Figure 2.20 (a) Beam-splitter and (b) directional coupler.

According to the symbolism introduced in Figure 2.20, neglecting losses, we have the following equations regarding a beam-splitter

$$P_2 = P_3 = \frac{P_1}{2} \qquad P_1' = P_2' + P_3' \tag{2.139}$$

and the following equations regarding a directional coupler, characterized by ratio R,

$$P_3 = RP_1 + (1 - R)P_2 \qquad P_4 = RP_2 + (1 - R)P_1$$
$$P_1' = RP_3' + (1 - R)P_4' \qquad P_2' = RP_4' + (1 - R)P_3' \tag{2.140}$$

Typical values of R range between 10^{-2} (weak coupling) to 0.5 (strong coupling).

An N-port device, with N/2 ports on the one side and N/2 ports on the other side, can be seen as a generalization of a directional coupler [49]. Typically, from any port of the first side we can obtain an equal subdivision of the incoming power towards all the ports of the second side, except losses.

Although all the fibers employed for the construction of such devices are single-mode, no effect of phase selection is exploited in order to ensure a certain behavior. Therefore no particular requirement is imposed on the coherence properties of the light in crossing the devices themselves. The choice of operating only on the powers of the various optical waves, in contrast to what occurs in similar components at microwave frequencies, is justified by the difficulties in reproducing an exact fraction of wavelength at these very high frequencies with fiber-based technology. Nevertheless, undesired wavelength selection properties can occur in such devices [50], to the point where particular care must be taken in order to have a wavelength-flattened behavior [51].

All-fiber devices are sometimes employed for WDM signal processing. WDM was originally conceived for enlarging the transmission capacity of a point-to-point optical link. Nevertheless, at present, a transmission capacity as high as about 10 Gbit/s can be ensured by a unique signal on a single-mode fiber, making such application of WDM unnecessary for point-to-point links with single services. More recently, WDM has been suggested for local area network applications or for distribution network applications. On the same transmission medium, several channels can be supported by means of as many source wavelengths separated by a few nm.

Insertion of such wavelengths into a single fiber from the light generated by suitable optical sources and their extraction close to the users are performed by means of wavelength-selective components, such as prisms or intereference filters. They are placed in beam-splitters, directional couplers, or N-port all-fiber devices, in particular at points where the various waveguides are fused together. In this way, such devices can be used only in one direction: from many inputs to one output, when the wavelengths have to be collected, and from one input to many outputs, when the wavelengths must be separated.

In these applications, the various wavelengths also form a spectrum not so dense as to require coherent processing of the signal. Furthermore, signal modulation is typically performed acting on the intensities of the light. This enables us to classify the entire family of all-fiber passive devices as components for incoherent light processing.

2.6.4 Micro-Optic Devices for Incoherent Light Processing

Traditionally, only lenses, gratings, prisms, and mechanical switches were considered to be micro-optic devices [52]. $LiNbO_3$ components [53] and SiO_2-substrate components [54] based on guided wave principles were considered integrated-optic devices, together with the first attempts at performing similar functions on GaAs substrate [55]. With the advent of single-mode fibers and the necessity of well-collimated beam processing, the requirement of integrating such functions over the same components able to generate light has become dominant. In this sense, GaAs and other semiconductor compounds appear preferable.

Micro-optic devices will be treated here for the applications regarding incoherent light processing, and in 2.7.4 for the applications in coherent light processing.

One of the most investigated functions in micro-optic components is optical switching. There are three principles of optical switching:

1. space division,
2. time division,
3. wavelength division.

An optical space division switch is applied in broadband video signal switching networks [56]. This generally requires a large number of optical switches and input/output connections, although it is characterized by not particularly high-speed control devices. Conversely, optical time division switch, which seems to be more suited for its use in conjunction with optical time division multiplexed transmission systems, requires high speed optical memories as well as optical write/read gates. Finally, wavelength division switch results from the combination of wavelength selectors/switches and wavelength converters.

In regards to the switches adopted in space and time division, those which better agree with the micro-optic components definition are the mechanical switches consisting of movable fibers [57], prisms, and mirrors. These devices are based on the deflection of an incident optical beam and are usually characterized by high switching times (on the order of 1 ms) and large sizes.

Two basic types of movable fiber switch have been proposed and developed up to now. The first, called *end-fire switch*, employs piezoelectric elements, whose bending is controlled by a suitable voltage, to move the input fibers relative to the selected output fibers. In a variant of this solution, a funnel arrangement is used to direct individual, electromechanically activated, fibers into the correct position for coupling with a fixed fiber. This scheme is then well suited for N-to-1 switching. The other fundamental type of movable fiber is called *remote connection switch*; it employs two facing arrays of fibers and lenses. In this case, piezoelectric elements are also used for deflecting the input fibers in such a way as to change the angle at which the collimated light is launched by the input lens. This lens directs the light towards the output lens, where the position of the output fiber has been deflected to provide maximum coupling efficiency.

Instead of moving the fibers, it is obviously possible to realize a switching operation by moving the light beams [58]. More precisely, the beams are collimated and focused by lenses, while prisms or mirrors selectively deflect the light, thus defining the switch routing. In a typical configuration, the input and output fibers, together with the corresponding lenses, are arranged along two sides of a rectangular matrix; prisms and mirrors are placed inside the matrix and electromechanically inserted into the beams [57, 59].

A quite different solution, but always classifiable as micro-optic component, is that based on liquid crystal switches [60]. The liquid crystal layer is sandwiched between glass supports together with transparent conductors, which allow the application of the control voltage, and alignment layers, which impose the direction of the molecules in the crystal. When a voltage is applied, the molecules tend to rotate according to the direction of the applied field. As a consequence, the refractive index seen by a light beam passing through the layer is varied. By a suitable choice of the angle formed by the incoming beam for a fixed-pair glass-liquid crystal, it is possible to have total internal reflection at the glass-liquid crystal interface when no voltage is applied or a complete transmission, at least in principle, under the action of a proper applied field. Because the liquid crystal is birefringent and then sensitive to the state of polarization of the incoming light, it is necessary to add a polarizer at its input. Some power is then lost, owing to the rejection of one polarization state, but this can be easily recovered by introducing some modified designs, arranged to separately switch the two polarization components.

For switching capacity expansion, multistage networks (for instance, of the type space-time-space) are advisable [56]. The number of stages is mainly limited by the optical insertion loss for the matrix in the time and space switches. High-speed operation for the time switch is, instead, limited by the read/write memory cycle time in a bistable laser diode. A value of 1 Gbit/s, for the optical memory operation of a bistable laser diode, has been, however, experimentally obtained since 1985.

Other applications of micro-optics include single optical elements like lenses, gratings, prisms, which can be assembled in different fashions depending on the function to be realized. Lenses are typically employed for focusing a light beam onto the core of a single-mode fiber [61] or onto that of a planar waveguide [62]. Circular and cylindrical lenses are used for changing properly the beam transverse geometry. Gratings or prisms separate different wavelength components in WDM multiplexing or demultiplexing [63].

2.6.5 Integrated-Optic Devices for Incoherent Light Processing

One of the most investigated applications for integrated-optic devices is optical switching. A change of direction of an optical guided beam can be accomplished by the application of several principles:

1. refractive index change induced by electric signals (electro-optic effect),
2. refractive index change induced by elastic waves (acousto-optic effect),
3. refractive index change directly induced by optical signals [64],
4. nonlinear interaction [65].

The most common electro-optic material is undoubtedly $LiNbO_3$. The fabrication technology of $LiNbO_3$ components is at present well consolidated and consists of diffusing a patterned layer of titanium thin-film into the polished surface of a lithium niobate substrate. In this way, channels of raised refractive index are produced that each forms an integrated-optical waveguide. An example of the application of such a technology is given by the so-called stepped $\Delta\beta$ directional coupler switch. This device is formed by two monomode waveguides that when brought together allow their modes to couple. Thus energy can couple back and forth periodically between the guides along the length of the coupling region. By applying an electric field, the period of this region can be varied so that the selection of the guide, in which optical energy remains at the end of the device, can be easily controlled. Switching times are typically on the order of some ns, but special structures can be designed in order to significantly reduce it.

In an acousto-optic switch, the variation of the refractive index (necessary for scattering the incident beam out of its original direction) is generally due to an acoustic wave launched into a block of fused silica. In comparison with electro-optic devices, the acousto-optic switches are generally much slower and provide inferior contrast. On the other hand, they require considerably smaller voltages and are more transparent as regards the maximum repetition rate that can be obtained.

The strong polarization dependence which is typical of the waveguide switches based on the electro-optic effect can be overcome by resorting to InGaAsP/InP optical switches, in which the refractive index variation is induced by a current injection. Such a structure is composed of two multi-mode InGaAsP/InP buried heterostructure ridge waveguides whose intersecting angle is on the order of some degrees. The carrier confinement area is then fabricated in the middle of the intersecting area. When carriers are injected from the *p-n* junction into the InGaAsP waveguide, the refractive index of this area decreases and a total internal reflection is produced. It can be seen that the induced refractive index variation increases almost linearly with the injection current density and reaches a value on the order of 10^{-2} for a current nearly equal to 5 kA/cm^2, when typical dimensions for the structure are considered.

By generalizing this mechanism, monolithic optical switches have been realized, based on bistable laser diodes, which can be triggered by an incoming optical pulse and can store this information bit until reset. In order to switch the state of the output optical signal, the optical input can also exploit the intrinsic nonlinear properties of the device material. More precisely, one of many fascinating features of nonlinear waveguides is the flux-dependent behavior of the field distribution. In other words, the local electric field strength is a function of the the total optical power as its intensity changes the refractive index value. With reference to a very simple slab waveguide with nonlinear core and linear cover and cladding, in the range of low power levels the strength at the center of the core is larger than in the cladding and both of them increase with increasing the power. However, in the range of high power levels, the dominance of the field strength is reversed and the strength at the center decreases with increasing the power whereas that in the cladding still increases. This characteristic indicates that when the optical flux density increases the wave evolves into a single interface surface polariton, that is, a field distribution with a maximum (bulge) of solitonic type in the cladding. Consequently, if the guide is used as an all-optical switch, its output passes from state 1 to state 0.

The electro-optic effect is, at present, the one more frequently employed in practical realizations. Electro-optic waveguide switches can be integrated on a single substrate to form optical switch arrays. To date, arrays of up to 16×16 in size have been reported with the array performance which is critically dependent on the type of crosspoint adopted and on the size of the array. Sometimes several N×N switch arrays, with N on the order of few units, are connected to each other in order to obtain multistage switch matrices [66].

In some particular situations it could be convenient to construct hybrid integrated-optical circuits, which combine GaAs optical sources and multiplexing elements mounted on a $LiNbO_3$ substrate [67]. The problem of interconnecting a series of stripe waveguides and associated devices formed in glass and lithium niobate substrate can be solved by using sandwich ribbon fiber coupling elements or resorting to jointing techniques which employ etched grooves. Even more effective, especially in order to satisfy the need of exact phase matching between the coupled mode, is the adoption of tapered waveguiding regions in lithium niobate guides.

Besides optical switching we can have a certain number of functions to be carried out by suitable integrated-optical devices, such as second- and third-harmonic generation, and four-wave mixing. All these functions are based on the strong nonlinearity properties that some media exhibit. More precisely, by exploiting second-order nonlinearities, it is possible to design efficient, low-power, frequency-conversion devices, while third-order nonlinearities are the basis for many all-optical signal processing systems.

For a better understanding of the significance of these statements, it is useful to note that the response of any dielectric to light becomes nonlinear for intense electromagnetic fields. The origin of such a behavior is related to a harmonic motion of bound electrons under the influence of an applied field. The nonlinear mixing of optical fields can then be described by an expansion of the material polarization, induced from the electric dipoles appearing in the material, in terms of the Fourier components of the input fields of frequency ω_1 and/or ω_2 and/or ω_3 as

$$P_i(\omega) = \varepsilon_0 \left\{ \chi^{(1)}{}_{ij} E_j(\omega_1) + \chi^{(2)}{}_{ijk} E_j(\omega_1) E_k(\omega_2) + \chi^{(3)}{}_{ijkl} E_j(\omega_1) E_k(\omega_2) E_l(\omega_3) + .. \right\} \tag{2.141}$$

where $\chi^{(1)}$, $\chi^{(2)}$ and $\chi^{(3)}$ are the first-, second- and third-order susceptibilities, respectively. When second-order nonlinearities are considered, $\omega = \omega_1 \pm \omega_2$, which always leads to frequency conversion. On the contrary, by assuming third-order nonlinearities we can have $\omega = \omega_1 \pm \omega_2 \pm \omega_3$, which can lead to frequency conversion or, for example, $\omega = \omega_1 - \omega_1 + \omega_1$, which gives rise to an output signal at the same frequency of the input signals. The latter case is well-known as degenerate four-wave mixing and has potential application to real-time signal processing operations involving the mixing of two or more waveforms, such as convolution and time inversion. In this interaction, the signal power is proportional to the product of the input powers of the three input waves at the same frequency. This result has been experimentally demonstrated [68] in a nonlinear waveguide in which all the beams were derived from the same laser.

Four-wave mixing can be also used for achieving optical phase conjugation. By this term we basically mean a technique which allows us to precisely reverse the direction of propagation and the overall phase factor of each plane wave in an arbitrary beam of laser light. In practice, a beam reflected by a phase conjugator retraces its original path and this property can be very useful in the problems associated with maintaining high optical quality after passing through unavoidable nonuniform or distorting media.

In order to generate a phase conjugate return by a four-wave mixing process, the optical Kerr effect, already introduced in Section 2.3.5, can be adopted. Kerr-like media are often called n_2-materials, in that the refractive index is the sum of the linear index n_0 and the term n_2E^2, where E is the electric field associated with the external light wave. If two strong counter-propagating beams at frequency ω, playing the role of pump signals, are directed to set up a standing wave in a Kerr-like material, a third beam, playing the role of probe signal at the same frequency, incident upon the material from any direction, will result in a fourth beam precisely retracing the third one.

Finally, it is important to note that the variation of the refractive index, occurring in a Kerr-like medium, is due to the real part of the third-order susceptibility in equation (2.141). When such a change is proportional to the external electric field instead of the square field, the so-called *Pockels effect* results. The latter is most commonly employed in order to obtain the field dependent changes in the refractive index which rule the electro-optic or magneto-optic effects.

2.7 DEVICES FOR COHERENT LIGHT GENERATION AND PROCESSING

As anticipated in Section 2.6, by the term *coherent light* source we substantially mean all those light generators whose radiation spectral width is smaller than 0.1 nm. Special solutions must be generally adopted in order to obtain such a result. Among them, it is useful to consider the usage of a semiconductor device inserted in an external longer cavity [69] and the adoption of *distributed feedback* (DFB) lasers [70] or *distributed Bragg reflection* (DBR) lasers [71].

More recently, an increasing interest has been devoted to quantum-well semiconductor devices [72]. With respect to conventional double-heterostructure lasers, whose active layer thickness is typically in the range 0.1–0.3 μm, they are characterized by a thickness of the active layer of the order of 10 nm or less. In these conditions, the carrier (electron or hole) motion normal to the active layer is greatly restricted and the kinetic energy relative to this motion is quantized in discrete energy levels, similar to the well-known quantum-mechanical problem of the one-dimensional potential well.

Obviously, the objective of realizing very narrow spectral linewidths can also be achieved, in principle, also by using more traditional sources such as gas or solid-state lasers. The most common types of gas lasers are the helium-neon laser, the argon and krypton ion lasers, the carbon dioxide laser, the rare-gas-halide excimer laser, and some chemical lasers (including, for example, the hydrogen-fluoride laser). In all these devices, population inversion is accomplished by applying a high voltage between electrodes located within the gaseous medium. In this way, high-energy electrons are created and their collisions with gas atoms provide the energy necessary for preparing stimulated emission.

Solid-state lasers, in turn, usually consist of transparent crystals or glasses, playing the role of host materials, within which ionic species of laser atoms are dispersed. Typical host materials are aluminum oxide, garnets, and various forms of glasses. The most common lasing species are, instead, neodymium ions and ruby ions. A separate treatment has to be made for color center lasers. In these devices the laser species is produced by irradiating the crystal with x-rays, with the aim of producing defects capable of attracting electrons. Such defects lead to energy levels that absorb and emit light, and can be inverted to produce gain. The host material is usually an alkali-halide crystal. In almost all the types of solid-state lasers, population inversion is forced by a light source that is focused into the crystal to excite the upper laser levels.

Regardless of the properties of the amplifying material, the laser includes an optical resonator or mirror arrangement that provides feedback of the amplified beam into the gain medium. Intrinsically associated with the selectivity properties of such an optical cavity, there is therefore the possibility of obtaining sources characterized by very narrow spectral linewidths. Nevertheless, most of the sources currently adopted in the framework of fiber optic systems are based on semiconductor devices. Thus, in Sections 2.7.1, 2.7.2 and 2.7.3, we will restrict our analysis to the discussion of the main properties of some semiconductor components characterized by good performance for coherent system application, while we refer to other, more specific, texts for a comprehensive description of the properties of gas and solid-state lasers (see, for example [73] and [74]).

As regards the proposal of active or passive components suitable for coherent light processing we can repeat what we said in Section 2.6 with regard to the devices for incoherent light. The basic functions, in fact, are substantially the same, though only in the coherent case can the large number of degrees of freedom made available by light be completely exploited. In this sense, a substantial impetus to the development of coherent systems has been given in the last few years by the progress made in the fabrication technology of some key components, such as modulators [75, 76], high-efficiency detectors [77], integrated-optic components (i.e., couplers, hybrid optical systems, etc.) [78-80] and high-speed electronic circuits [81].

2.7.1 DFB Laser Sources

We have already stressed in Sections 2.3.1 and 2.3.4 that the possibility of accomplishing coherent transmission systems is related to the availability of laser sources characterized by very narrow spectral linewidths (less than a few MHz in some applications). Particular laser structures have been developed to this goal; among them, a preeminent place is occupied by the DFB laser.

Besides the reduced spectral linewidths, other advantages may also result from the use of such devices. In fact, most semiconductor lasers have a number of modes operating simultaneously as both longitudinal and transverse modes. The mode whose central frequency is the nearest to the peak of the gain curve is the most intense one, but other modes within the gain bandwidth but farther from the peak can also contribute to the output optical power. Though these modes are in general greatly attenuated in continuous wave operation, their power contents can be remarkably increased in the case of pulsed operation. Since direct modulation of semiconductor lasers at frequencies on the order of 1 GHz is, at present, often realized, the presence of undesired lateral modes, combined with a residual chromatic dispersion always introduced by the fiber, can severely limit the transmission rate. Therefore, the adoption of single-mode lasers such as the DFB ones seems preferable also from this point of view, especially for high modulation rates.

In these devices, the modal selectivity is improved by a nonuniform feedback, which is made dependent on the frequency in such a way that the cavity losses are not the same for different modes. This result is obtained by including a corrugated region in one of the laser claddings. More precisely, the grating structure is distributed along the entire length of the laser, just above the gain region.

This corrugation causes a periodic perturbation of the refractive index along the structure, which is responsible of a coupling between the forward and backward waves, according to a phenomenon known as *Bragg scattering*. For this purpose, the following condition between the corrugated period and the laser emission wavelength has to be satisfied

$$m\Lambda = \frac{\lambda}{2n} \tag{2.142}$$

where Λ is the grating period, λ the laser wavelength, n the refractive index, and m an integer which states the order of Bragg scattering. By means of a proper choice of Λ, an accurate selection of the frequency at which feedback takes place is possible. Moreover, as the modal selectivity is produced by an internal variation of the refractive index, single-mode operation is ensured for a wide range of working conditions and is not very sensitive to variations of temperature and driving current, for example.

Other attractive features of the DFB laser are a great mechanical stiffness, a reduced size, and the fact that the cavity is defined by the grating structure and not by cleaved facets. This allows the DFB laser to be better suited for direct incorporation into an integrated optoelectronic device.

In manufacturing the laser, it is generally preferable to avoid a direct etching of the active region because such a method could lead to the introduction of some defects which, in turn, could increase the nonradiative recombination mechanism and, consequently, the threshold current value. Proper choice of the corrugation position and thickness is also very important in this sense. The grating is then realized in the upper cladding, where a final adjustment, after testing the emission wavelength, is possible before the device is made commercially available.

The value of the coupling coefficient κ, which determines the device performance, is strongly influenced by the grating parameters (i.e., the shape, the thickness and the period of the corrugation) as well as by the width and composition of the active and cladding regions. For a fixed length of the device, the value of κ has to be maximized, in that it implies an improvement of modal selectivity, a reduction of the threshold current, a narrowing of the spectral linewidth, and a decrease of the device sensitivity to external undesired reflections. In particular, an optimum value of the active region width, d, exists, which allows to minimize the threshold current density J_{th}. For values of d lower than this optimum, J_{th} rapidly increases, owing to the reduction of the field confinement factor Γ.

A comparison between the values of J_{th} for the TE and TM modes supported by the structure shows that, similarly to more conventional lasers, the fundamental TM mode reaches threshold at a higher value of the injected current, but the threshold margin between TE and TM modes is much smaller than that, for example, of a Fabry-Perot laser. This suggest the possibility that a DFB laser is susceptible to a kink in its light-current characteristics, for relatively low power levels. The assumption of d ≈ 0.1 mm can be assumed as typical in order to have a threshold margin sufficiently high without significantly increasing J_{th} for the fundamental TE mode. Furthermore, it must be taken into account that the presence of an internal stress normal to the active layer can reduce threshold for the TM mode and increase it for the TE mode. When this stress is particularly heavy (equal or greater than 100 atm), the TM mode may even reach threshold first. An important role, in this sense, can be played also by the waveguide roughness. Finally, the threshold current density is also influenced by the cladding thickness, h, and may change by more than 10% for a variation of h by the same amount.

As regards the spectral linewidth $\Delta\nu$, it is possible to show [82] that, in case of DFB laser with no phase-shifters and no mirror facets, the following expression holds

$$\Delta\nu \propto \frac{\alpha_o}{\kappa^2 L^3} \tag{2.143}$$

where α_o is the sum of the absorption loss and the scattering loss while L is the device length. Therefore, $\Delta\nu$ rapidly decreases with increasing κ and/or L. Linewidths on the order of 1 MHz (that is, two orders of magnitude narrower than the conventional Fabry-Perot laser structures) can be achieved without particular technological efforts. Thus, these devices seem quite suitable for their application in coherent transmission systems.

The basic principle governing the operation of the DFB laser is applied also in the DBR laser. This device, however, uses a passive waveguide region to contain the corrugated structure. The presence of a discontinuity between the active and passive regions penalizes the coupling efficiency which is lower than in the DFB case. As a consequence, the threshold current density is higher.

2.7.2 External Cavity Stabilization

When even narrower linewidths than those provided intrinsically by DFB lasers are required, the adoption of an external cavity is advisable. The laser emission can be collected and collimated by an objective lens while a reflecting diffraction grating or etalon can be usefully adopted for improving frequency selection. In this case, linewidths as narrow as 1 kHz have been achieved [83].

As the optical feedback from the external cavity is, in general, strong, an antireflection coating has to be employed on the laser facet inside the external cavity and facing the frequency selectivity element. In this way it is possible to eliminate the longitudinal modes of the Fabry-Perot structure, avoiding competition between the longitudinal modes defined by the external cavity and the cleaved facets.

The main disadvantage of external cavity lasers lies in their relatively large dimensions, typically on the order of many centimeters, which make these devices ill-suited for direct integration in optoelectronic systems. On the other hand, they offer the significant advantage of allowing wavelength tunability - more precisely, the possibility of tuning the laser emission wavelength over a certain (at least discrete) range, by changing the external cavity length. This can be done by acting on the device temperature or by using a piezoelectric transducer in such a way to realize an electronic control. Alternatively, for a fixed external cavity length, tuning can be obtained by changing the temperature of the laser. In any case, a thermal variation induces a change in the refractive index distribution. This, in turn, causes a variation in the optical length, thus shifting the modes sustained by the laser. For a 250–μm–long laser whose coupled external cavity has length 150 μm, a tuning range of 6 nm can be easily achieved [84].

The spectral linewidth can also be varied by acting on the external cavity dimensions. In particular, it can be reduced by increasing the external cavity length, which has the effect of increasing the photon lifetime. For a given output power, this mechanism enhances the number of intracavity photons and, as one can demonstrate, this increase leads to a linewidth decrease.

Before concluding the present subsection, it is worthwhile to stress that the external cavity can be itself pumped to provide gain, thus realizing an active-active scheme. If both sections are independently pumped, we obtain an additional degree of freedom that can be used to control the behavior of the device. The most common choice consists of using identical active materials for both cavities or, even better, in creating the two cavity sections by means of a gap formed in a conventional semiconductor laser; in the latter, alignment between the active regions is automatically satisfied [85].

In the active-active solution, the modes of the external cavity can be electronically shifted by independently controlling the current in the two cavities. If one of the cavities operates below laser threshold, a change in its drive current leads to a significant variation of the carrier density inside the active region. Thus the refractive index changes as well and, consequently, different modes of the laser cavity can be selected. Nevertheless, when both sections operate above threshold, the device is able to maintain the same longitudinal mode over a wide range of current. This mode of operation is then useful in order to achieve wavelength stability, which is of considerable importance for some applications such as wavelength multiplexing.

2.7.3 Quantum-Well Laser Sources

When the active layer thickness becomes comparable with the de Broglie electron wavelength, some quantum-mechanical effects can occur. These effects are perceivable either in the absorption and emission characteristics or in the transport characteristics of the laser itself. More than one active region can be present in the structure. Lasers with one active region are usually called *single-quantum-well* lasers, while those with multiple active regions are named *multi-quantum-well* lasers.

The laser spectral width, $\Delta\nu$, in these devices is a function of the number of wells and of their thickness, besides the total internal loss. Generally speaking, $\Delta\nu$ is reduced greatly in the region of thin active layer and, for a fixed value of the active layer thickness, it increases monotonically when the number of wells increases. Best results are expected for thin single-quantum-well lasers, where a reduction of $\Delta\nu$ by a factor of ~1/10 is possible [86]. A further reduction of the laser spectral width can be obtained by exerting compressive strain to the wells [87] or combining the wavelength selction properties of a grating structure with the advantages of the quantum-mechanical effects. In the latter case, by introducing a multi-quantum-well structure in a long-cavity DFB laser, $\Delta\nu$ of 250 kHz has been experimentally achieved [88]. Nevertheless, even narrower spectral linewidths at high output power are expected by applying a corrugation-pitch modulated structure to a long-cavity multi-quantum-well DFB laser [89]. In these devices, the dependence of the spectral width on the number of wells is no longer monotonic, so that an optimum value of such a number can be determined in order to minimize $\Delta\nu$. In [90], for example, a minimum 170-kHz spectral linewidth was obtained in the case of a four-well structure.

The introduction of quantum-well devices seems promising also for obtaining very high transmission rates. The adoption of external cavity lasers, in fact, is often unsatisfactory because of their low modulation efficiency. On the contrary, the adoption of distributed Bragg reflection quantum-well lasers, for example, operating at $\lambda = 1.55$ μm allows a very high modulation efficiency to be obtained while maintaining a spectral linewidth on the order of 4 MHz [91].

2.7.4 Micro-Optic Devices for Coherent Light Processing

In order to achieve optimum coherent detection, it is necessary that the state of polarization of the incoming signal at the receiver is perfectly matched to that of the local oscillator. On the other hand, in optical communications using single-mode fibers, the state of polarization is generally eliptical and fluctuates even when the transmitted light is stably linearly polarized. In this case the introduction of a state of polarization control scheme is indispensable. Several solutions have been proposed in this sense, including the use of polarization-maintaining fibers, the adoption of polarization-diversity receivers or the introduction of a proper polarization control on output, acting on the signal, or local oscillator light. All these solutions have been introduced in Section 2.5.3. However, as most of the polarization control devices make use of lenses, gratings, prisms, and other mechanical components, they can be considered as micro-optic devices, although they frequently they include, for realizing the control, $LiNbO_3$ components as well.

A vast body of literature exists on the design of polarization control devices. Thus it is very difficult to do individual justice to all authors that have worked in this area over the last 15 years. Be it enough to quote the following basic references where others can be found [92–95].

In order to have an idea of the complexity of such systems, we can fix our attention on some typical solutions. In [96], for example, four different kinds of phenomena or devices have been collected for achieving polarization control: the linear birefringence in a fiber induced by lateral stress, the birefringence of an electro-optic crystal, the polarization-plane rotation in a fiber induced by twist, and rotable phase plates. The poposed method is based on two fiber-type Faraday rotators used as control elements and on the "gradient method" as the optimization principle. The scheme exhibits negligible insertion loss and the receiver sensitivity is experimentally proved to be kept within ±2.5% even when the state of polarization changes drastically.

An automatic electro-optic polarization control system of the light output transmitted through a long single-mode fiber driven by a 1.62–μm–wavelength GaInAsP/InP laser has been described in [97]. The scheme is based on an electric feedback acting on two $LiNbO_3$ modulators tilted 45 deg relative to each other. The feedback signal detection consists of two loops. In the first, a part of the light passing through the second modulator is separated by a half-mirror and then passes through a Rochon prism, which is rotated by ±45 deg from the desired final linear orientation. Two outputs are then detected by Ge solar cells; these are proportional to the components of the output light oriented ±45 deg from the desired axis. In this way, the electric feedback circuit generates the modulation voltage applied to the first modulator until the equality between the outputs of the two cells is satisfied.

A further solution is proposed in [98]. Light from signal and local oscillator are coupled through an optical isolator into a 2×2 directional coupler. More precisely, the signal light enters directly one port of the directional coupler while the light emitted from the local oscillator is firstly split into two beams using a polarization beam-splitter. The beams outgoing from this device are two orthogonally polarized beams of nominally equal intensity. One of these beams is then frequency-shifted by using an acousto-optic modulator and recombined with the other beam using a second polarization beam-splitter. This composite wave is then launched into the second input port of the directional coupler and detected by an Si APD at one of the coupler otput ports.

In addition, it is interesting to mention the possibility of using a technique based on polarization scrambling [99]. This solution is similar in some respects to the more conventional polarization diversity receivers, but, in this case, diversity is achieved by suitable operations realized at the transmitter instead of at the receiver. More precisely, the transmitted optical field is stepped between orthogonal polarization states by means of a polarization scrambler comprising a fiber coupler, an $LiNbO_3$ waveguide phase modulator, a bulk optic polarization selective coupler, and several manual polarization controllers. The performance of such a system is then based on the property that the orthogonality between the states outgoing from the scrambler is unaffected by any absolute polarization changes that may occur during propagation through the fiber. The advantages of this technique are that it is particularly suited for local network applications (because the receiver design is very simple and cheap) and for wavelength-multiplexed systems, where one scrambler per transmission fiber is all that is required.

Some of the polarization control devices described above make use of an isolator for protecting the frequency stabilized source from spurious back-reflections originated, for example, by fiber splices and material scattering. Such an isolator can be realized by micro-optic technology, using an *yttrium iron garnet* (YIG) crystal, as Faraday rotator, and GRIN lenses, for collimating the laser beam. YIG is a material characterized by a high value of the Verdet constant [100]. This last parameter expresses the proportionality coefficient which relates the rotation of the plane of polarization to the product between the magnetic flux parallel to the direction of propagation and the path length, in a material subjected to the Faraday effect. By GRIN lenses, instead, we mean a particular kind of lens characterized by a parabolic distribution of the refractive index. A micro-optic isolator directly incorporated in a laser diode package [101] is able to meet the specifications of isolation (40 dB) and insertion loss (< 1 dB) required in coherent communication applications.

Finally, it is useful to note the results of some experiments conceived with the goal of obtaining a nonmechanical laser beam deflection. This is an operation frequently required, for example, by advanced optical space instrumentation to perform scanning functions of various kinds. The most direct solution, which is based on the adoption of mirrors mounted on pivots or gimbals with position read-out by means of angular encoders, is characterized by low reliability, and the moving parts may have negative effects on the spacecraft stabilization. A quite different solution is that based on dynamic holograms [102]. A hologram is the physical record of an interference pattern. It contains phase and amplitude information about the wavefronts that produce it. The proposed scheme employs the diffraction of a signal laser beam, positioned at Bragg angle, from dynamic holographic gratings recorded in a phorefractive crystal at different wavelengths. When the recording wavelength changes, the grating spacing varies as well; as a consequence, the diffracted beam changes its direction, that is to say it is deflected. With the configuration relative to a grating spacing of about 3.3 μm, for example, and the adoption of a laser diode emitting at 840 nm, a maximum deflection angle of 1.8 deg has been measured.

2.7.5 Integrated-Optic Devices for Coherent Light Processing

Integrated-optical devices are essential for the development of advanced coherent optical systems. The most consolidated technology available now is that based on $LiNbO_3$, although some realizations on GaAs, InP and SiO_2 substrate have been recently proposed. Currently, however, the performance of semiconductor electro-optic waveguide components is still inferior to comparable components fabricated on $LiNbO_3$ substrates, so that almost all the integrated-optics components used in coherent optical systems prototypes and studied today have been fabricated from $LiNbO_3$.

The basic devices, in some cases commercially available in a fully packaged form, include electro-optic waveguide modulators, frequency translators, switches, and those polarization controllers that, because of their size, technology, and elementary components, are not classified as micro-optic devices.

In regards to electro-optic waveguide modulators, traveling wave Mach-Zehnder interferometric modulators have been developed to allow high-speed amplitude modulation of the output from 1.3– and 1.55–μm lasers. The necessary drive voltages scales with bandwidth. In particular, a 18–GHz component has been proposed [103], which requires a drive voltage of 11 V. An evolution of this device consists in removing the input fiber and integrate a DFB narrow linewidth laser and modulator into one package to realize a compact modulated source.

$LiNbO_3$ electro-optic waveguide phase modulators consist essentially of a monomode waveguide, often formed by titanium diffusion, and two appropriately oriented electrodes, typically aluminium or gold [104]. The electrode separation is on the order of some μm while active lengths longer than 5 mm are required for producing a phase shift of π with a ~ 5 V drive signal. The operating voltage can be reduced by incresing the active length value. At the same time, however, the maximum frequency of operation is reduced, owing to the corresponding increase of the electrode capacitance. For very high speed operation, it is necessary to use devices employing traveling wave electrodes. In such a solution, the electrodes are designed to form part of a transmission line so that the frequency limiting factor becomes dependent on the difference between the propagation velocity of the electrical and optical signals.

The phase modulator structure can operate as a frequency translator if appropriately modulated. More precisely, a saw-toothed voltage waveform has to be applied to the electrodes in order to generate a single frequency shifted optical signal. The frequency shift entity is proportional to the rate of change of the applied voltage. This fact, together with the possibility of appearance of undesired frequency components, limits the use of such a device to the applications in which a constant and small frequency shift is required. More effective solutions, consisting of two sets of interleaved mode converters, modulated by sinusoidal voltages in quadrature, are based on TE-TM conversion [105]. The excellent performance predictable by these devices, in terms of suppression of the carrier and elimination of unwanted optical signals, can be worsened by the appearance of some parasitic electric fields, whose elimination requires a very careful design.

Single Mach-Zehnder interferometric Y junction modulators are able to generate frequency translated optical signals when modulated by a sinusoidal voltage waveform. In this case the frequency shift is proportional to the frequency of the electric modulating signal [106]. Single sideband signals can also be generated by combining the outputs of two similar interferometers [107]. This solution is however rather involved, requiring more complex electronics and limiting broad band tunability. Thus when the existence of two sidebands is not a problem, the simpler Y junction interferometer is much more attractive, especially for those applications which require large frequency translations.

In a coherent optical receiver, optical components are necessary for performing the functions of optical combining, polarization control, frequency translation, and phase control. All these functions can be implemented by using $LiNbO_3$ technology, and it is clear that the insertion loss and system cost can be greatly reduced by integrating a number of components on a single substrate. A typical $LiNbO_3$ integrated-optic coherent receiver device is designed for operating on a single TM mode. Therefore, it consists of [108]:

1. a single phase modulator and TE-TM mode converter for transforming the incoming signal into the required TM mode;
2. a single Mach-Zehnder device, appropriately modulated, for realizing frequency translation;
3. a directional coupler for optical combining;
4. a pair of semiconductor detectors for detecting the resulting optical output.

The conversion efficiency of such a device is smaller than – 4.7 dB [108], but the use of sufficiently powerful local oscillator laser, ensures limited detection of the shot noise. The signal-to-noise ratio can then be made extremely high, according to the analysis developed in Section 2.3.4.

When a multiport coherent detection system is considered [109], the receiver performance is made independent of the relative phases of the signal and local oscillator by the adoption of suitable signal processing and multiple detectors. Micro-optic components can also be used for this purpose, but the adoption of electro-optic waveguide technology leads to simpler and more reliable system.

As an example of polarization controller manufactured by integrated-optic technology, we can consider the classical device based on $LiNbO_3$ and comprising two polarization control elements in series to ensure that continuous polarization control is achieved [92]. A device of this type typically is 40 mm long, has an insertion loss of 5 dB, and requires control voltages of the order of 150 V.

Before concluding the present section, it is worthwhile to mention the role of some grating components for integrated-optics and optical interconnections. Diffraction gratings have been designed and experimentally demonstrated for a very large number of integrated-optical circuits [110]: input-output couplers for optical waveguides, filters, beam-splitters and beam reflectors, beam expanders, distributed reflectors for semiconductor lasers, dispersive components for multiplexers and demultiplexers, and focusing and imaging elements.

Particular emphasis has to be given to Bragg and Fresnel lenses. These may be considered as thick and thin gratings, respectively, so that the former type has usually higher efficiency but is also more critically dependent on the angle of incidence. In turn, geodesic lenses, in which the focusing effect is obtained by exploiting the different path lengths of rays guided in a film of constant thickness on a substrate of proper curvature, can be used for fabricating guided-wave geodesic multiplexers and demultiplexers.

Finally, regarding interconnections of optical fibers, the most common solution, adopted for many years, employs the diffraction properties of transmission "bulk" gratings. Now the trend moves towards the use of holographic components, with the goal of increasing the system speed, which is currently limited by the bandwidth and crosstalk of electronic interconnections and by the connectivity of planar chips. Furthermore, this technique is characterized by a remarkable flexibility. A holographic coupler, for example, allows one to couple optical components in an almost arbitrary predesigned way, offers the possibility of unlimited (at least in principle) reconfigurability and is intrinsically bidirectional.

REFERENCES

[1] Ramo, S., J. R. Whinnery, and T. Van Duzer, *Fields and waves in communication electronics*, Wiley, New York, 1965.

[2] Henry, C., H., "Theory of the linewidth of semiconductor lasers", *IEEE J. Quantum Electron.*, Vol. QE-18, 1982, pp. 259–264.

[3] Stein, R., and J. J. Jones, *Modern communication principles*, McGraw Hill, New York, 1967.

[4] Born, M., and E. Wolf, *Principles of optics*, Pergamon Press, Edinburgh, 1964.

[5] Stratton, J. A., *Electromagnetic theory*, McGraw Hill, New York, 1941.

[6] Hariharan, P., *Optical interferometry*, Academic Press, Sydney, 1985.

[7] Michelson, A. A., and H. G. Gale, "The effect of the earth's rotation on the velocity of light", *Nature*, Vol. 115, 1925, pp. 566.

[8] Booth, R. C., K. H. Cameron, and A. E. Glaccum, "Optical components for telecommunications", *Proc. EFOC 91*, London, 1991, pp. 211–215.

[9] Olshansky, R., et al., "Frequency response of 1.3 μm InGaAsP high speed semiconductor lasers", *IEEE J. Quantum Electron.*, Vol. QE-23, 1987, pp. 1410–1418.

[10] Pearsall, T. P., "Photodetectors for optical communication", *J. Opt. Commun.*, Vol. 2, 1981, pp. 42–48.

[11] Cancellieri, G., and U. Ravaioli, *Measurements of optical fibers and devices. Theory and experiments*, Artech House, Norwood, MA, 1984.

[12] Salz, J., "Coherent lightwave communications", *AT&T Tech. J.*, Vol. 10, 1985, pp. 2153–2209.

[13] Levin, B. F., and C. G. Bethea, "Single-photon detection at 1.3 μm using a gated avalanche photodiode", *Appl. Phys. Lett.*, Vol. 44, 1984, pp. 553–555.

[14] Okoshi, T., A. Hirose, and K. Kikuchi, "Bit-error rate in photon-counting communications", *Proc. ECOC '87*, Helsinki, 1987, pp. 329–332.

[15] Pierce, J. R., "Optical channels, practical limits with photon counting", *IEEE Trans. Commun.*, Vol. COM-26, 1978, pp. 1819–1821.

[16] Oliver, B. M., " Thermal and quantum noise", *Proc. IEEE*, Vol. 53, 1965, pp. 436–454.

[17] Sugiyama, H., and K. Nosu, "MPPM: a method for improving the band-utilization efficiency in optical PPM", *IEEE J. Lightwave Technol.*, Vol. LT-7, 1989, pp. 465–472.

[18] Kanaya, F., and K. Nakagawa, "Theoretical studies on the performance of lossy photon-channels", *IEEE Trans. Commun.*, Vol. COM-37, 1989, pp. 39–45.

[19] Cancellieri, G., F. Chiaraluce, and E. Gambi, "A new method for improving the bandwidth utilization efficiency in optical PPM", Lecture Notes in Physics, Vol. 378, C. Bendjaballah, O. Hirota, S. Reynaud (Eds.), in *Quantum Aspects of Optical Communications*, Springer-Verlag, Berlin, 1991.

[20] Shannon, C. E., "A mathematical theory of communication", *Bell Sys. Tech. J.*, Vol. 27, 1948, pp. 379–423 (Part I), pp. 623–656 (Part II). Reprinted in book form with postscript by W. Weaver, Univ. Illinois Press, Urbana, 1949.

[21] McEliece, R. J. "Practical codes for photon communication", *IEEE Trans. Inf. Theory*, Vol. IT-27, 1981, pp. 393–398.

[22] Cancellieri, G., "Transmission efficiency in photon counting channels", *IEEE Trans. Commun.*, Vol. COM-37, 1989, pp. 183–187.

[23] Imoto, N., H. A. Haus, and Y. Yamamoto, "Quantum nondemolition measurement of the photon number via the optical Kerr effect", *Phys. Rev. A.*, Vol. 32, 1985, pp. 2287–2292.

[24] Snyder, A. W., and J. D. Love, *Optical waveguide theory*, Chapman & Hall, London, 1984.

[25] Gloge, D., "Weakly guiding fibers", *Appl. Opt.*, Vol. 10, 1971, pp. 2252–2258.

[26] Urbach, F., "The long wavelength edge of photographic sensitivity and of the electronic absorption of solids", *Phys. Rev.*, Vol. 92, 1953, pp. 1324–1324.

[27] Agrawal, G. P., *Nonlinear fiber optics*, Academic Press, San Diego, 1989.

[28] Petermann, K., "Fundamental mode microbending loss in graded index and W fibres", *Opt. Quantum Electron.*, Vol. 9, 1977, pp. 167–175.

[29] Cancellieri, G., and A. Orfei, "Asymptotic effective cut-off condition in single-mode optical fibers", *IEEE J. Lightwave Technol.*, Vol. LT-5, 1987, pp. 199–205.

[30] Cancellieri, G., *Single-mode optical fibres*, Pergamon Press, London, 1991.

[31] Mears, R. J., L. Reekie, S. B. Poole, and D. N. Payne, "Neodymium-doped silica single-mode fibre lasers", *Electron. Lett.*, Vol. 21, 1985, pp. 738–740.

[32] Meli, F., et al., "Saturated Erbium doped fiber amplifier", *Proc. OSA Topical Meeting on Optical Amplifiers*, Monterrey, 1990, paper MB3.

[33] Cancellieri, G., F. Chiaraluce, and E. Gambi, "An accurate model for the analysis of erbium-doped single-mode fiber optical amplifiers", *Proc. ISFOC '91*, Leningrad, 1991, pp. 154–159.

[34] Kaminow, I. P., "Polarization in optical fibers", *IEEE J. Quantum Electron.*, Vol. QE-17, 1981, pp. 15–22.

[35] Okoshi, T., et al., "Polarization-diversity receiver for heterodyne coherent optical fiber communications", *Proc. IOOC 83*, Tokyo, 1983, pp. 386–387.

[36] Okoshi, T., "Polarization-state control schemes for heterodyne or homodyne optical fiber communications", *IEEE J. Lightwave Technol.*, Vol. LT-3, 1985, pp. 1232–1237.

[37] Noé, R., "Endless polarisation control in coherent optical communications", *Electron. Lett.*, Vol. 22, 1986, pp. 772–773.
[38] Walker, N. G., and G. R. Walker, "Endless polarisation control using four fiber squeezers", *Electron. Lett.*, Vol. 23, 1987, pp. 290–292.
[39] Okoshi, T., N. Fukaya, and K. Kikuchi, "New polarisation-state control device: rotable fiber cranks", *Electron. Lett.*, Vol. 21, 1985, pp. 895–896.
[40] Matsumoto, T., and H. Kano, "Endlessly rotable fractional-wave devices for single-mode-fiber optics", *Electron. Lett.*, Vol. 22, 1986, pp. 78–79.
[41] Betti, S., F. Curti, G. De Marchis, and E. Iannone, "Phase noise and polarization state insensitive optical coherent systems", *IEEE J. Lightwave Technol.*, Vol. LT-8, 1990, pp. 756–767.
[42] Okoshi, T., and K. Oyamada, "Single-polarization single-mode optical fibre with refractive index pits on both sides of the core", *Electron. Lett.*, Vol. 16, 1980, pp. 712–713.
[43] Okoshi, T., T. Aihara, and K. Kikuchi, "Prediction of the ultimate performance of side-tunnel single-polarization fibre", *Electron. Lett.*, Vol. 19, 1983, pp. 1080–1082.
[44] Jeunhomme, L., and M. Monerie, "Polarization-maintaining single-mode fiber cable design", *Electron. Lett.*, Vol. 16, 1980, pp. 921–922.
[45] Meyer-Arendt, J. R., *Introduction to classical and modern optics*, Prentice-Hall, Englewood Cliffs, NJ, 1972.
[46] Schafer, F. P. (ed.), "Dye lasers", Vol. 1 of *Topics in Applied Physics*, 2nd revised ed., Springer-Verlag, New York, 1977.
[47] Minelly, J. D., and C. D. Hussey, "Single-mode fibre Y-junction beam-splitter", *Electron. Lett.*, Vol. 23, 1987, pp. 1087–1088.
[48] Bricheno, T., and A. Fielding, "Stable low-loss single-mode couplers", *Electron. Lett.*, Vol. 20, 1984, pp. 230–232.
[49] Varshney, R. K., and A. Kumar, "Design of a 4x4 fiber coupler with equal power splitting at two different wavelengths", *Optics Commun.*, Vol. 81, 1992, pp. 167–170.
[50] Wright, J. V., "Wavelength dependence of fused couplers", *Electron. Lett.*, Vol. 22, 1986, pp. 320–321.
[51] Mortimore, D. B., "Wavelength-flattened fused couplers", *Electron. Lett.*, Vol. 21, 1985, pp. 742–743.
[52] Rosemberg, D., "Micro-optic passive devices for multimode optical fiber communication systems", *Siemens Res. and Dev. Rept.*, Vol. 8, 1979 pp. 125–129.
[53] Syms, R. R. A., "Advances in channel waveguide lithium niobate integrated optics", *Opt. Quantum Electron.*, Vol. 20, 1988, pp. 189–213.
[54] Kawachi, M., "Silica waveguides on silicon and their application to integrated optic components", *Opt. Quantum Electron.*, Vol. 22, 1990, pp. 391–416.
[55] Carenco, A., L. Menigaux, and P. Delpech, "GaAs electro-optical directional coupler switch", *Proc. 4-th ECOC*, Genova, 1978, pp. 404–409.

[56] Sakaguchi, M., and H. Goto, "High speed optical time-division and space-division switching", *Proc. IOOC-ECOC 85*, Venice, 1985, pp. 81–88.
[57] *Proc. Optical Communication Conference*, Amsterdam, 1979.
[58] Tomlinson, W. J., et al., "Multiposition optical fiber switch", *Electron. Lett.*, Vol. 15, 1979, pp. 192–194.
[59] Minowa, J, et al., "Nonblocking 8x8 optical matrix switch for fiber-optic communications", *Electron. Lett.*, Vol. 16, 1980, pp. 422–423.
[60] Gratze, S. C., "Optical switching", in *Encyclopedia of lasers and optical technology*, R. A. Meyers ed., pp. 633–643, Academic Press, San Diego, 1991.
[61] Kawano, K., M. Saruwatari, and O. Mitomi, "A new confocal combination lens method", *IEEE J. Lightwave Technol.*, Vol. LT-3, 1985, pp. 739–745.
[62] Tamir, T., *Integrated optics*, Springer Verlag, Berlin, 1979.
[63] Howes, M. J., and D. V. Morgan (Eds.) "Optical fibre communications", Section 6.8.2, Wiley, New York, 1980.
[64] Ishida, K., H. Nakamura, H. Inoue, S. Tsuji, and H. Matsumura, "InGaAsP/InP multimode optical switches operated by a carrier induced refractive index change", *Proc. IOOC-ECOC 85*, Venice, 1985, pp. 357–360.
[65] Stegeman, G. I., and C. T. Seaton, "Nonlinear integrated optics", *Proc. IOOC-ECOC 85*, Venice, 1985, pp. 55–60.
[66] Kondo, M., N. Takado, K. Komatsu, and Y. Ohta, "32 switch-elements integrated low-crosstalk $LiNbO_3$ 4x4 optical matrix switch", *Proc. IOOC-ECOC 85*, Venice, 1985, pp. 361–364.
[67] Stewart, G., R. H. Hutchins, P. J. R. Laybourn, and C. A. MacRae, "Construction of hybrid integrated optical circuits", *Proc. IOOC-ECOC 85*, Venice, 1985, pp. 39–42.
[68] Karaguleff, and al., " Degenerate four wave mixing in planar CS_2 covered waveguides", *Appl. Phys. Lett.*, Vol. 46, 1985, pp. 621–623.
[69] Mellis, J., et al., "Miniature packaged external cavity semiconductor laser with 50 GHz continuous electrical tuning range", *Electron. Lett.*, Vol. 24, 1988, pp. 988–989.
[70] Ogita, S., et al., "Dependence of spectral linewidth on cavity length and coupling coefficient in DFB laser", *Electron. Lett.*, Vol. 24, 1988, pp. 613–614.
[71] Kotsai, Y., et al., "Tunable DBR laser with wide tuning range", *Electron. Lett.*, Vol. 24, 1988, pp. 503–505.
[72] Holonyak, Jr., N., R. M. Kolbas, R. D. Dupuis, and P. D. Dapkus, "Quantum-well heterostructure lasers", *IEEE J. Quantum Electron.*, Vol. QE-16, 1980, pp. 170–186.
[73] Thyagarajan, K., and A. K Ghatak, *Lasers – Theory and applications*, Plenum, New York, 1981.
[74] Yariv, A., *Quantum electronics*, Wiley, New York, 1989.
[75] Tench, R., et al., "Performance evaluation of waveguide phase modulators for coherent systems at 1.3 and 1.5 μm", *IEEE J. Lightwave Technol.*, Vol. LT-5, 1987, pp. 492–501.

[76] Wood, T. H., "Multiple quantum well (MQW) waveguide modulators", *IEEE J. Lightwave Technol.*, Vol. LT-6, 1988, pp. 743–752.

[77] Bowers, J. E., and C. A. Burrus, "Ultrawide-band long-wavelength p-i-n photodetectors", *IEEE J. Lightwave Technol.*, Vol. LT-5, 1987, pp. 1339–1350.

[78] Thylèn, L., "Integrated optics in $LiNbO_3$: recent developments in devices for telecommunications", *IEEE J. Lightwave Technol.*, Vol. LT-6, 1988, pp. 847–861.

[79] Somekh, S., et al., "Channel optical waveguide directional couplers", *Appl. Phys. Lett.*, Vol. 2, 1973, pp. 46–47.

[80] Belager, M., G. L. Yip, and M. Haruna, "Passive planar multibranch optical power divider: some design considerations", *Appl. Opt.*, Vol. 22, 1983, pp. 2383–2389.

[81] Sobol, H., "The application of microwave techniques in lightwave systems", *IEEE J. Lightwave Technol.*, Vol. LT-5, 1987, pp. 293–300.

[82] Kojima, K., K. Kyuma, and T. Nakayama, "Analysis of the spectral linewidth of distributed feedback laser diodes", *IEEE J. Lightwave Technol.*, Vol. LT-3, 1985, pp. 1048–1055.

[83] Delfyett, P. J., and Chang-Hee Lee, "Semiconductor injection laser", in *Encyclopedia of lasers and optical technology*, R. A. Meyers ed., pp. 587–601, Academic Press, San Diego, 1991.

[84] Van der Ziel, J. P., and R. Mikulyak, "Single-mode operation of 1.3 μm InGaAsP/InP buried crescent lasers using a short external optical cavity", *IEEE J. Quantum Electron.*, Vol. QE-20, 1984, pp. 223–229.

[85] Agrawal, G. P, and N. K. Dutta, *Long-wavelength semiconductor lasers*, Van Nostrand Reinhold, New York, 1986.

[86] Arakawa, Y., and A. Yariv, "Theory of gain, modulation response, and spectral linewidth in AlGaAs quantum well lasers", *IEEE J. Quantum Electron.*, Vol. QE-21, 1985, pp. 1666–1674.

[87] Dutta, N. K., H. Temkin, T. Tanbun-Ek, and R. A. Logan, "Linewidth enhancement factor for InGaAs/InP strained quantum well lasers", *Appl. Phys. Lett.*, Vol. 57, 1990, pp. 1390–1391.

[88] Yamazaki, H., T. Sasaki, N. Kida, M. Kitamura, and I. Mito, "250 kHz linewidth operation in long cavity 1.5 μm multiple quantum well DFB-LDs with reduced linewidth enhancement factor", *Proc. Int. Conf. Opt. Fiber Commun.*, San Francisco, CA, 1990, postdeadline paper PD33–1.

[89] Okai, M., N. Chinone, H. Taira, and T. Harada, "Corrugation-pitch modulated phase-shifted DFB laser", *IEEE Photon. Technol. Lett.*, Vol. 1, 1989, pp. 200–201.

[90] Okai, M., T. Tsuchiya, K. Uomi, N. Chinone, and T. Harada, "Corrugation-pitch modulated DFB lasers with narrow spectral linewidth", *IEEE J. Quantum Electron.*, Vol. QE-27, 1991, pp. 1767–1772.

[91] Glance, B., et al., "Densely spaced FDM coherent optical star network using monolithic widely frequency tunable lasers", *Electron. Lett.*, Vol. 25, 1989, pp. 672–673.

[92] Brain, M. C., et al., "Progress towards the field deployment of coherent optical fibre systems", *IEEE J. Lightwave Technol.*, Vol. LT-8, 1990, pp. 423–437.

[93] Honmou, H., et al., "Stabilization of heterodyne receiver sensitivity with automatic polarisation control system", *Electron. Lett.*, Vol. 22, 1986, pp. 1181–1182.

[94] Frigo, N. J., A. Danridge, and A. B. Tveten, "Technique for elimination of polarisation fading in fibre interferometres", *Electron. Lett.*, Vol. 20, 1984, pp. 319–320.

[95] Rashleigh, S. C., "Origins and control of polarisation effects in single-mode fibres", *IEEE J. Lightwave Technol.*, Vol. LT-1, 1983, pp. 312–331.

[96] Okoshi, T., K. Kikuchi, and Y. Cheng, "A new polarization-control scheme for optical heterodyne receiver", *Proc. IOOC-ECOC 85*, Venice, 1985, pp. 405–408.

[97] Kidoh, Y., Y. Suematsu, and K. Furuya, "Polarization control on output of single-mode optical fibers", *IEEE J. Quantum Electron.*, Vol. QE-17, 1981, pp. 991–994.

[98] Kersey, A. D., A. M. Yurek, A. Danridge, and J. F. Weller, "New polarisation-insensitive detection technique for coherent optical fibre heterodyne communications", *Electron. Lett.*, Vol. 23, 1987, pp. 924–926.

[99] Hodgkinson, T. G., R. A. Harmon, and D. W. Smith, "Polarisation-insensitive heterodyne detection using polarisation scrambling", *Electron. Lett.*, Vol. 23, 1987, pp. 513–514.

[100] Kobayashi, K., and M. Seki, "Microoptic grating multiplexers and optical isolators for fiber-optic communications", *IEEE J. Quantum Electron.*, Vol. QE-16, 1980, pp. 11–22.

[101] Bendelli, G., S. Donati, and T. Tambosso, "Optical isolators for coherent fiber optic communications", *Proc. OCTIMA*, Rome, 1989, pp. 184–189.

[102] Delvò, P., M. L. Rizzi, and G. Visin, "Non-mechanical laser beam deflection by dynamic holograms", *Proc. OCTIMA*, Rome, 1989, pp. 108–112.

[103] Donaldson, et al., "$LiNbO_3$ integrated optical components for coherent optical systems", *Proc. OCTIMA*, Rome, 1989, pp. 174–182.

[104] Alferness, R. C., "Waveguide electrooptic modulators", *IEEE Trans. Microwave Theory and Tech.*, Vol. MTT-30, 1982, pp. 1121–1137.

[105] Heismann, F., and R. Ulrich, "Integrated-optical frequency translator with stride waveguide", *Appl. Phys. Lett.*, Vol. 45, 1984, pp. 490–492.

[106] Auracher, F., and R. Keil, "Method for measuring the RF modulation characteristics of Mach-Zehnder-type modulators", *Appl. Phys. Lett.*, Vol. 36, 1980, pp. 626–629.

[107] Izutsu, M., S. Shikama, and T. Suetu, "Integrated optical SSB modulator/frequency shifter", *IEEE J. Quantum Electron.*, Vol. QE-17, 1981, pp. 2225–2227.

[108] Booth, R. C., "Integrated optic devices for coherent transmission", *Proc. IOOC-ECOC 85*, Venice, 1985, pp. 89–96.

[109] Walker, N. G., T., and J. E. Carroll, *Proc. IEE Colloquium on* "Advances in coherent optic devices and technologies", Digest No. 1985/30, London, 1985, pp. 10/1–10/6.

[110] Suhara, T., and H. Nishihara, "Integrated optics components and devices using periodic structures", *IEEE J. Quantum Electron.*, Vol. QE-22, 1986, pp. 845–867.

Chapter 3

Measurements of Fiber Optics in Incoherent Light

A. Galtarossa

E. Nava

G. Valentini

It is necessary, first of all, to subdivide the measurement techniques on fiber optics for telecommunications into two main classes:

1. those performed on the transmission medium,
2. those performed on optical components (passive and active) placed in particular sites along an optical link or externally to it.

The measurements of the first class concern silica-based single-mode fibers and are defined by standards. Among the measurements of the second class we can list:

1. measurements on fibers and fiber-based devices,
2. measurements on opto-electronic active devices.

In this chapter and in the following one, these measurements will be discussed. The present chapter is devoted to the description of measurements performed in incoherent light. They represent the greater part of such measurements, especially when repeatability and simplicity of operation are required.

The chapter is organized as follows. Measurements on the transmission medium, including those on fiber-to-fiber joints, all-fiber passive devices, and active fibers, are collected in Section 3.1. The subsequent section is devoted to describing measurements of active optical components when they are not required to have a particularly high degree of light temporal coherence. Among them, great attention is given to semiconductor LEDs and conventional laser diodes. In Section 3.3, measurements on photodiodes are described.

3.1 INCOHERENT LIGHT FIBER CHARACTERIZATION

Single-mode fiber characterization concerns the evaluation of geometrical and transmission parameters. Among the first of these are: mode field diameter and geometrical dimensions, refractive index behavior, and cut-off wavelength. With reference to the definitions introduced in Chapter 1, the term "mode field diameter" appears as more general than the "spot-size" one, especially in the case of fibers whose field is far from the Gaussian shape. So, it will be widely used in the following. The two main transmission parameters are attenuation and time dispersion. Geometrical parameters are tested after fiber drawing or cable fabrication, frequently with semiautomatic or completely automatic procedures. Time dispersion is usually tested at the end of the cable fabrication process as well. Attenuation measurements are performed, instead, various times during cable assembling and also after cable installation. In fact, several possible defects in the fiber can be revealed from an attenuation measurement. Correspondingly, there are different procedures for attenuation measurements; some of them are conceived for an industrial environment, while others are more suitable for operating in the field.

After the description or these types of measuring the geometrical and transmission parameters of the fiber, the experimental characterization of fiber-to-fiber joints is discussed. Measurements of all-fiber devices are then outlined. Finally some possible measurements on active fibers are presented. Such relatively new optical components are particularly attractive as on-line signal repeaters. The reason for introducing here the characterization procedures for the fiber-optic components previously mentioned is that they must exhibit a very good compatibility with the fiber employed in the transmission link. Such compatibility includes geometrical parameters (e.g., core size, refractive index shape) but also transmission parameters, like attenuation and even time dispersion.

3.1.1 Mode Field Diameter and Geometrical Characteristics

Modal parameters like field diameters and cut-off wavelength are closely related to the characteristics of the electromagnetic field propagating along the optical fiber. They depend on geometrical properties (refractive index shape, core and cladding diameters) as well as on transmission properties (wavelength, presence of microbending and macrobending, etc.).

All modal parameters may be analytically and exactly calculated starting from Maxwell equations, as demonstrated in Sections 2.4 and 2.5. On the other hand, real situations show that there are some uncertainties in measuring them. Moreover, modal parameter values may vary as functions of environmental conditions and these variations affect strongly the system behavior.

The spot-size or mode field diameter depends on the radial width of optical transverse power distribution. Sometimes, its experimental evaluation is related to the inverse of the r.m.s. width of the fiber far-field distribution [1]. The value of such a parameter indicates how high the transverse concentration of the electromagnetic energy is in a single-mode fiber. It is used to estimate joint losses in splicing, as well as bending and microbending effects.

Typical values of W_{rms}, as defined by (1.4), are 9–10 μm at 1300 nm, with a maximum deviation of ± 10% from the nominal value. These depend also on the refractive index behavior. For example, matched-cladding fibers have a nominal value close to 10 μm, while depressed-cladding fibers are characterized by $W_{rms} \approx 9$ μm.

There are some suitable techniques to measure mode field diameter, such as:

1. far-field scan,
2. variable-aperture technique,
3. knife-edge scan,
4. near-field domain.

The transverse offset technique, which was one of the most diffuse techniques for such measurement, has been nearly abandoned because of its setup complexity [2].

There are transform relationships among mathematical definitions on which all the techniques described above are based [3], as shown in Figure 3.1. Recommendations and standard definitions can be found in [4, 5]. Measurements have to be done with great care in particular in regards to launching conditions, since higher order modes may strongly affect the results.

The far-field intensity distribution $I^2(x)$ is plotted as a function of the far-field angle Φ. The mode field diameter, $\bar{W}$, in turn, is given by

$$\bar{W} = \frac{2}{\pi} \frac{1}{\left[2\int_0^\infty x^3 I^2(x)dx \Big/ \int_0^\infty x I^2(x)dx \right]^{1/2}} \tag{3.1}$$

where $x = \sin\Phi/\lambda$ and λ is the operating wavelength. As a consequence, $\bar{W}$ is wavelength dependent.

When the variable-aperture technique is used, the complementary aperture power transmission quantity a(x) is measured as a function of the aperture radius x. More precisely, a(x) is defined as

$$a(x) = [1 - P(x)/P_{max}] \tag{3.2}$$

where P(x) is the power transmitted by each aperture, while P_{max} corresponds to the largest aperture. The aperture plane is located at a distance D from the fiber and the mode field diameter is calculated as

$$\bar{W} = \frac{2}{\pi} \frac{1}{\left[4\int_0^{\infty} xa(y)dx\right]^{1/2}} \tag{3.3}$$

where y = DtanΦ.

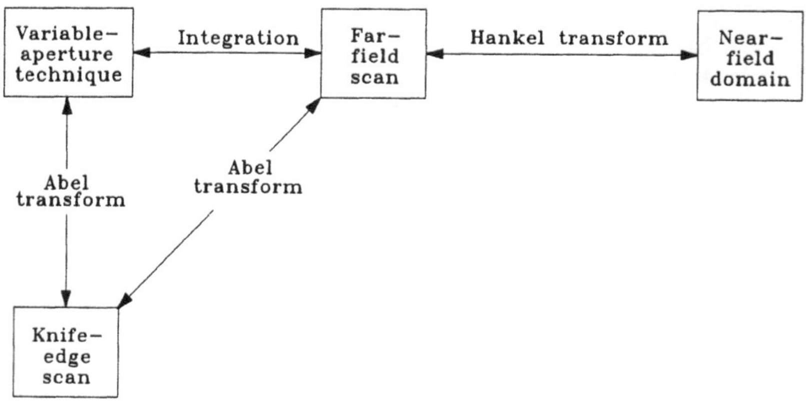

Figure 3.1 Mathematical relationships among mode field diameter measurement techniques (adapted from [3]).

The knife-edge power transmission quantity K(y) is measured as a function of the lateral offset y, with the plane of the knife-edge separated by a distance D from the fiber, and the mode field diameter is evaluated as

$$\bar{W} = \frac{2}{\pi} \frac{1}{\left[4\int_0^\infty K'(y)x^2dx \,/ \int_0^\infty K'(y)dx \right]^{1/2}}$$

$$K'(y) = \frac{dK(y)}{dy} \tag{3.4}$$

where x and y have the same meaning as in (3.3).

When the measurement is that called near-field domain the evaluation of the near-field intensity distribution f(r) is performed as a function of the radial coordinate r and the mode field diameter becomes

$$\bar{W} = 2\left[2\int_0^\infty r\, f^2(r)dr \,/ \int_0^\infty r \left[\frac{df(r)}{dr}\right]^2 dr \right]^{1/2} \tag{3.5}$$

Electromagnetic field distribution may be expressed as a Gaussian function, according to (1.2). So near-field and far-field have approximately a Gaussian shape, and the spot-size or mode field diameter corresponds to the diameter where electromagnetic field value reduces by a factor 1/e. This leads to the definition of W_0 given by (1.2). In all the previously described techniques, fiber-end face has to be prepared with care, because deviations in mode field diameter measurements may occur due to face roughness.

Fiber geometry is characterized by mode field diameter and mode field concentricity error, as well as by cladding diameter and cladding noncircularity. Nevertheless, other quantities, like mode propagation constant and static parameters of a perturbation, when present, are necessary for low-loss single-mode fiber design [6]. In addition, other geometric characteristics, such as core-cladding concentricity errors and primary and secondary coating diameter, are also to be taken into account. All these measurements are performed only in factory tests during optical fiber and cables manufacturing. However, geometrical fiber quality is a crucial parameter in order to guarantee a suitable fiber to fiber coupling, in particular for minimizing losses in mechanical splices and connectors. Birefringence, stress, and microbending are also influenced by geometrical characteristics. Thus, system performances, like dynamic range or bit rate, are related to geometric quantities.

The reference test method directly related to geometric quantity definitions is the transmitted near-field technique (see Section 3.1.2). This method suffers from large reproducibility errors as demonstrated by round robin tests [7]. Consequently, some geometrical parameters have a reproducibility error that is comparable to the recommended tolerance range. On the other hand, equipments based on the refracted near-field method with better reproducibility have become recently commercially available. They are suitable for factory test use and can be easily recalibrated. The National Institute of Standards and Technology (NIST) in the U.S. and the National Physical Laboratory (NPL) in the U.K. have worked on calibration improvements and new measurement methods [8, 9] using interferometric and micrometric techniques. All methods are based on a longitudinal invariant fiber diameter because the reference fibers must be recleaved several times during tests to calibrate equipment.

Investigations on diameter uniformity along the fiber length have been performed [10] with a total uncertainty in calibration accuracy estimated to be ±0.23 μm for the diameter and ±0.1% for ovality, with no correlation between the two parameters. Sometimes, periodic diameter fluctuations versus fiber length have been observed, with a period of about 0.5–2 m. Comparison of five different fibers has given a maximum variation in cladding diameters of 2.7 μm on peak-peak difference and 0.55 μm on ovality.

Contact micrometry [8] may be an attractive alternative. The fixed part is the anvil (Figure 3.2) while the moving part is the spindle, which is pressed against the anvil by means of a well-determined and known force. Measurements are performed inserting and taking away the fiber between the spindle and the anvil. The difference between the two spindle positions is the fiber diameter. It is suggested to perform measurements a few times to ensure repeatability. A total uncertainty of any measurement of ±40 nm has been obtained. Comparison between contact micrometry and scanning confocal microscope [11, 12] has demonstrated a difference of less than 0.03 μm in fiber diameter measurement.

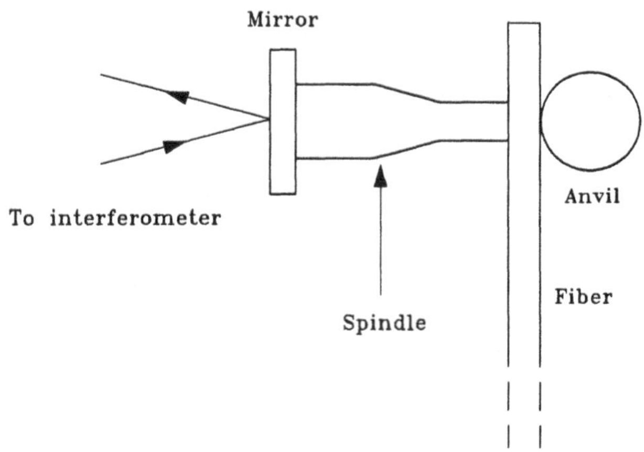

Figure 3.2 Contact micrometry technique (adapted from [8]).

Another solution particularly devoted to measure and to control fiber diameter during the drawing process is laser micrometry [13]. In the solution shown in Figure 3.3, there is a reference arm to self-calibrate the setup, which is able to compensate temperature fluctuations. The laser beam is scanned by the deflector that oscillates at a constant frequency. A first beam-splitter divides the light into two beams; one is sent to the reference rectangular grating and the other one is injected into the test fiber through a second beam-splitter. Optical signals are then detected by means of *photodiodes* (PDs). The diameter of the fiber under test is derived by the comparison of the time for the interception of light signal from the test fiber and the rectangular grating. The temperature dependence coefficient of the measured fiber diameter is smaller than 0.002 μm/°C (at 20–40 °C) and the diameter fluctuation is less than 0.1 μm in 150 hours. However, also higher frequency components, determining a period of about 1 mm, have been observed [7, 10].

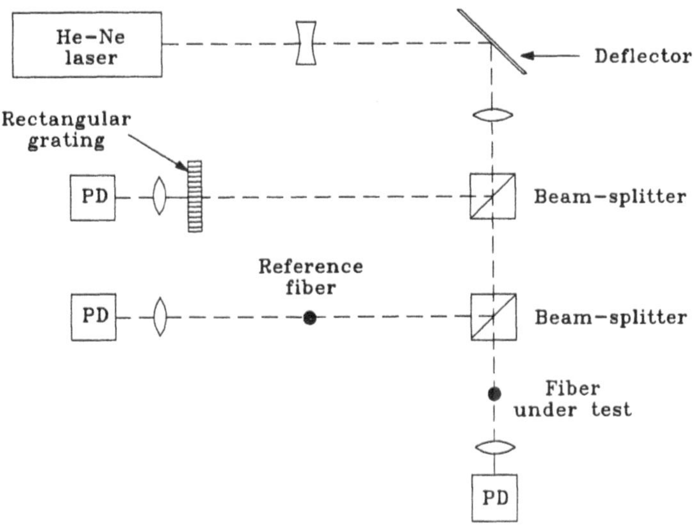

Figure 3.3 Laser micrometry technique (adapted from [13]).

Methods based on lateral measurement instead of end face ones avoid the necessity of a well-cleaved fiber head. This is one of the advantages of the side view method [14] where the fiber acts as a lens. A para-axial light beam, perpendicular to the fiber axis without primary coating, is refracted by the fiber. Refracted rays direction gives the index fiber profile, and the light intensity distribution, observed by means of a CCD-camera array, determines the fiber geometry. By rotating the fiber around its axis, it is possible to evaluate cladding diameter and noncircularity as a function of the azimuthal angle. Repeating this measurement along the fiber length, enables us to measure longitudinal variation of the geometrical parameter. This method is largely used also in fusion-splice equipment to realize high-quality field splices for single fibers or for ribbon fibers [15]. Comparison with refracted near-field technique on a reference fiber made by different research groups confirms [16] that side-view method is an excellent alternative for geometrical measurement. The accuracy of side view method is about ±0.1 μm for cladding diameter, ±0.1% for cladding noncircularity, and ±0.15 μm for core eccentricity.

The critical point for all geometrical methods is that a reference fiber for the setup calibration is needed. Actually, inaccuracy on calibration of standard fiber diameter is about 0.19 μm.

3.1.2 Refractive Index

Refractive index measurements have a fundamental role only in factory tests, particularly to check optical fiber quality after drawing and before cabling.

It is useful to distinguish between measurement techniques based on the fiber guidance properties and methods that use an external light source. Among the former we have:

1. the refracted near-field technique,
2. the near-field distribution technique.

The main methods adopting an external source are:

3. the slice interferometry,
4. the reflection method,
5. the transverse interferometry,
6. the transverse focusing technique.

Finally, not belonging to any of the classes listed above, there are also the so-called:

7. far-field scanning techniques,
8. scattering methods,
9. diffraction methods.

Despite this wide range of possible solutions, we will focus our attention here only on the methods based on the fiber guidance properties. These, in fact, seem to be the most appropriate to the application in a factory environment, by virtue of their low cost and reduced time commitment.

In the refracted near-field technique [3, 4], the refractive index shape is obtained by detecting the light that is not trapped in the core. A 0.5–m optical fiber whose primary coating has been removed, is immersed in a liquid (Figure 3.4(a)) characterized by a refractive index greater than the cladding one. A stable light source, like an HeNe laser (λ = 632.8 nm) is typically employed (Figure 3.4(b)); the refractive index behavior at longer wavelengths, on the other hand, may be easily extrapolated. The laser beam passes through a quarter-wave plate, changing its polarization from linear to circular, since the reflectivity on air-glass interface depends strongly on polarization.

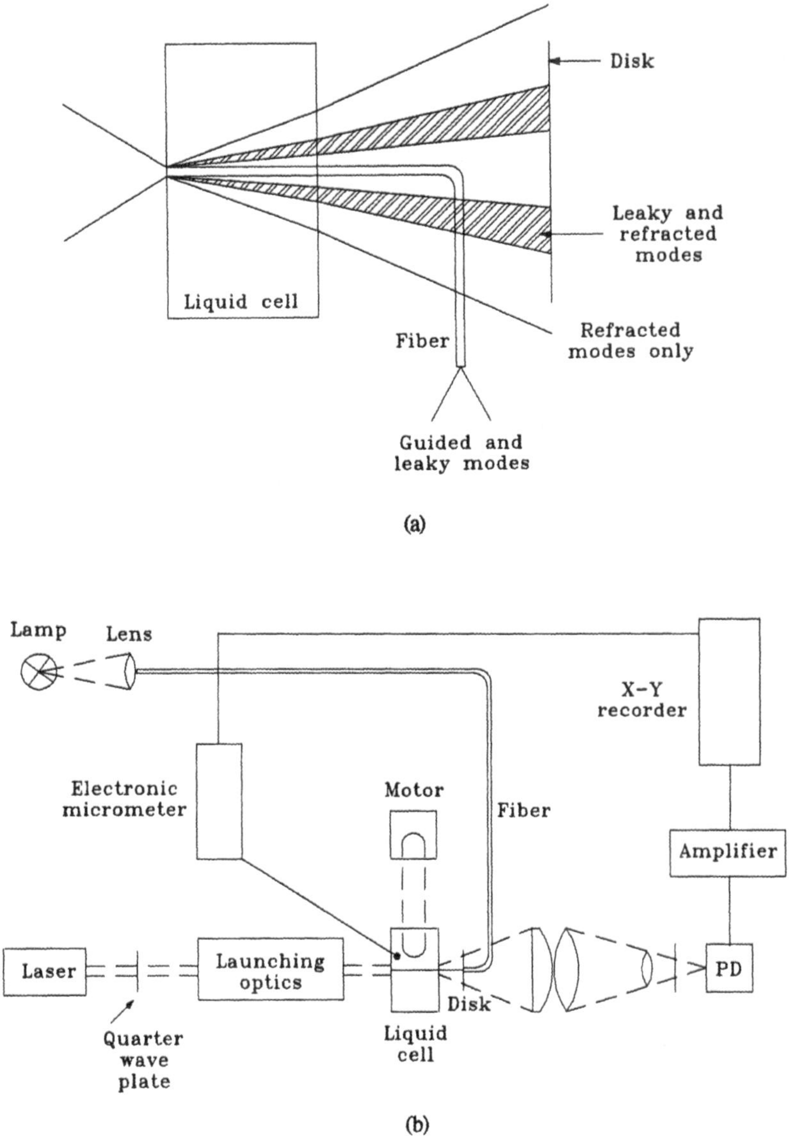

Figure 3.4 Refracted near-field technique: (a) schematic diagram and (b) measurement setup.

A lens with NA larger than that of the fiber is located between the light source and the fiber-end face. Consequently, part of the light is transmitted through the fiber and the remaining part is radiated outside it. The fiber is also back-illuminated by light from a tungsten lamp.

The measurement spatial resolution depends on mode field diameter dimension; the smaller the mode field diameter dimension, the better the spatial resolution. A disk is centered on the output cone and is properly placed in order to obtain optimum spatial resolution. Refracted modes passing the disk are collected and focused onto a photodiode; refractive index variation is then directly plotted. The equipment calibration is performed removing the fiber from the liquid cell. Absolute index values can be obtained only when cladding or liquid index are accurately known at the measurement wavelength and temperature.

The near-field distribution technique permits to analyze the fiber-end cross section, provided that it is perpendicular to the fiber axis. Fiber-end face is magnified using a microscope or another suitable optic system and examined by means of a digital video analyzer. Inspection can be also performed simply using direct view, scanning detection, or a photographic camera. Amplifying optics and detection device shown in Figure 3.5 are usually included in the equipment. Two light sources are sometimes used, for core and cladding illumination, respectively. The light launched into the core is incoherent and intensity controlled. As described earlier, fiber-end face is laid in front of a microscope with a calibrated magnification of 600 times or more. A video analyzer is also used as the detection system for its simple use. It is generally a CCD or a scanning vidicon and the fiber amplified image is analyzed and processed using a computer. A photographic camera or direct visual inspection represent less precise solutions, while scanning detection requires particular attention and a more expert approach.

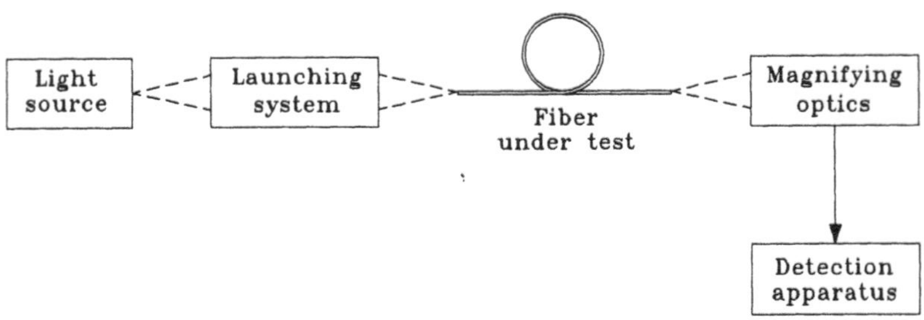

Figure 3.5 Near-field distribution technique.

3.1.3 Cut-Off Wavelength

In an ideal single-mode fiber, the cut-off wavelength λ_c of the LP_{11} mode has a precise definition and indicates that the fiber is in a multi-mode regime for wavelengths shorter than λ_c. Theoretical cut-off definition for a step-index fiber is given by (2.122). Nevertheless, index shape often departs significantly from the step distribution, either intentionally (dispersion-optimized fibers) or not. This implies the necessity of an experimental evaluation of the cut-off wavelength λ_c.

Furthermore, the establishment of a cut-off wavelength definition, which is unambiguous and useful in practice, is not simple. The LP_{11} mode can be present in the fiber for cut-off wavelengths greater than the theoretical one, though with a remarkable attenuation coefficient. Thus, the introduction of an effective cut-off wavelength, defined as the wavelength at which the LP_{11} mode really extinguishes, is necessary [17]. This more practical parameter is typically 100–200 nm smaller than the theoretical one. However, it depends on several factors, such as cable length, presence of curvatures, mechanical stresses on the fiber due to the cable, and so on. A compromise solution, often applied, consists in assuming as effective cut-off wavelength the value of λ for which the LP_{11} mode does not propagate along a fiber 2 meters long, wound on a mandrel with a radius of 140 mm and in the absence of any external stress. In any case the result is heavily influenced by the measurement technique. As will be shown, the most reliable solution, in this sense, exploits the noticeable sensitivity to curvature losses of the LP_{11} mode when the wavelength is close to the cut-off condition. Thus the latter can be appreciated by observing the transmitted power changes, as a function of wavelength, when a single-mode fiber is subjected to a known curvature amount. According to CCITT definition, cut-off wavelength is then the wavelength above which the ratio between the total power and the fundamental mode power decreases by less than 0.1 dB.

From another point of view, there are two different cut-off wavelength definitions, for uncabled (λ_c) and cabled (λ_{cc}) fibers. The former parameter is easier to measure, because the fiber can be arranged in such a way as to match a precise curvature radius when it is uncabled. The latter definition is important as well, since λ_{cc} is tightly related to the system performances. For this reason, it is sometimes advisable to know the exact cut-off wavelength in an installed link. International standardization groups are still working to establish unambiguous and reliable relations between the two measurements.

The reference test method for the determination of λ_c (but also for that of λ_{cc}, according to the procedure we will discuss later) is the transmitted power technique. Alternative methods are those called *mode field diameter versus wavelength* and *split-mandrel*.

The transmitted power technique (Figure 3.6) is based on the comparison between the transmitted power on a short length of fiber under test, which is wavelength dependent, and a reference power.

The light source has to be stable in position, intensity, and wavelength over a time period long enough to perform the complete measurement. A tunable optical source whose linewidth is smaller than 10 nm must be arranged, for instance by means of a lamp and a monochromator. Wavelength scanning is necessary over a spectral range of about 300 nm. The experimental setup resembles that used for a spectral attenuation evaluation through cut-back technique, which will be better described in Section 3.1.4. The light source is usually a halogen lamp, and the emitted light is on-off modulated by means of a light-chopper. A lock-in amplifier is employed in order to improve the signal-to-noise ratio at the receiver. Both the LP_{01} and the LP_{11} mode are launched into the fiber through a joint with a multi-mode fiber or a suitable optical system.

A *cladding mode stripper* (CMS) forces the power conversion from cladding guided modes into cladding radiation modes, which are, consequently, stripped out from the fiber. The test fiber, about 2 m long, is inserted into the apparatus and loosely bent with a uniform radius of nominally 140 mm. Bending with a smaller radius or external stresses acting on the fiber may influence λ_c, lowering its value. The output power $P_1(\lambda)$ is recorded over a large spectral range around the expected cut-off wavelength. Primary coating usually does not affect λ_c, while the secondary one is removed because its presence may lower the value of the wavelength under measurement.

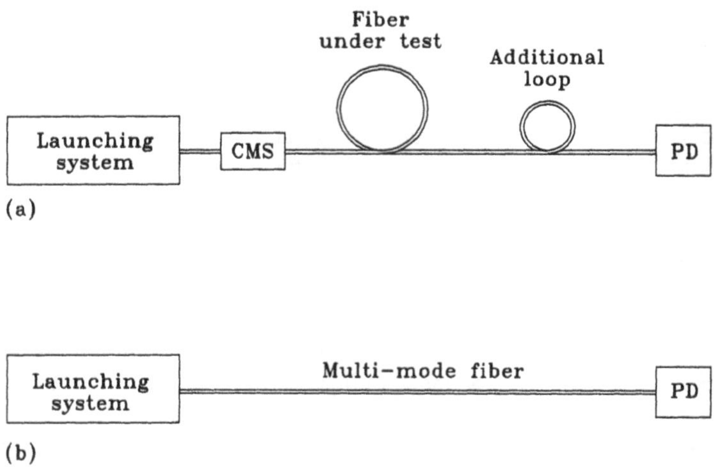

Figure 3.6 Transmitted power technique.

The reference power can be detected in two different ways [3, 4]:

a) using the same test sample with at least one loop with a small radius to filter out the LP_{11} mode and recording the output power $P_2(\lambda)$;
b) using a sample of multi-mode fiber (1–2 m long) and measuring the output power $P_3(\lambda)$ over the same spectral range.

With the former method, which is directly related to cut-off wavelength definition, typical loop radius is about 30 mm. With the latter method, which appears more empirical, the output signal may be affected by ripples due to leaky modes. Choosing suitable launching conditions (i.e., a mode filter or a restriction of multi-mode core diameter and NA to about 70%) can limit the effects due to leaky modes. A small difference between the results obtained using the two techniques on the same test fiber sometimes occurs.

The ratio between the two transmitted powers is expressed (in dB) as follows

$$R(\lambda) = 10 \log_{10}\left[\frac{P_1(\lambda)}{P_i(\lambda)}\right] \tag{3.6}$$

where i = 2 or 3 depending on which method (a or b) has been used.

With method a, λ_c is determined as the highest value for which $R(\lambda) = 0.1$ dB. With method b, λ_c is evaluated as the intersection of $R(\lambda)$ with a straight line displaced by 0.1 dB and parallel to the $R(\lambda)$ line, fitted to the long wavelength part of $R(\lambda)$ itself. These two definitions are described in Figure 3.7.

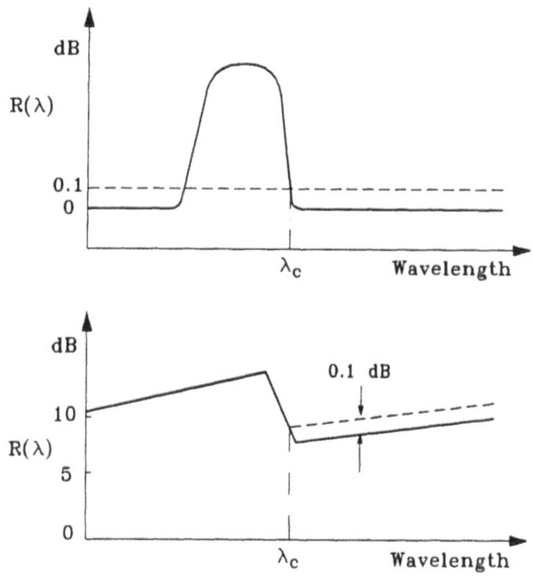

Figure 3.7 Description of the methods for determining the λ_c value.

Using the same setup, the attenuation coefficient of the LP_{11} mode can be also evaluated in two different ways. Under the assumption of uniform mode excitation, the attenuation coefficient per unitary length (in dB/m) of such mode is [3]

$$A_{s11}(\lambda) = \frac{10}{L} \log_{10}\left[\frac{P_1(\lambda) - P_2(\lambda)}{2P_2(\lambda)}\right] \tag{3.7}$$

where L is the curved fiber length with a nominal radius of 140 mm.

Alternatively [4] we can cut the fiber at a distance of about 0.5 m and measure the output power $P_4(\lambda)$, maintaining the same launching conditions. The attenuation coefficient (expressed in dB/m) of the LP_{11} mode is now given as

$$A_{s11}(\lambda) = \frac{10}{L} \log_{10}\left[\frac{P_1 - P_2}{P_4 - P_2}\right] \tag{3.8}$$

where L is the loop circumference (in meters) whose nominal radius is 140 mm. The attenuation coefficient of the LP_{11} mode, as a function of the wavelength, permits to estimate bend sensitivity in cabled fibers [18].

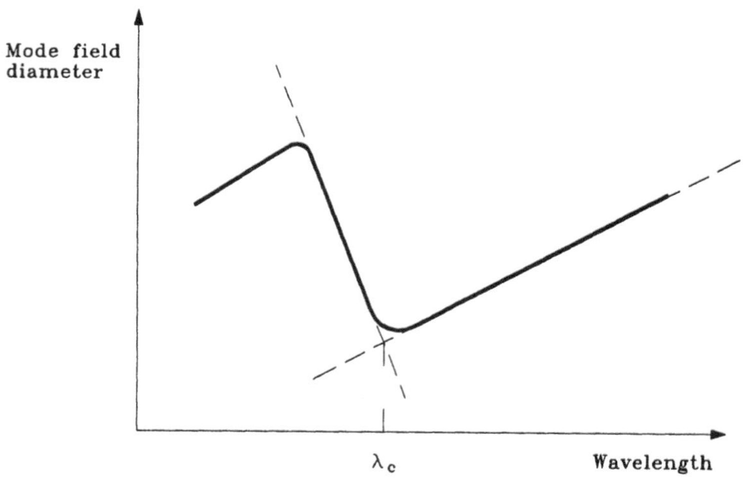

Figure 3.8 λ_c evaluation using mode field diameter.

The first alternative technique for cut-off wavelength measurement, called *mode field diameter versus wavelength*, is based on a mode field diameter evaluation as a function of the wavelength in order to derive indirectly the cut-off wavelength. At wavelengths far from cut-off point, the fundamental mode field diameter increases almost linearly with increasing the wavelength. At the cut-off point there is a large variation, due to the appearance of the LP_{11} mode. Mode field diameter behavior, linearly extrapolated and plotted, permits to define λ_c as the intersection of linear extrapolations (Figure 3.8).

The second alternative technique for cut-off wavelength measurement, that called the *split-mandrel method* [3], is based on the setup shown in Figure 3.9. Measurement is performed employing a loosely bent sample, 2 m long. Loop is composed by two arcs (180 deg each) of 140–mm radius connected by straight segments, while external stresses on the fiber are to be avoided. The output power $P_1(\lambda)$ is recorded as function of the wavelength. The lower semicircular mandrel moves away to remove any slack from the fiber loop, keeping completely stable the launching and receiving systems. All the successive steps for λ_c determination are the same as in the transmitted power technique.

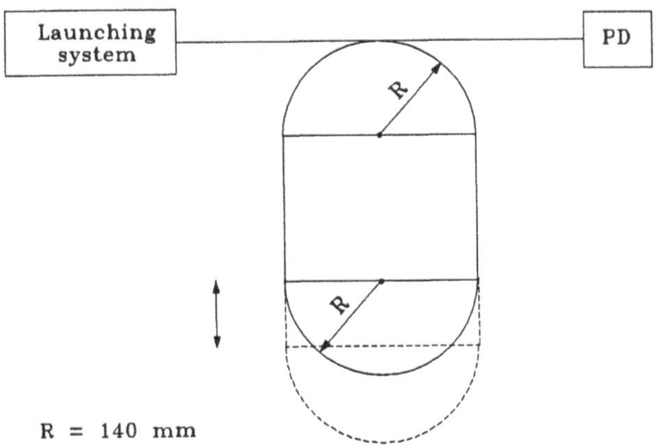

Figure 3.9 Split-mandrel setup.

Completely automatic instruments for the evaluation of λ_c are at present available in many laboratories. An example of the output given by such an apparatus is reported in Figure 3.10. It is worthwhile to stress that the same instrument is also normally able to operate the measurement of many other fiber parameters, such as mode field diameter.

The measurement of cut-off wavelength is performed on cabled single-mode fibers (λ_{cc}) simulating conditions similar to those previously described for outside plant features. Effective single-mode region is defined only for real installed fibers. In general $\lambda_{cc} < \lambda_c$, but it is difficult to establish an unambiguous correlation between such two quantities. Typical values recommended for fibers employed in the second window region are 1100 < λ_c < 1280 nm and λ_{cc} < 1270 nm.

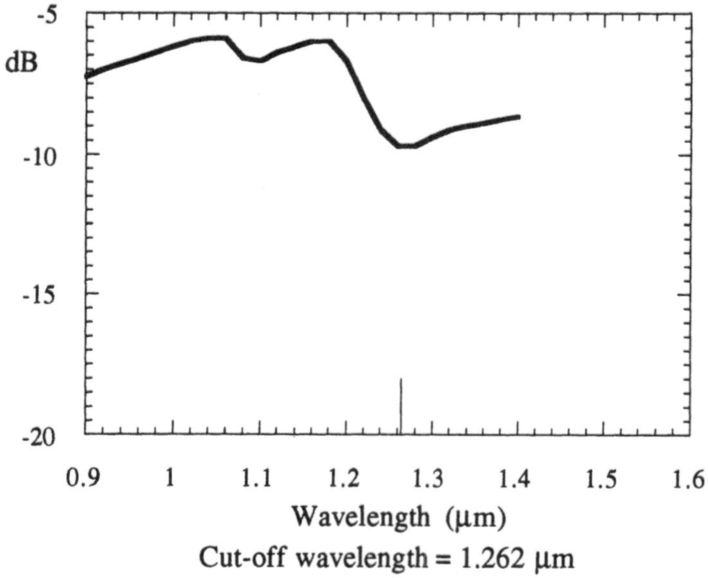

Figure 3.10 Example of output from a completely automatic instrument for λ_c evaluation (courtesy of Teleco Cavi SpA).

The reference test method for λ_{cc} determination is the transmitted power technique, which uses one of the two options described before. In this case the test sample is a piece of cable 22 m long. One meter of uncabled fiber is exposed at both cable ends; the remaining part of cable (20 m) is laid with no bends. Uncabled fibers are then bent in a loop, whose radius is R, expressed in mm, to simulate splice organizer effects (CCITT suggests R = 40 mm). The cut-off wavelength determination is then realized with the same steps as for uncabled fibers. It is evident that this method is heavily destructive. A more practical alternative is the replacement of the cable with uncabled fiber having the same length and wound with a suitable radius.

Anomalous behavior of spectral loss for the LP_{11} mode has been observed due to whispering gallery modes interference or to mode coupling between coating modes and higher order modes. Sometimes, one or more humps in spectral loss curve have been observed. This effect may introduce a large uncertainty in cut-off wavelength evaluation, in particular when it is strongly dependent on curvature.

Interlaboratory comparison [19] has demonstrated that reproducibility is acceptable inside the individual research group, but it is poor among different research groups, both for cabled and uncabled fiber cut-off wavelength measurements. This uncertainty was larger for dispersion-shifted than for matched cladding fibers. The same round robin test demonstrated a small difference also in the standard deviation between λ_c and λ_{cc}. The large uncertainty was strongly dependent on the loop radius.

3.1.4 General Considerations on Attenuation Measurements

First of all it is useful to recall that, in facing measurement problems, gain and attenuation, as well as noise figure and signal-to-noise ratio, are usually expressed in logarithmic units. In this way we obtain values in dB. In particular, gains and attenuations can be expressed either as ratios between signal amplitudes or as ratios between signal powers. In the first case, denoting by A_1 and A_2 the amplitude levels to be compared, we have

$$\left(\frac{A_1}{A_2}\right)_{dB} = 20 \log_{10}\left(\frac{A_1}{A_2}\right) \tag{3.9}$$

while, in the second case, denoting by P_1 and P_2 the power levels to be compared, the following expression holds

$$\left(\frac{P_1}{P_2}\right)_{dB} = 10 \log_{10}\left(\frac{P_1}{P_2}\right) \tag{3.10}$$

That being stated, three methods for attenuation measurements are well-defined: cut-back (which is a reference method), insertion loss, and backscattering. Setup complexity, resolution, accuracy, and the amount of information on attenuation properties differ from one method to another. We will analyze the advantages and disadvantages of all methods. The evaluation of losses is in fact probably the element that best characterizes the performance of a fiber during the fabrication process as well as in cable manufacturing, installation, and service. The first two techniques, previously mentioned, give a value of attenuation between two sections. The third technique, instead, is able to follow local fluctuation of the loss because it is based on a reflectometer principle. Attenuation along the fiber has an approximately linear behavior in logarithmic scale. In order to test this longitudinal uniformity a backscattering measurement is to be performed.

The experimental setup of a cut-back technique is shown in Figure 3.11. A direct comparison between the two measured levels of optical power under the same excitation is taken to evaluate the insertion loss between the two fiber sections. The results achievable are extremely accurate and the measurement can be taken over a wide spectral range, provided that a tunable optical source is used. A typical solution consists in employing a lamp followed by a monochromator.

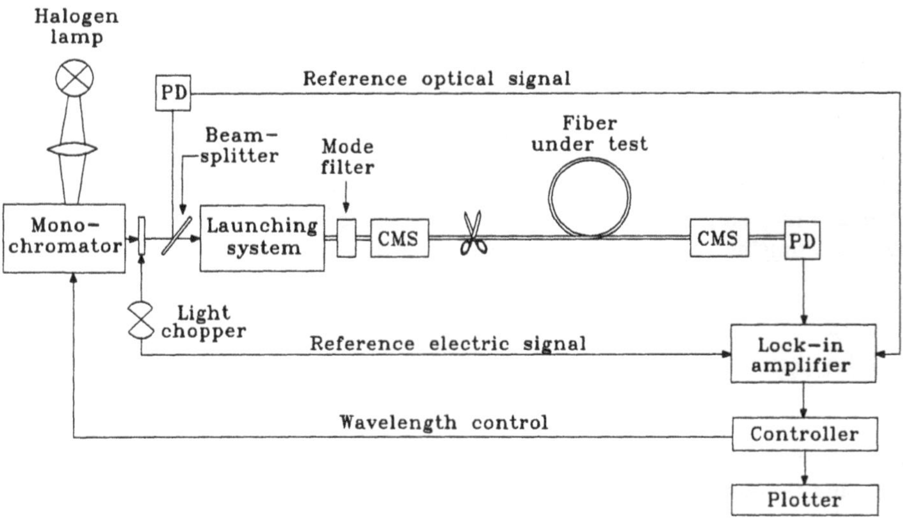

Figure 3.11 Cut-back technique.

Access to both ends of the fiber is required. This fact, together with a certain setup complexity, limits cut-back technique to factory tests. The attenuation A (expressed in dB, according to (3.10)) is defined as

$$A = 10 \log_{10}\left(\frac{P_1}{P_2}\right) \tag{3.11}$$

where P_1 and P_2 are the power levels measured at two end points without any change in the input conditions; generally P_2 is the power at the end of the fiber, and P_1 is the power at a point close to the input head of the fiber (about 2 to 3 m from it). If the fiber is uniform along its length L, it is possible to determine the attenuation per unitary length A_s defined as

$$A_s = \frac{A}{L} \tag{3.12}$$

which is expressed in dB/km.

The source must be stabilized in intensity and emission wavelength for a time period long enough to complete the measurement, and any movement has to be accurately avoided. When the measurement is performed at a fixed wavelength, an LED or a conventional laser diode is preferable as optical source. Only the fundamental mode has to be excited. Suitable launching techniques [3] consist of jointing the test fiber with a launch fiber or employing a proper optical system in front of the test fiber. Care must be taken against the possible influence of cladding or higher order modes. These modes usually have no effect on the far head of the fiber, because they are completely attenuated, but they can be still present close to the input head. As a consequence, attenuation coefficient becomes higher than the true one. Mode filters, like a bend, followed by cladding mode strippers, are commonly used to resolve these problems.

Optical detectors must have a large and uniform active area to capture all the optical signal coming out from the head of the fiber. They must also exhibit a good linear characteristic. The measurement procedure, even when repeated at different wavelengths, is rather simple. The fiber under test is placed in the measurement setup with the fiber-head cleaned and perpendicularly cut; the fiber holder shall be aligned to the launch cone or connected coaxially to a launch fiber. The output power P_2 is recorded at different wavelengths. The same measure is repeated for evaluating P_1 after fiber cutting (2 or 3 meters far from input head) and keeping all the launching conditions fixed. Care must be devoted to reproduce exactly the same wavelengths as before, especially when the attenuation spectral behavior is rapidly varying. The ratio between P_1 and P_2, which is wavelength dependent, represents the attenuation due to the fiber cut away, at the selected wavelength.

An alternative optical fiber attenuation measurement is that called the *insertion loss technique* [4], which is a two-point method (like cut-back). The experimental setup is arranged as shown in Figure 3.12. Measurements are generally performed at one fixed wavelength, though they can be repeated at different wavelengths. The light source is usually an LED or a conventional laser diode. The optical source, detector, and launching system have the same technical characteristics as in the cut-back method. Furthermore, in the insertion loss measurement, a very precise fiber-to-fiber coupling device is used in such a way as to ensure repeatability. Generally, it is a special connector or a mechanically adjustable device (like a V-groove). The result of the measurement is not as precise as that obtained by the cut-back technique, because an uncertainty remains in the estimate of the loss introduced by such launching system. Nevertheless, when the overall attenuation to be measured exceeds a few dB, this uncertainty becomes as low as some percentage.

This method is suitable for field tests and maintenance, in particular to analyze installed cables or in-service lines. The measurement procedure is quite simple. A reference power P_1 is obtained by connecting the transmitting apparatus (composed by an optical source and the launching system) directly to the receiving one (formed by a detector and a power-meter). It is advisable to adjust coupling for maximizing the detected optical power. The same operation is then repeated including the fiber under test, in order to obtain an output power P_2. The overall attenuation A between the two points (in dB) is still given by expression (3.11). This attenuation value summarizes the one due to the test fiber and the loss produced by the launching system, whose contribution to A, however, as already stressed, is significant only for short links. Also in this case, longitudinal uniformity enables us to obtain the attenuation per unit length A_s, by means of (3.12).

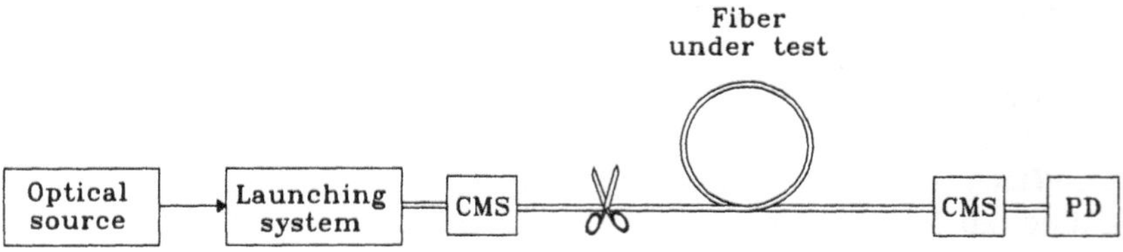

Figure 3.12 Insertion loss technique.

3.1.5 Attenuation Measurements by OTDR

In comparison with cut-back and insertion loss techniques, backscattering measurement has some advantages. Among them we can list:

1. access to only one fiber-head is necessary;
2. the experimental setup is simpler;
3. the measurement can be simply performed either in the factory or during installation and maintenance;
4. some additional information is available about longitudinal uniformity.

Property 2 has allowed the possibility of designing commercial instruments particularly suited for field tests. Among the data about possible longitudinal nonuniformities, we can appreciate splice and connector losses, faults, defects, and fiber segments with anomalous attenuation. Furthermore, the lengths of the various fiber pieces and the exact positions of connectors, splices, and faults can be easily determined. A laser diode is employed as the optical source. Therefore measures are generally performed at fixed wavelengths (1310 and 1550 nm). The accuracy of the results obtained, in terms of attenuation per unit length, are only slightly inferior to that achieved in cut-back or insertion loss techniques, and an uncertainty not greater than 0.05 dB/km is typically present also for commercially available portable instruments. Backscattering attenuation measurement is generally performed using *optical time domain reflectometry* (OTDR). A modern instrument for optical fiber characterization [20] is not very different from the first demonstrations of this technique [21, 22].

The same equipment can be used for the evaluation of connector insertion loss [23], splice loss [24], microbending loss [25], diameter fluctuations [26] and fiber and cable length [27]. The arrangement of a test equipment is depicted in Figure 3.13. Index matching compounds can be used to limit Fresnel reflections and the light beam is coaxially launched into the fiber input face. A coupling device, generally acoustooptic [28], must be added to realize coupling between the laser diode LD and the fiber, and between the fiber and the detector. Isolation between laser diode and detector is requested to be larger than 50 dB. Couplers based on polarization effects are usually less efficient. Pulse width and repetition rate are chosen as a trade off between spatial resolution and dynamic range, as we will discuss later. In reception, there is a cascade composed by an optical detector and an amplifier with a suitable bandwidth; the output signal is then processed to improve the signal-to-noise ratio and the curve of backscattered signal is presented in logarithmic scale.

A simple way for understanding the OTDR operating principle is as follows. If we have an input optical pulse, with peak power P_0 and width W, the Rayleigh backscattered power P_R observed at the input fiber-head at the time T is [29]

$$P_R = \frac{P_0 S \alpha_s W c}{2n} e^{-2\alpha[cT/(2n)]} \tag{3.13}$$

where S is the fraction of scattered signal traveling back into the fiber, α_s the field attenuation constant due to Rayleigh scattering, given by (2.125), $\alpha[cT/(2n)]$ is the cumulative attenuation (in nepers) up to a distance cT/(2n), c is the light velocity in the free space, n the core refractive index (considered for simplicity as a constant, so that c/n represents approximately the light velocity in the fiber), and the time T corresponds to the position $z = cT/(2n)$ in the fiber. In order to have a correct operation, it should be $W \ll [2n/(\alpha c)]$.

Rayleigh backscattered signal is about five to six orders of magnitude lower than forward optical signal and three to four orders smaller than Fresnel reflections due to air-fiber interface, if care is not taken to lower it. However, backscattered trace due to the residual Fresnel reflection cannot be appreciable. It means that there is a "dead zone" whose longitudinal amplitude depends on the pulse width.

The main features characterizing an OTDR instrument are spatial resolution, dynamic range, and noise reduction. Spatial resolution is limited by pulse width; information referred to two points in the fiber may be distinguishable when there is a distance larger than cT/(2n) between them. For example, a pulse width of 3 ns means about 30 cm of uncertainty. However, there is a trade-off between dynamic range and resolution [30, 31]. In fact a reduction of pulse width by a factor of N implies a one-way dynamic range smaller than 7.5 $\log_{10}N$ dB [29], due to a change in power pulse by a factor of N and in minimum detectable optical power by a factor of $N^{1/2}$.

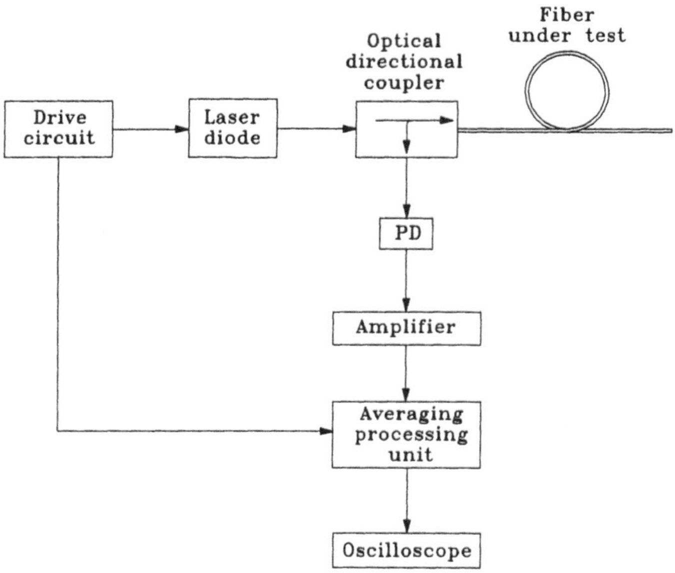

Figure 3.13 OTDR equipment composition.

Other techniques have been proposed to enhance resolution, like nonlinear detection [32] or low coherence interferometry [33]. Two kinds of tools have been suggested to overcome the problems related to dynamic range limitation, especially for single-mode fibers where the backscattered signal has a very low amplitude: increasing of input optical power or decreasing the minimum detectable optical power [29]. Enhancement of optical power had been obtained using high power light source like Q-switched Nd:YAG or $Er^{(3+)}$:glass lasers [34] for wavelengths of 1300 and 1550 nm, respectively. Another solution is OTDR from stimulated Raman scattering where a measurement range from 1200 to 1700 nm is achieved [35, 36]. OTDRs using Nd:YAG lasers or $Er^{(3+)}$:glass lasers appear as the solution with the largest dynamic range. Nevertheless it is impossible to realize portable equipment for field tests when the light source is a high-power laser because of large size and high electric power consumption. Correlation OTDR [37] appears as a more practical method, based on enhancing the receiver sensitivity.

This technique (Figure 3.14) attains a typical improvement of about 10 dB in the dynamic range using a pseudorandom bit sequence as modulating signal. A cross correlator circuit, composed by a multiplier and an integrator, performs time correlation between the electrical reference signal (delayed) and the optical backscattered signal. The measurement point of the fiber is varied by changing the delay.

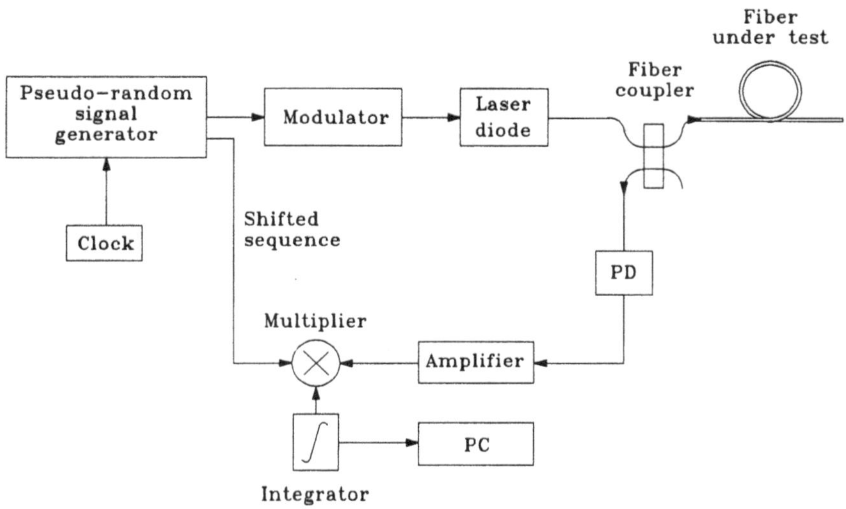

Figure 3.14 Correlation OTDR (adapted from [37]).

Another very attractive solution in order to reduce the minimum detectable power is based on the amplification of the backscattered signal by means of an optical preamplifier. The first proposal of this technique was arranged using a semiconductor laser [38] with an improvement of 4.5 dB compared to direct detection. Recently, better performances have been achieved employing erbium-doped fiber amplification [39, 40] of both transmitted pulses and backscattered signal. A net improvement of almost 10 dB and a potential increase of 20 dB in one-way measurements have been demonstrated.

Other proposals exploit a photon counting method [41] or semiconductor single-photon detectors [42]. Finally, interesting solutions are those based on heterodyne detection [43] and *optical frequency domain reflectometry* (OFDR) [44]. In the former solution, local oscillator and optical signals are mixed in the photodetector. In the latter, a CW optical carrier is modulated by an RF signal constant in amplitude and with a periodic linear frequency sweep. Correlation between the backscattered optical signal and the reference electric signal is converted in the time domain using a spectrum analyzer. More details on coherent backscattering will be done in the next chapter.

Instability of laser central frequency and variations of the scattering points along the optical fiber due to environmental variations (mechanical and thermal fluctuations) are the main sources of undesired fluctuations for the measurement signal. They can be reduced using a square-law detection [29] or by other nonlinear averaging of optical backscattered signal.

A large number of characteristics are included in the technical data of commercial OTDR instruments. Some of these are: transmitter pulse energy and repetition rate, logarithmic display accuracy, receiver sensitivity (in dB/m), span length, dynamic range (one-way), signal averaging, trace linearity [45].

Repetition rate is of the order of some kHz, whereas span length and longitudinal resolution depend by use. In *local area network* (LAN) applications optical cable is only a few km long, whereas long-haul connections of the public network exceed 100 km without repeaters. So pulse width varies between 1 ns to 1 μs. The spatial resolution in correspondence of the former extremum can reach a value as good as some centimeters. In long-haul connections, a spatial resolution a thousand times larger is still acceptable.

In (3.13) the backscattered optical signal versus time is expressed, for a measurement performed in the time domain (OTDR). Conversion of time into distance along the fiber can be done taking into account that $z = cT/(2n)$. Therefore n must be a well-known parameter, and it is usually set by the operator. A typical value for single-mode fibers is about 1.467, but it may change slightly from one cable to another. Determination of time and its conversion into distance can be affected by systematic errors; another source of uncertainty is the rise time of the forward pulses. A periodic calibration is required to guarantee precise OTDR performances. We will show later some active and passive techniques for instrument calibration.

Spatial resolution is usually defined as $\Delta z = Wv/2$, W being the pulse width and $v = c/n$. This expression is exact [46] when the pulse has a rectangular shape and its width only determines the minimum longitudinal resolution. However, contributions to improve the resolution could arise also from instrument characteristics, such as postdetection bandwidth or digital sampling intervals. It is helpful to introduce a new definition of spatial resolution [47] as $\Delta z' = W_m c/(2n)$ where W_m is the full width at half-maximum (FWHM) of a discrete reflection as observed on the OTDR display and there are no fiber dispersion contributions on pulse width. In logarithmic scale this corresponds to the pulse width at – 3 dB (optical) level. Fresnel reflections due to perfect fiber cuts or connectors can be so large to saturate the photodetector, especially when they are close to input fiber head. Measurements cannot be exactly performed under these conditions. On the other hand, in all equipment a suitable threshold level is included to limit Fresnel reflection effects, so as to avoid the photodiode saturation.

The total fiber loss between $z = 0$ and $z = L = cT/(2n)$ is given by

$$A(L) = -5 \log_{10} \frac{P_R(L)}{P_R(0)} \tag{3.14}$$

A(L) is commonly defined as one-way loss (in dB). The backscattered optical signal, expressed in (optical) dB due to the same fiber sample between $z = 0$ and $z = L$, is

$$A_R = -10 \log_{10} \frac{P_R(L)}{P_R(0)} \tag{3.15}$$

Expressions of A(L) and A_R differ for a factor of 2. This is because the electromagnetic field must travel first in forward direction and then in backward direction to the photodetector, so that backscattered signal is attenuated twice.

The signal-to-noise ratio (S/N) at the detector output, refers to electrical signals which are proportional, via a square law due to photodetector, to the optical power. This is expressed as

$$S/N = 20 \log_{10} \frac{P_R}{P_N} \tag{3.16}$$

where P_N is the power noise, and P_R is a length function, with a maximum for $L = 0$.

A dynamic range definition [47] may be given in terms of the highest measurable one-way loss as

$$D = 5 \log_{10} \frac{P_R(0)}{P_N} \tag{3.17}$$

where $P_R(0)$ is the backscattered input signal due to a reference fiber and P_N is the power noise after averaging. Using this definition, directivity is easily related to the S/N ratio. Therefore, no direct measurements on internal parameters of the OTDR are necessary [47].

Signal averaging decreases noise power and is useful to improve the signal-to-noise ratio on OTDR display. In particular, in high-resolution measurements, probe pulses and, consequently, backscattered signals have very low levels. Commercial OTDRs realize a time averaging of up to 2^{16} values in a few seconds with an improvement on noise power suppression of about 20 dB. A suitable electric amplifier is included before the processing unit to increase the signal level. Bandwidth amplifier is a trade-off between time resolution and noise reduction.

Nonlinearity of the OTDR power response affects loss measurement repeatability. This is the case of measurements performed at different power levels, i.e., short fiber pieces installed inside long fiber links. Typical manufacturers' specification for nonlinearity is of the order of 0.05 dB/km, but this can be reduced using correction functions [48].

Now we are going to give some more details on backscattering trace (Figure 3.15) where we can see the input-head Fresnel reflection (point A), internal defects (C), local discontinuities (D), end-fiber Fresnel reflection (E). Portion B of the curve is linear as it is typical for an homogeneous fiber. Thus it has a constant slope, i.e., the fiber attenuation per unitary length A_s (expressed in dB/km) is constant. The measurement of A_s can be performed in one way only using two points chosen out of the dead zone. They are characterized by a difference between the power backscattered signals $\Delta P = P_1 - P_2$ and by a distance $L = L_2 - L_1$ between them. The two points must be taken as far as possible. Sharp peaks in the backscattering trace (C) may be measured as in the previous section. Multi-Fresnel-reflections can give a small peak in the fiber trace when a large pulse width is chosen and/or the fiber length is small. Thus Fresnel reflections due to fiber-end (or connectors) travel backward and forward a few times. Consequently, there are peaks in the trace, with a constant distance between them.

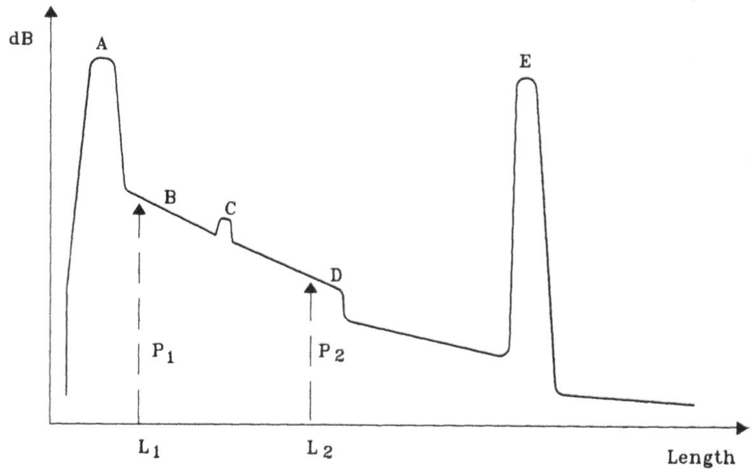

Figure 3.15 Example of OTDR trace.

There is a large variation in the backscattered signal around point D due to a splice or other external causes. If referred to a splice, the step in backscattered signal may be positive or negative, since it is the result of splice losses and different backscattered behaviors of the two fibers (see Section 3.1.9). Therefore the measurement has to be made from both fiber-end sides and splice loss is estimated as the mean value of the two measurements. Longitudinal resolution depends on pulse width and suitable optical pulse has to be chosen, in particular, when there are more splices. In case of very close splices it may be impossible to measure the two single attenuation values.

A large positive reflection indicates a change of refractive index like that occurring at an air-fiber interface. It is typical of fiber-end (with a good fiber cut) or connectors. Fiber-end may be also revealed by a sharp decrease in backscattered trace when the fiber output head has a poor quality.

Sometimes there are ripples inside the trace. They are due to measurement equipment limitations, in particular when the fiber length approaches the dynamic range of the backscattered equipment. Such a problem can be avoided using a two-way measurement and a least-mean square approximation.

Other causes of changes in the slope behavior may occur, resulting in a bow trace with concave or convex shape [3]. Bows are caused by internal or external influences and are the result of cladding mode propagation.

Internal effects (absorption losses excepted) may be due to a continuous change in mode field diameter. External effects could derive from an increasing force on fiber related to cable installation or fiber coatings.

The great detail presented for OTDR equipment may be simply explained with the large introduction of backscattering technique as an automatic diagnostic tool for network supervisioning, in particular for LANs with a great number of branches. OTDRs are surely suitable for maintenance measurements with a minimum operator contribution. Furthermore, since fiber geometrical and physical parameters are well standardized, one-way measurement may be automatically realized using OTDR and optical switches. It permits the analysis of hundreds of fibers and comparison of their loss evolution in time.

There are several demonstrations of in-service supervisory systems using OTDR [49–52] with WDM technique. Optical wavelength selective couplers and bandpass filters permit insertion of OTDRs on in-service telecommunication networks with no degradation on the BER. There are also solutions [53] that avoid the use of WDM techniques. A first example of these may be the transmission of OTDR signal using the same laser diode of the telecommunication link in a time slot available between periods of normal traffic. Another proposal consists of a pseudo-random baseband sequence superimposed, optically or electrically, upon the traffic signal.

Since backscattering signal is directly related to link parameters (loss, length, discontinuities, etc.), OTDR tests must be accurately calibrated. Several calibrating methods have been proposed. They can be subdivided into active and passive methods [54].

A first simple active calibration method is realized using a cascade composed by two variable attenuators and a mirror-ended fiber. An alternative setup for an active calibration method is shown in Figure 3.16. The OTDR output optical probe pulse passes through a directional coupler and is converted into an electric signal by a photodetector (O/E converter). Electric signal is delayed and subsequently reconverted into an optical pulse which comes back to OTDR via a cascade composed by two attenuators and a directional coupler. Changing the delay it is possible to calibrate the distance scale.

Attenuation scale can be simply calibrated, comparing the attenuation changing (in dB) measured by an optical calibrated power meter (A_{PM}) and by the OTDR (A_{OTDR}). The loss error can be written as

$$\Delta A = \frac{2A_{ODTR} - A_{PM}}{A_{PM}} \tag{3.18}$$

and reproducibility is usually better than 0.01 dB/dB.

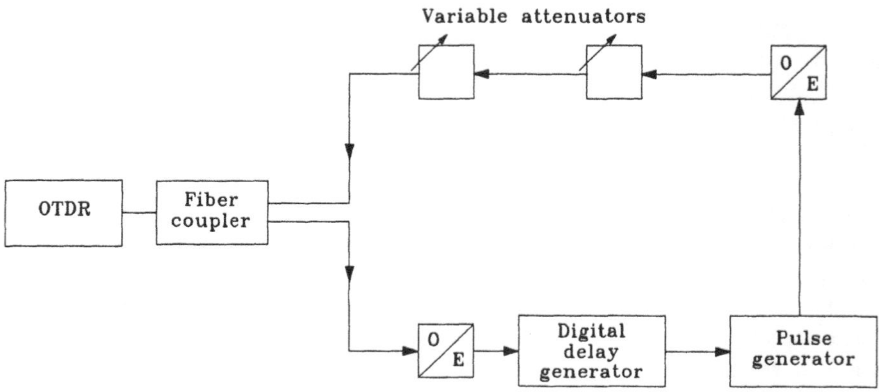

Figure 3.16 Active OTDR calibration method (adapted from [54]).

Passive methods employ a calibrated or reference fiber. A first proposal is shown in Figure 3.17 [48]. Two points are fixed and selected on the fiber length. Attenuation between them does not depend on different values of attenuation due to the attenuator. OTDR linearity can be tested changing attenuator contribution. Absolute calibration can be done using a reference uniform fiber with a known attenuation coefficient.

Another calibration technique [55] is based on a coupler test device. It is very simple to carry out because only attenuators and power meters are needed. Calibration uncertainty in such a method is typically very low (0.011 dB/dB). A further proposal for instrument calibration is based on a reference device consisting of a recirculating optical delay line [56]. Such a device can operate both in active and passive manner. Optical fiber length can be measured with an accuracy of better than 1 part in 10^4. Using the same device, it is possible to precisely measure attenuation and nonlinearity.

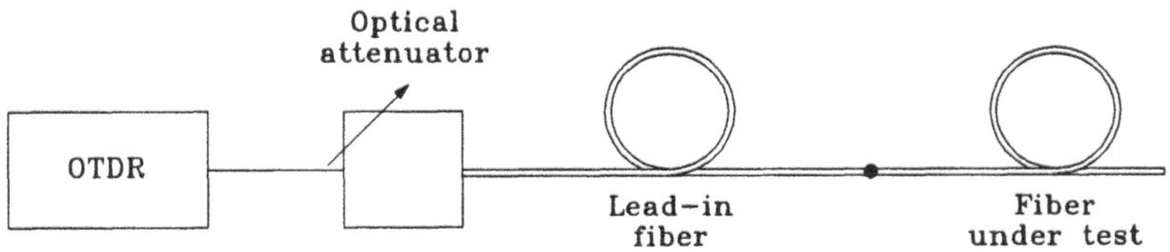

Figure 3.17 Passive OTDR calibration method (adapted from [48]).

3.1.6 Chromatic Dispersion

Optical pulse broadening occurs because of nonperfect monochromaticity of the laser source. Thus different wavelengths, starting at the same instant, travel the fiber with their own velocities and arrive at the receiver at different times, determining pulse broadening. Material and waveguide effects are the main sources of signal distortion in single-mode fibers, as explained in Section 2.4.3. We will treat polarization mode dispersion in Sections 4.1.2 and 4.1.3. This additional cause of signal distortion, not related to chromatic effects, is due to birefringence that always affects real fibers. When material and waveguide contribution are the opposite of the other, fibers approximately free of second-order chromatic dispersion can be realized. The wavelength value at which an exact compensation of such two effects occurs is called the *zero dispersion wavelength*. Overall chromatic dispersion is well characterized through the knowledge of the chromatic dispersion coefficient $\tau'(\lambda)$ already introduced in Section 2.4.3, the zero-dispersion wavelength λ_0, and the zero-dispersion slope (that is the slope of $\tau'(\lambda)$ at $\lambda = \lambda_0$).

There are three main techniques for an experimental evaluation of such parameters:

1. pulse delay,
2. phase shift,
3. interferometric method.

The last is the most accurate method but it can be used only for short pieces of fiber. The other two techniques are closely related, as they are considered two different versions of the same method. The chromatic dispersion coefficient is obtained by the measurement of the spectral time delay using particular functions for experimental data fitting. These are:

three-term Sellmeyer equation for dispersion-unshifted fibers and the parabolic equation for dispersion-shifted fibers. However, problems arised for dispersion-flattened fibers and approximate methods are still in progress [2, 57].

The three-term Sellmeyer equation allows us to represent the wavelength dependence of the refractive index n in a dielectric medium at optical frequency, according to

$$n^2 = 1 + \sum_{j=1}^{3} \frac{B_j \lambda^2}{\lambda^2 - \lambda_j^2} \tag{3.19}$$

where B_j are suitable constants and λ_j correspond to the fundamental resonances of the considered material. The values of B_j can be evaluated by the knowledge of some characteristic points, also experimentally determined, in the refractive index distribution (for example, the values of n for certain prefixed wavelengths and the value of λ at which $d^2n/d\lambda^2$ vanishes).

An alternative expression, based on the experimental determination of three energies, and then more significant from a physical point of view, is

$$n^2 = 1 + \frac{E_d E_o}{E_o^2 - \left(\frac{hc}{\lambda}\right)^2} - \frac{E_l^2}{\left(\frac{hc}{\lambda}\right)^2} \tag{3.20}$$

where E_d is the electronic oscillation energy, E_o is the average electronic energy gap, E_l is the lattice oscillation energy, h is Planck's constant and c is the velocity of light in free space.

Double differentiation of (3.19) or (3.20), with respect to λ, allows us to determine the second order material dispersion, whose effect has to be compensated by the waveguide dispersion in order to obtain fibers characterized by zero total chromatic dispersion at the transmission wavelength.

Chromatic dispersion characterization can be performed measuring time delays among optical pulses at different wavelengths propagating along a fiber whose length must be higher than 0.5 km [58]. One optical source (or more) generates narrow pulses at selectable wavelengths in the spectral range included in the region between 1.1 and 1.6 μm. A simplified experimental setup is shown in Figure 3.18. A fiber Raman laser source can be employed, for example, to obtain an almost continuous wavelength spectrum. The pump is a mode-locked and Q-switched Nd:YAG laser with a peak power of several kW [59] at a wavelength of 1.06 μm. In a single-mode fiber about 100 m long so excited, a large number of wavelengths are generated because of the stimulated Raman scattering. This arrangement permits to obtain a broad spectrum light source. Pulses are then selected by a wavelength selector, typically a monochromator, and launched into the fiber under test. Time delays at the fiber output are compared for the various wavelengths with respect to those produced by a reference fiber. The light signals are detected by a fast photodiode, displayed using a sampling oscilloscope with proper time resolution and elaborated using a personal computer (PC). An optical attenuator is sometimes included between the Raman fiber and the monochromator to adjust power level.

Alternatively, an array of laser diodes may be employed in order to obtain a source whose wavelengths are a discrete number, although properly distributed in the wavelength range of interest [60]. Laser diode driver is a picosecond electrical pulses generator. Wavelengths are selected by optical switches inserted between light sources and test fiber.

A comparison of the two solutions here presented for wavelength generation shows how the one based on the fiber Raman laser is probably more accurate and hence appears particularly suited for an industrial environment. The one consisting of an array of laser diodes, on the other hand, appears simpler, lighter, and easy employable for field tests. In both solutions, delay time is measured at different wavelengths by an oscilloscope. The curve obtained fitting delay time values is then differentiated with respect to the wavelength for calculating the total chromatic dispersion coefficient.

The so-called phase shift technique is largely used because of its simplicity, which follows from the possibility of operation in the frequency domain. The experimental setup appears less sophisticated than the previous one. Measurement results are very accurate, leading to a time resolution as good as 1 ps. The dynamic range is large, thus making possible a very precise evaluation of the chromatic dispersion coefficient [61–63].

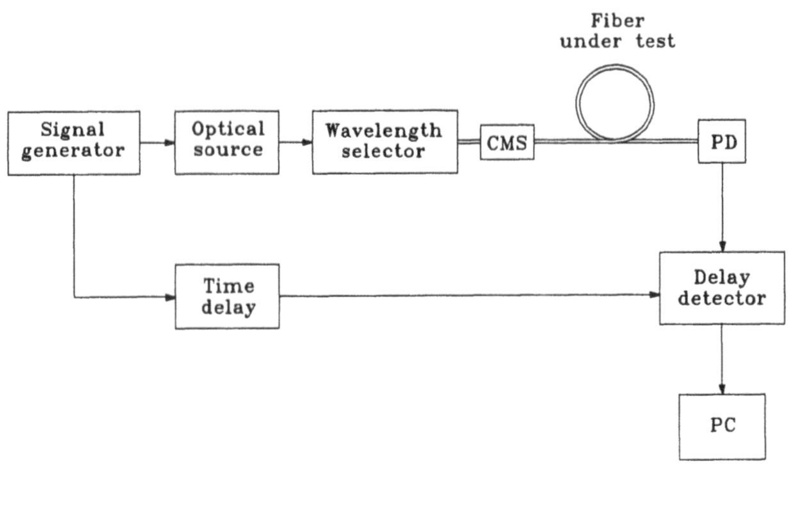

Figure 3.18 Chromatic dispersion measurement by means of pulse delay method.

This method is based on the phase shift of a sinusoidal signal at frequency f_0 on the test fiber as a function of wavelength. The resulting phase-shift $\Phi(\lambda)$ is then related to the modal group delay $\tau(\lambda)$ as $\Phi(\lambda) = 2\pi f_0 \tau(\lambda)$. The optical source is usually composed by one (or more) LED or one (or more) laser diode. The wavelengths emitted belong to the spectral range of interest. A unique oscillator drives such optical source giving also a reference signal for the vector voltmeter by which the two waveforms are compared (Figure 3.19). Wavelength selection is performed using a monochromator inserted at the input or at the output of the fiber under test, with about 10 nm linewidth. Dynamic range is very high at the LED central wavelengths, but it decreases at the outer part of LED spectrum. The chromatic coefficient is measured on fiber pieces typically 10 km long. The results are in good agreement with those obtained in the time domain with Raman lasers. A vector voltmeter characterized by a resolution of ±0.1 deg determines an uncertainty of about ±10 ps on the pulse delay measurement when frequency modulation f_0 is only 30 MHz.

In comparison with time delay and phase-shift techniques, the interferometric method is certainly the most accurate one, since it exhibits a time resolution of about 0.1 ps (two orders of magnitude better). Yet the measurements can be performed here only on pieces of fiber a few meters long. Chromatic coefficient extrapolation on long fiber section requires the assumption of longitudinal fiber uniformity, which is not always verified.

Interferometric measurements are based either on Michelson [64] (see Section 2.2.2) or on Mach-Zehnder [65, 66] (see Section 2.2.3) configurations. A high-resolution measurement is easily achieved because a small length change in one arm of the interferometer corresponds to a fringe shift in the interferogram itself.

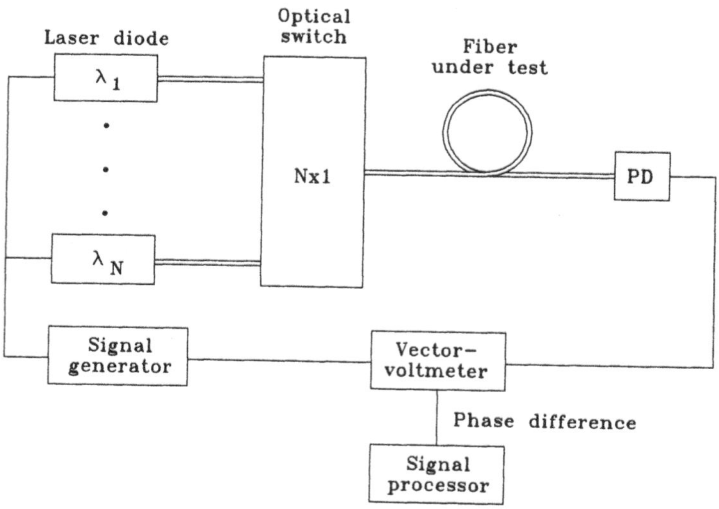

Figure 3.19 Phase shift method for chromatic dispersion measurement (adapted from [62]).

The wavelength-dependent delay is obtained comparing the test sample with a reference arm, generally in air or through a single-mode fiber with a known spectral group delay [58, 67] (Figures 3.20 and 3.21).

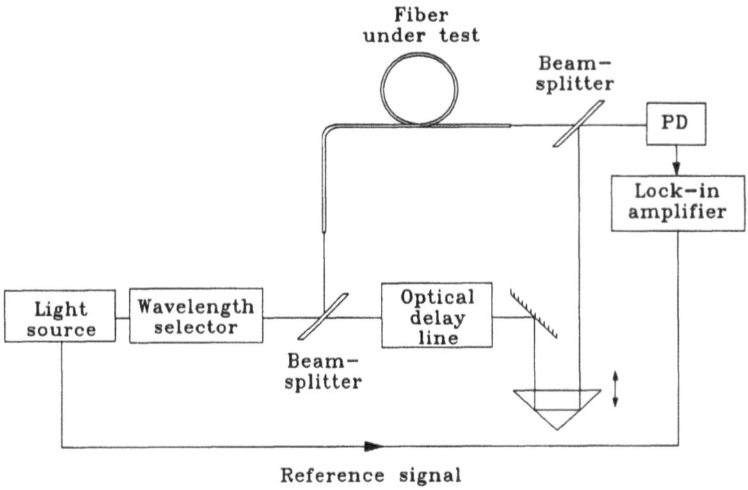

Figure 3.20 Interferometric technique: reference arm in air (adapted from [3]).

Optical sources, stable in position, intensity and wavelength, and characterized by a wide emission spectrum, can be a Raman laser, a halogen lamp or an LED. The adoption of a lock-in amplifier permits to improve the signal-to-noise ratio. Wavelength selection is performed by means of a monochromator or an interference filter with a linewidth on the order of few nm. Balancing of optical path into two arms of the interferometer is realized using a linear positioning device having suitable accuracy, stability, and uniformity. Linear translation stages exhibit traveling lengths varying from 20 to 100 mm with an accuracy of about 2 μm [3].

Measurement procedures are performed as follows. The fiber under test is placed in the measurement setup and the fiber-end face, carefully prepared, is accurately set in front of the photodiode active area by means of a three-axis linear translator. The linear positioning device included in the reference path permits balancing of the two optical paths corresponding to the two arms of the interferometer. Considering now the position x_j, in which maximum intensity of the interference pattern is reached at the wavelength λ_j, and the position x_o, where the maximum of interference pattern is reached at the wavelength λ_o, the difference group delay $\Delta T_g(\lambda_j)$ between the reference path and the test path, is [3]

$$\Delta T_g(\lambda_j) = \frac{x_o - x_j}{c} \tag{3.21}$$

where c is the velocity of light in a vacuum. The resulting group delay of the fiber under test T_{ris} is derived by the group delay of the reference path plus ΔT_g. If L is the fiber length, the group delay per unit length $\tau_g(\lambda)$ of the test fiber is calculated from

$$\tau_g(\lambda) = \frac{T_{ris}}{L} \tag{3.22}$$

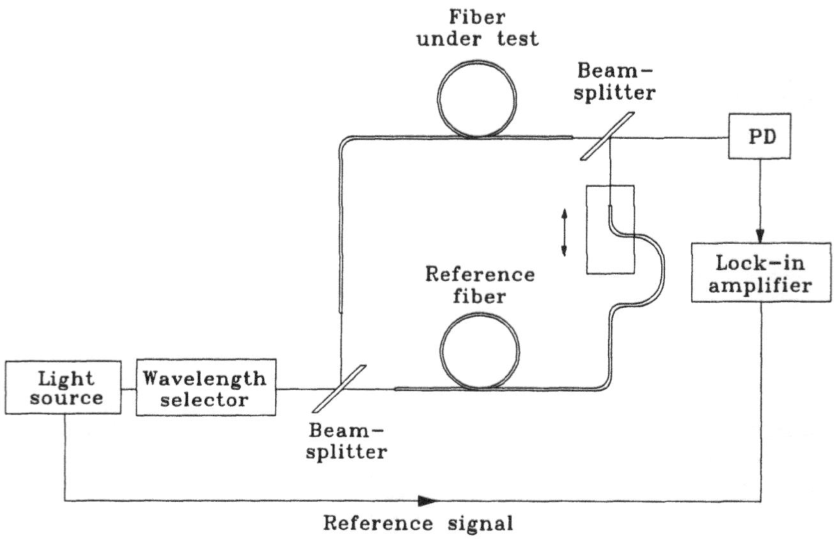

Figure 3.21 Interferometric technique: reference arm in fiber (adapted from [3]).

The chromatic dispersion coefficient is then obtained, according to (2.129), as

$$\tau_g'(\lambda) = \frac{d}{d\lambda}\tau_g(\lambda) \tag{3.23}$$

Finally, if we define λ_o as the zero dispersion wavelength, an important parameter is the zero dispersion slope defined as

$$S_o = S(\lambda_o) = \frac{d}{d\lambda} \tau_g'(\lambda) \Bigg|_{\lambda = \lambda_o} \tag{3.24}$$

This permits evaluation of the extent of a spectral region around the zero dispersion wavelength, in which fiber dispersion is less than a preset value.

3.1.7 Outline of Optical Cables for Telecommunication

Optical cables for telecommunication can be subdivided depending on their application. We can have:

1. indoor applications, where typically one or two fibers are to be assembled into a unique cable and a soft protection is set against environmental impact forces which could damage the fibers;
2. outdoor applications, where a cable unit, in which a higher number of fibers are usually assembled, is typically surrounded by suitable sheaths to ensure mechanical protection.

Outdoor cables must be properly designed also to avoid that longitudinal strains, due to the installation procedure, act on the fiber. A first solution, in this sense, consists in the adoption of a loose secondary coating, made of nylon or other material characterized by a high value of Young's modulus, which forms a tube able to host the fiber in a floating way. It is just this loose disposition of the fiber which ensures good protection against external mechanical stresses or temperature variations. In addition, a central reinforcing member is present, made of steel or fiberglass and surrounded by polyurethane. When more than one fiber is placed within the same cable, they are stranded around the central member. A plastic binding tape, a metal sleeve and a polyethylene outer sheath are also present to complete the structure of a typical buried cable.

An alternative solution, plotted in Figure 3.22, is based on a polyurethane grooved cylinder directly incorporating the reinforcing member. One or more fibers are placed within each groove, where they can float inside a tamponing jelly solution.The latter is introduced in order to reduce water permeability in case of cable sheaths damaging. Obviously, either the presence of several fibers in the same groove or the filling with jelly contribute to reduce fiber mobility with respect to the case in which the fiber is unique and completely free in its floating movement. This reduces the effectiveness of the loose solution and in some situations an enlarged size of the housing slots has to be foreseen.

The grooved structure can be also adopted for realizing ribbon cables. By this term we mean a particular type of cable, suitable for underground laying, whose main characteristic is the presence of a double primary coating. In fact, in addition to their normal protection, the fibers are assembled in ribbons and then housed inside the cable (an example is shown in Figure 3.23). More precisely, each ribbon consists of a group of fibers (whose number usually ranges between 4 and 16) arranged in a planar array and showing defined and constant spatial sequence in all their length. The fibers, each covered with its own primary coating, are held together by means of an additional common coating made of a layer of polymeric ultraviolet curable resins and having characteristics similar to those of the fiber primary coating itself. The central reinforcing element is, in this case, made of a dielectric material.

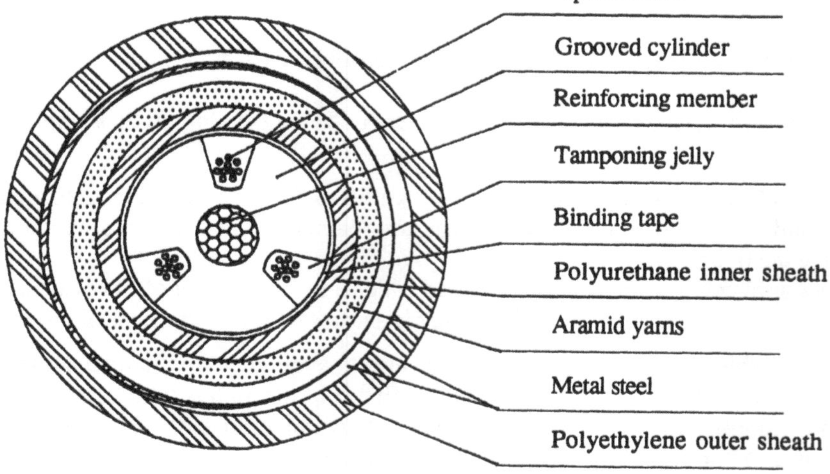

Figure 3.22 Sketch of a grooved cable (courtesy of Pirelli Cavi SpA - Divisione Italia).

3.1.8 Splices and Connectors

Joints between subsequent fiber pieces can be subdivided into two different classes: permanent joints are usually called *splices*, while in case of demountable joints the term *connectors* is preferred.

Splicing between two optical fiber heads is inevitably affected by power losses due to intrinsic and extrinsic causes [68]. The former ones are related to differences in the physical parameters, like refractive index and geometrical dimensions, and they can be estimated analytically when fiber parameters are well-known. In particular we can have differences in core and cladding refractive indices, core and cladding dimensions, ellipticity, and so on. Such differences limit theoretically the minimum splice loss value. Fiber parameters influence all together the spot-size figure. This suggests that a rough prediction for the intrinsic loss value can be determined considering a Gaussian behavior for the mode field distribution and calculating the cross section overlapping between such propagating fields. The splice loss in this case is

$$A_{SL} = 10 \log_{10}\left[\left(\frac{W_1^2 + W_2^2}{2W_1W_2}\right)^2\right] \tag{3.25}$$

where W_1 and W_2 are the two Gaussian spot-size values.

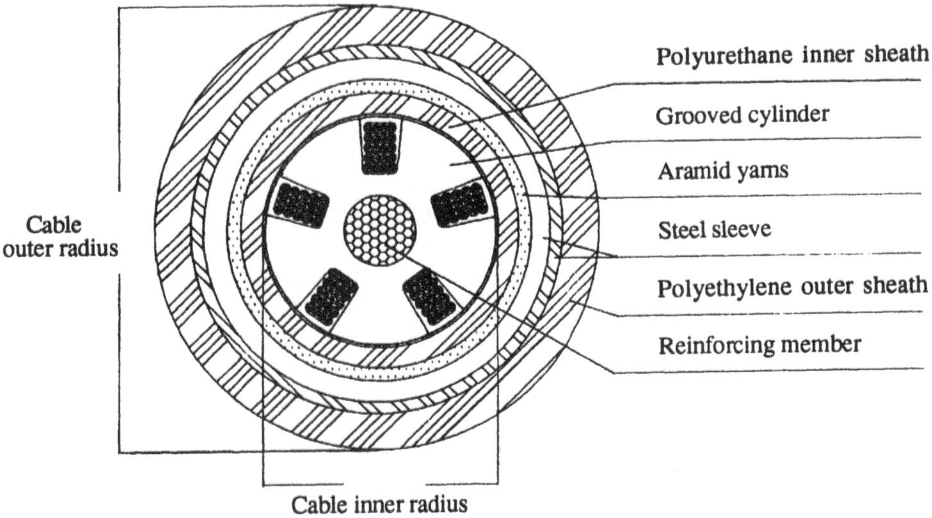

Figure 3.23 Sketch of a ribbon cable (courtesy of Pirelli Cavi SpA - Divisione Italia).

Offset, tilt, and fiber end-to-end separation due to face roughness are the sources of extrinsic losses for a splice. Effects of these parameters on splice loss can be limited using high-quality equipment such as a fiber cleaver and a fusion splicer. It can be estimated [69] that a loss of about 0.1 dB is the result of an offset of 0.8 μm or of a tilt angle of 0.3 deg. Joints between fibers are made using mechanical support or arc fusion splicer.

Mechanical splices are sometimes employed because they are cheap and easy installable in the field. They can be realized in a permanent manner or not. Durable splices are made by means of ultraviolet curable adhesive, while mechanical supports are V-groove, precision-made glass terminals or round rods [70–74].

All these joints are affected by a new source of loss due to the reflection at the air gap between the two fiber-ends. Such additional cause of loss can be estimated of about 0.32 dB. Oil is sometimes used as index matching to limit reflection at the glass-air interfaces.

Mechanical splice performances are also affected by vibration and temperature fluctuations. Care must be taken against water penetration and performance variations in time. A tool to minimize optical reflection consists in cleaving the fiber at angles of 6 to 12 deg. Suitable equipment has been proposed [75] and an average splice loss of about 0.11 dB (using index-matching) and a reflectance < 70 dB have been achieved.

Arc fusion splicer gives rise to a permanent joint by heating the two fiber-heads and fusing them together. Other heat sources, like CO_2 laser or H_2 flame, are more difficult to organize in transportable equipment. Losses are smaller than for mechanical splices, but rather sophisticated equipment is needed. The last generation of arc fusion splicer is a fully automatic equipment that obtains core alignment by means of piezo-electric transducer, and fiber cleaning by means of well-controlled electrical discharge. Arc fusion splicers are largely diffused and permit reliable performances.

Splices are made following a series of operations which can be described as listed below. First of all, fiber-heads, accurately cleaned and perpendicularly cut, are inserted into a V-groove and axially aligned to reduce possible offsets or tilts. The two fiber-heads are about 15–20 μm apart from each other. An electric discharge cleans and melts fiber-heads in a short time (this step is called *prefusion*). Fibers are then moved and coupling is enhanced as possible; the two surfaces are then brought in contact and a well-controlled (in time and intensity) discharge completes fusion operation. In this step, the two fibers are forced to align to each other because of surface tension. A crucial step is coupling optimization, which also gives information about splice loss value. Various solutions have been proposed to achieve this goal.

Some equipments are based on *local injection and detection* (LID). Input and output fibers are bent before and after the splice point to permit launching and radiation of optical signal using an LED and a photedetector included in the fusion splicer. In this way, the overall coupling is maximized and splice loss can be estimated by considering the detected power level before and after splicing. Because locally launched power is distributed over the whole fiber section, the fundamental mode, but also some higher order modes, are excited in the fiber. This may imply that splice loss becomes higher than splicer estimation. Such behavior is particularly favored with fibers having different parameters or whose eccentricities are quite large.

Moreover, LID methods introduce microbends in the fiber, which can determine surface cracks, and injection or detection of the local optical signal is difficult with some type of colored jacket.

Side view [76, 77] and image processing of warm fibers [78] are more suitable techniques for obtaining good alignment and very low loss splices. Both such techniques permit profile alignment system through image processing.

Side-view methods are based on a lateral light source that illuminates orthogonally the fiber without coating. Fiber itself is used as a convex lens and its focused image is then detected by a video camera permitting the alignment. Dark zones of focused images correspond to boundaries of different refractive indices inside the fibers, i.e., core-cladding and air-cladding interfaces. It is also possible to estimate splice losses comparing data from the two fiber images. The same procedure allows one to compare the estimation loss before splicing with the splice loss measured after fusion splicing. It has been demonstrated [79] that this method can also be easily adopted for carbon-coated fiber, because this strength layer is removed using a cleaning discharge. Image processing of warm fiber permits core alignment as well as the side-view method using warm fiber image during prefusion and fusion. As described before, prefusion is basically used for fiber cleaning. During the arc discharge in GeO_2-doped fibers, the core exhibits a higher absorbing loss and hence a stronger thermal radiation because the core has an higher germanium-doped concentration than does the cladding. By virtue of this property, it is possible to measure core offset and fiber center offset by processing images of the two warm head fibers. Such a method also permits fiber monitoring during fiber fusion and splice loss estimation can be done using microbending theory [80]. Since splice losses depend strongly on tilt angle and offset, estimation of fiber quality end face is automatically performed by the machine before beginning the splicing procedure.

Arc fusion splicers are largely used also for massive splicing, for instance, those proposed with fiber ribbons. Massive splicing of ribbons including a large number of fibers reduces time and costs, and the splice quality can be considered as being not particularly degraded [81]. Nevertheless, the uniformity and intensity of discharge must be accurately controlled. Furthermore, a similar accuracy is devoted to fiber positioning between the electrodes because splicing conditions must be the same for all the fibers.

Also when splicing equipments give a direct loss estimation, it is advisable to verify splice loss value by means of proper instruments.

3.1.9 Measurements of Splices

Measurement of splice loss can be performed using some matching oil before arc discharge. Because of surface tension, matching oil behaves as an optical waveguide between the two fiber-ends. Fibers are then brought in contact close to each other, with almost no space between the two heads, so that coupling can be maximized. The splice loss can be estimated as the difference between the output power levels measured with matching oil and after splicing [82]. This method, however, is affected by some uncertainties and results fundamentally unpractical.

Much more frequently, splice loss measurement is performed with OTDR. Some further information such as the exact splice position, the total number of splices, whether there are more than one, besides the simpler splice loss can be obtained from such a one-end measurement [83].

Nevertheless, one-end measurement may be affected by some errors. If there are two fiber pieces called AB and CD, there are two backscattered signals, V_B and V_C, corresponding to power levels just before and just behind the splice BC. Assuming that both fibers have constant but different properties along their length, we can write

$$5 \log_{10}\left[\frac{V_B}{V_C}\right] = \frac{1}{2}[\alpha_{BC} + \alpha_{CB}] + 5 \log_{10}\left[\frac{\rho_{AB}}{\rho_{CD}}\right] \tag{3.26}$$

in which α corresponds to the insertion loss, while the backscattering coefficient ρ is defined as the ratio of the backscattered power per unit length of fiber and the local mean power [84]. This measurement reproduces exactly the splice loss when $\rho_{AB} = \rho_{BA}$ and $\rho_{CD} = \rho_{DC}$. This occurs when the backscattering coefficient is independent of direction of propagation. If attenuation loss and backscattering contribution have opposite sign, we should have a negative one way splice loss, which means an apparent gain. This is obviously false. The backscattering coefficient is a function of the Rayleigh scattering attenuation, the refractive index difference, the V value, and the mode field diameter, as discussed in Section 3.1.2. It is more correct to assume as splice loss the average between the two measurements performed by the two fiber-ends. In fact, contributions due to backscattering coefficients have the same amplitude but different signs when measurements are realized from reverse directions, and hence are cancelled.

The same result can be obtained from one-end fiber measurement using a mirror on the far fiber-end [85]. The forward optical pulse propagating through the fiber is completely reflected by the mirror, so generating a backward pulse. If the splice is placed at a distance L_s and the total fiber length is L, the two optical backscattered pulses referred to the splice under measurement arrive at the instants $2L_sn/c$ and $2(2L - L_s)n/c$, respectively. Therefore they can be considered separately. Splice loss is assumed as the average between such two splice measurements.

Another one-way splice measurement is based on splice measurements performed at two different wavelengths. It is possible in that OTDR can be equipped with two optical sources, one at 1.3 μm and the other one at 1.55 μm. In this way we can separate the contributions due to the backscattering coefficients from the intrinsic loss of the splice [86, 87].

3.1.10 Measurements of Connectors

While splice quality is fully characterized by optical loss, connectors need some more details. The most popular connectors for telecommunication applications are the Biconic, FC, SC, and ST types. All these connectors have performed well. They are generally pull-proof, preferably mounted in laboratory. The SC connector [88] has the advantage of being a push-pull type. Furthermore it can be fabricated in a multiple structure to obtain higher packaging density. Multifiber connectors [89–92] properly designed for fiber ribbons are also available.

Air gap between two connectors may give some problems. In particular, fiber-end faces are very flat and parallel. Therefore, the transmitted power exhibits fluctuations depending on such fiber separation, which acts as a Fabry-Perot cavity. It can be verified [92] that the maximum amplitude modulation on the transmitted power is

$$A = -10 \log_{10} (1 - 4R) \tag{3.27}$$

where R is the reflectivity at the air-fiber interfaces expressed by

$$R = \left[\frac{n - 1}{n + 1}\right]^2 \tag{3.28}$$

For a fiber refractive index n of 1.46, theory predicts for A a value of 0.65 dB, which is very close to the measured value. An even larger amplitude modulation is sometimes present because the refractive index of the glass surface increases because of polishing. Modulation in transmitted power is particularly severe when the light exhibits a high degree of temporal coherence. Physical contact connectors reduce interference effects as well as index matching oil or end face angle.

Because connectors are preferably factory mounted, it is advisable to assign the losses for field operation. A statistical method has been proposed [93] able to predict the maximum allowable factory insertion loss based on the specified field insertion loss.

Insertion loss of a couple of connectors can be experimentally evaluated by OTDR technique, in a manner analogous to that previously described for splices. Connector loss performances are tested in regard to repeatability, temperature fluctuations, vibrations, and humidity. Thus an average value and a standard deviation or a maximum range of variation is typically given. Insertion losses can be made smaller by using index matching oil, but this solution is impractical and performances are less stable.

A parameter with increasing importance for high-quality laser diodes used in optical telecommunication systems is the optical return loss, defined as the ratio (expressed in dB) between the injected and the reflected optical powers. This will become more and more significant especially for very high bit rate or coherent systems, in which backreflected optical signals may degrade source linewidth and, consequently, system performance. Optical return loss is directly determined measuring the backward signal at the fiber input head [94]. Some attentions may be devoted to avoiding any spurious reflection. Their experimental evaluation can be done using backscattering technique, both in the time and in the frequency domain [95]. The measured reflection from a reflective fault is due to Rayleigh backscattered signal and to fault reflection. It is necessary to separate these two terms, which can be done by comparing reflected and backscattered signals in the OTDR trace.

The measured reflection R_m is the sum of the Rayleigh backscattered contribution R_{bs} and of the reflection due to the fault R_f. If we set a first cursor on the left-hand side of the Fresnel reflected light waveform and a second one on the peak of the Fresnel waveform, we determine the difference (in dB), noted by H, between such two power levels. Optical return loss (in dB) can be calculated as [51]

$$O_{RL} = -R_1 - 10\log_{10}[10^{0.2H} - 1] - 10\log_{10}W \tag{3.29}$$

where W is the pulse width (normalized by 1 μs) and R_1 is the backscattered signal corresponding to 1 μs pulse width. Because almost all commercial OTDR equipments are dedicated to measure small reflections, they can saturate. In this case it is necessary to reduce ODTR gain or to insert an attenuator.

Optical return loss can be measured also using *optical continuous wave reflectometry* (OCWR) [51, 95]. OCWR can be employed when the total reflected power (including the fiber backscattering) has to be appreciated. OTDR is better suited, instead, when splices and connectors are widely spaced on a long fiber link.

A simple method often used for connector return loss measurement is shown in Figure 3.24. We can define three steps in the measurement procedure:

1. the measurement of the incident power P_I;
2. the measurement of the intrinsic reflection due to the setup P_{RO};
3. the measurement of the total reflection power P_{RI}.

Particular care must be taken during the determination of P_{RO} and P_{RI} so as to avoid the reflection due to the far fiber-end.

Thus, the return loss is given by

$$O_{RL} = 10\log_{10}\left[\frac{P_I}{P_{RI} - P_{RO}}\right] + 10\log_{10}R_{23} \tag{3.30}$$

in which R_{23} represents the transmission coefficient between the ports 2 and 3 of the directional coupler. Typical return loss values for telecommunication single-mode connectors are 30 dB or higher.

Other more sophisticated techniques are sometimes used to characterize connectors during its certification. For example, a microfocus radiographic technique [96] permits to obtain an x-ray enhanced image to visualize possible fractures inside the connectors. Investigation on roughness in the front face of single-mode connectors are performed using far-field measurements [97]. A rough surface typically excites, at the same time, the fundamental mode and a spurious field. In the far-field domain such a field can be easily shown because it has a wider angular distribution.

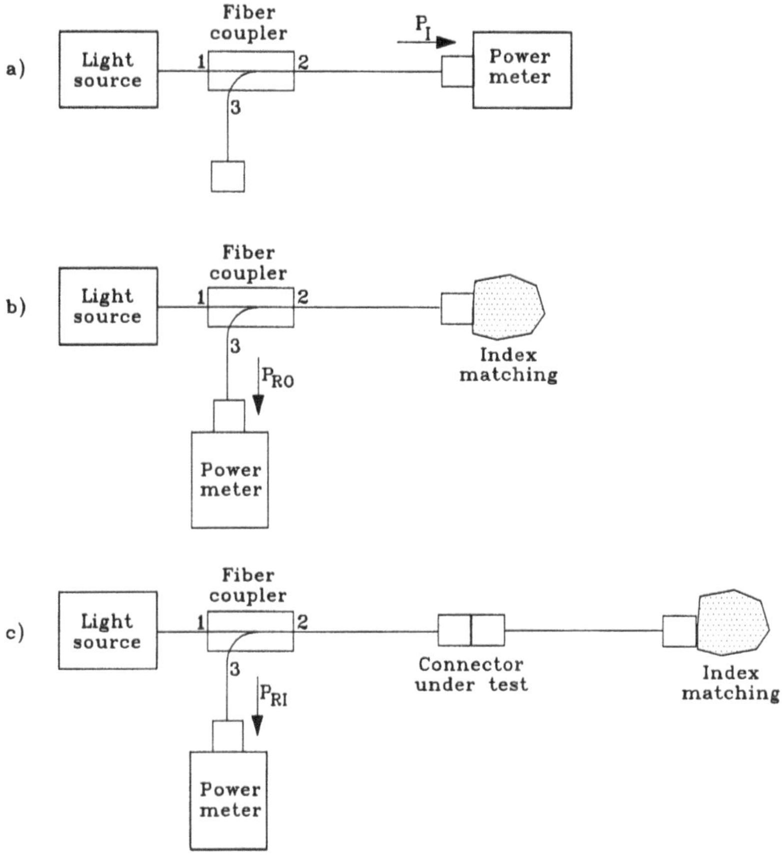

Figure 3.24 Return loss measurement of a connector; measurement a) of the incident power, b) of the intrinsic reflection due to the setup, and c) of the total reflection power.

3.1.11 Measurements of All-Fiber Passive Devices

Optical bulk-devices, like beam-splitters, couplers, filters, and so on, are usually based on a planar technology and are largely used for measurements on optical benchs. On the other hand, their geometrical mismatch with optical fibers characterized by a circular cross section makes their use impractical in telecommunication systems. This limitation has been overcome thanks to the introduction of all-fiber devices. In a general configuration, such devices are multiport couplers realized using only optical fibers as basic elements. Up to now, all-fiber couplers are a crucial tool in order to enhance networks interconnectivity. This is particularly attractive for broadcast networks or for LANs, where a common transmission medium is shared. Couplers are also employed for multiplexing and demultiplexing optical signals in WDM systems.

The success of all-fiber devices has been determined by their simplicity in connectivity, small dimension, and low-cost.

The main features of a multiport optical coupler are:

1. performances practically independent of the light polarization and launching conditions;
2. wavelength independence around the selected operation wavelength;
3. low losses;
4. availability of components characterized by various values of the coupling ratio R, whose definition has been introduced in Section 2.6.3.

A multiport coupler consists of N-input ports and M-output ports. Devices having up to 64×64 ports are commercially available. The most diffused all-fiber devices are: wavelength multi-demultiplexers, couplers, Y-junctions, splitters, and attenuators. Other devices based on the same technology, but more sophisticated, include isolators, polarization splitters, polarization maintaining couplers, and polarizers. Polarization-independent devices are used for distribution networks and bidirectional transmission systems, while polarization-dependent ones are preferred in coherent transmission systems.

All-fiber devices are typically reciprocal (although some significant exceptions, like that constituted by isolator components, should be taken into account), so that input and output ports can be exchanged between them. This behavior occurs because they are completely made of silica, that is, a reciprocal material. Thus any port may be considered as either an input or an output port.

With reference to Fig. 2.20(b), where a directional coupler 2×2 device is depicted, the insertion loss is defined as [98]

$$L_{13} = -\ 10\ \log_{10} \frac{P_3}{P_1} \tag{3.31}$$

Directivity of the same device is then called the parameter

$$L_{12} = -\ 10\ \log_{10} \frac{P_2}{P_1} \tag{3.32}$$

Due to the symmetry and reciprocity characteristics of the directional coupler, it gives the fraction of the optical power entering one port of the device, which is extracted on the other port at the same side.

Finally, the power splitting ratio is defined as the parameter

$$R_{34} = \frac{P_3}{P_4} \tag{3.33}$$

It describes the way in which the power entering one port (for example port 1) is subdivided among the ports at the opposite side.

An index of coupler quality is the excess loss, defined as

$$L_{ex} = -\ 10 \log_{10} \frac{P_3 + P_4}{P_1} \tag{3.34}$$

Commercial components have L_{12} ranging between 30 and 60 dB, and L_{ex} between 0.1 and 1 dB. Other quantities determine the performance of a directional coupler, including geometrical and modal parameters of fiber pigtails, environmental sensitivity (to temperature, humidity, etc.), and relative delay between input and output optical signals.

The general setup employed to characterize the behavior of a directional coupler of this type is represented in Figure 3.25. Measurements are generally performed over a large spectral range. A suitable light source consists of a lamp followed by a monochromator (700–1600 nm). Alternatively we can employ a tunable laser (with 80 nm of spectral range around the central emission wavelength) or an LED.

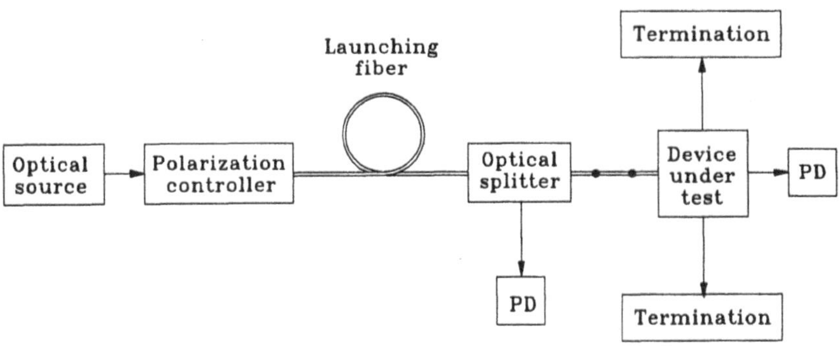

Figure 3.25 Fiber coupler characterization setup (adapted from [99]).

A polarization controller can be used either as a polarizing element or as a polarization scrambler, depending on the coupler parameter. In other words its function is different when polarized or unpolarized light is considered. Polarization scan is realized in two different ways. A first solution is based on an unpolarized light source, a monochromator and a Glan-Thompson polarizer as polarization controller. The second solution employs a linearly polarized light obtained by a laser source, while polarization control is performed by means of three fiber loops.

The experimental setup includes an optical splitter only when the return loss has to be measured. It consists of a 1×2 high-quality fused fiber coupler. Some conditions are to be satisfied for correct return-loss measurement. In particular, fiber outputs have to be terminated to avoid reflections. Such terminations are realized by means of index matching liquid, a fiber loop with a very small radius (5 mm or less), or by polishing an angle on the fiber-end face. A laser diode is generally preferred as a light source because of its very large dynamic range.

The results of all these measurements can be summarized in a transfer matrix, whose elements represent the power transmission coefficients of the directional coupler under test. In this measurement only a single wavelength is tested.

Another suitable way to represent the coupler behavior is the wavelength scanning of the output power level. Polarization sensitivity of the coupler can be verified by changing the state of polarization of the light entering the device. By the same procedure, we can verify also output power stability against environmental changes, such as vibrations or temperature fluctuations. They determine the range of uncertainty of the coupler transfer matrix elements.

3.1.12 Measurements of Active Fibers

As anticipated in Section 2.4.4, an active fiber is generally made starting from a SiO_2- or GeO_2-based fiber and adding a low concentration of rare-earth dopants, like Er^{3+} or Nd^{3+}.

Up to now, definitions on fiber amplifier characteristics, dimension, testing, and so on are still in progress. In particular, they concern active fiber geometrical and modal parameters, configuration and performances of pump lasers, characteristics and dimension of optical components (couplers, isolators, connectors).

The main measurements on active fibers concern spectral attenuation, excited-state absorption, and fluorescence [100]. Spectral attenuation is experimentally obtained by means of a multiple cutback method [101] in which several measurements are performed with different fiber lengths over various spectral ranges and the separate results are summarized.

Spectral attenuation measurements are conditioned by longitudinal and transversal variations in the rare-earth dopant concentration. Longitudinal inhomogeneity has almost negligible effects on spectral attenuation. This fact can be verified by using wavelength-tunable OTDR. The undesired effects due to geometrical and refractive index variations can be avoided using a two-end measurement. Transversal variations in the rare-earth concentration influences the modal behavior of the electromagnetic field and, in turn, the spectral attenuation. Measurements are generally performed on preform samples or large core fibers since resolution is about 2 μm.

There are two types of excited-state absorption: the one referred to the signal and the one referred to the pump. That acting on the signal is due to absorption from the metastable level towards a higher level whose wavelength corresponds to the emission one. The excited-state absorption acting on the pump causes depletion of the metastable level by the absorption of pump photons in a transition from the metastable towards a higher energy level. The experimental setup for excited-state absorption measurements is shown in Figure 3.26 [102]. A pump signal, generated by a laser diode, is injected into the fiber through a coupling device (shown as a beam-splitter in the figure). A white-light source generates a counter-propagating signal that is detected by a system composed by a monochromator and a photodiode permitting measurements of induced gain and excited-state absorption [100].

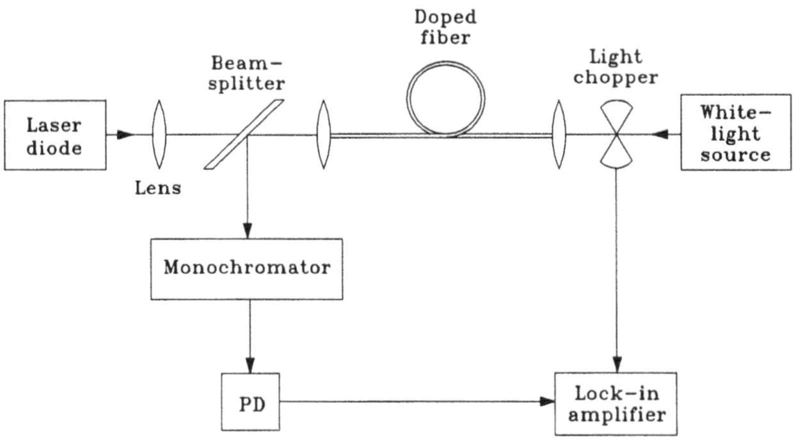

Figure 3.26 Excited-state absorption setup (adapted from [102]).

The experimental setup for fluorescence measurements (spectral output, decay-time and so on) [101] is composed by a pump laser in transmission and by a monochromator, a photodetector, and a signal processing unit in reception [100]. Output spectrum characterization can be influenced by reabsorption of emission due to long interaction lengths. This effect is reduced by using pieces of fiber whose length is shorter than the interaction length.

Nondestructive measurement of fiber gain is performed employing the setup shown in Figure 3.27 [103], which is based on macrobending loss in the active fiber. The fiber bend is moved along the whole active fiber length, detecting the scattered light by means of a large area detector. From this method, we can obtain information on amplifier gain, output power, pump efficiency, and optimal length in a nondestructive manner.

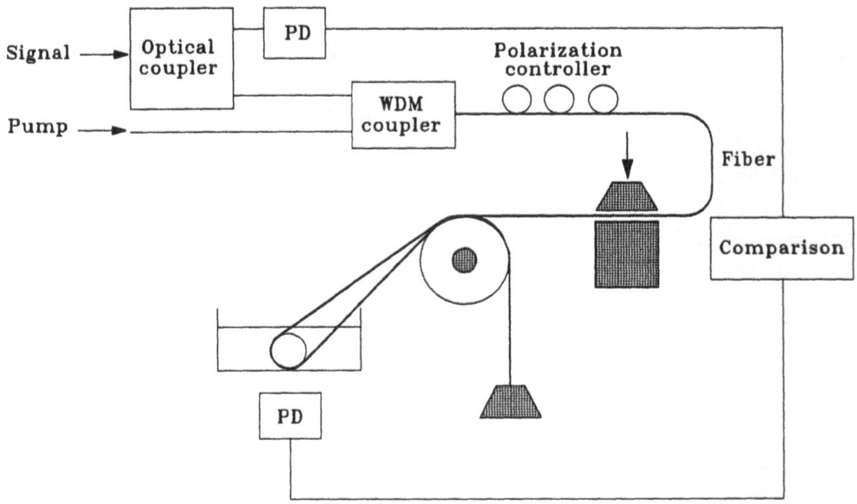

Figure 3.27 Nondestructive method for fiber gain measurement (adapted from [103]).

Active fibers are usually included into optical amplifiers in which pump power is coupled via a directional coupler. The use of optical amplifiers, which are now commercially available, on long-span fiber transmission systems, requires measurements on in-line equipment. The main testing tools are still OTDRs [104, 105], in particular for amplifier gain, polarization sensitivity, and fault location. Broadband amplified spontaneous emission and backward amplified reflections are to be controlled during such a measurement by means of optical bandpass filters and isolators. When isolators are included into optical amplifiers, OTDR cannot be used directly because backward signal is highly attenuated. A suitable technique [106] has been proposed for overcoming such problem employing optical circulators and return transmission lines.

Suitable parameters characterizing erbium amplifiers are gain, pumping efficiency, output saturation power, and amplifier noise [2]. Internal gain is defined as the ratio of the output signal power to the input signal power. Pumping efficiency is called the maximum slope of the gain versus pump power curve; the best values obtained are 10.2 dB/mW and 6.3 dB/mW for pumping wavelength of 980 nm and 1480 nm respectively. Output saturation power is defined as the output signal power corresponding to a 3 dB decrease from maximum gain value; the best performance reported is 22.3 dBm.

Finally, noise figure is defined as the ratio between the output and the input noise-to-signal ratios. The best values obtained range between 3 dB and 4.1 dB depending on the pumping wavelengths and the mechanisms originating such noise involve coherent properties of the amplified light. For this reason the measurement techniques devoted to an experimental evaluation of the noise figure of an active fiber amplifier will be discussed, with more detail, in Section 4.3. There, other aspects, regarding the technological characterization of such fibers, will be also presented.

3.2 MEASUREMENTS OF INCOHERENT LIGHT OPTICAL SOURCES

The light sources used in fiber optics are mainly LEDs and laser diodes. In recent years LEDs have lost importance in comparison with laser diodes in optical communication and fiber sensor applications. Nevertheless they still play a significant role especially for LAN transmitter modules. In fact, in low-rate and short-distance communications they are competitive with lasers. The wide emission spectrum and the low-frequency response of LEDs, which prevents their use in high-speed communications, is compensated by the insensitivity to optical feedback, lower intrinsic noise, and fairly high output power. Moreover LEDs can be efficiently fiber coupled.

LEDs used for fiber optics fall into three categories: *edge-emitting diodes* (ELEDs), *surface-emitting diodes* (SLEDs) and *superluminescent diodes* (SLDs). The last devices exhibit intermediate behavior between LEDs and laser diodes. These present an output power comparable to that of laser diodes together with a low temporal and spatial coherence. Due to these properties, they are specially suitable for OTDR and optical fiber gyroscopes [107]. A detailed description of devices used for incoherent optical fiber applications is reported in Section 2.6. In this section we describe several measurement techniques for the characterization of LEDs and partially coherent laser diodes from the optical and electrical standpoints. The output power as a function of the driving current, the emission spectrum, the pulse response, and the mode partition noise of laser diodes are addressed. The methods proposed to evaluate the electrical properties of *multi-longitudinal mode* (MLM) lasers are applicable also to test the highly coherent *single longitudinal mode* (SLM) laser diodes, which will be considered in Section 4.2.

3.2.1 Electro-Optic Characteristics

In this subsection we consider the measurements of electrical characteristics of light-emitting semiconductor devices. Whenever not specifically indicated, the methods discussed can be applied both for LEDs and laser diodes. During the electrical test of these devices the maximum ratings specified by the manufacturer must not be exceeded;

moreover, special care has to be taken in avoiding electrostatic discharge, large reverse biasing, or current overshoots since all can lead to an irreversible damage of the device. The first and simplest measurement is that of emitted light power versus driving current. From the plot of the P-I curve, important information on efficiency can be extracted and the device operating point for rated optical power can be determined. In the case of laser diodes, this characteristic allows one to evaluate the lasing threshold [108, 109]. This measurement can be performed in CW or pulsed regimes by using a stabilized current driver and a radiometer, taking into account the portion of the emitted power that is collected by the detector. In general, an optical system with high numerical aperture will be used. For pigtailed diodes it is necessary to consider the fiber type and the characteristics of the diode to fiber coupling optics (proximity, sphere, GRIN lens, etc.). Obviously, the useful power is only the fraction propagating into the fiber core, so that an appropriate mode stripper before the optical power meter is recommended.

Since laser diodes, unlike LEDs, are strongly sensitive to optical feedback, the P-I curve of lasers can be significantly perturbed in presence of spurious reflections from any optical surface. To avoid this trouble it is necessary to insert an optical isolator immediately behind the laser diode. As the laser threshold is strongly affected by the operating temperature, the measurement requires thermal stabilization of the device within 1 °C around the selected temperature (usually in the range 0–30 °C). A typical behavior of the CW P-I characteristic for different temperatures is shown in Figure 3.28.

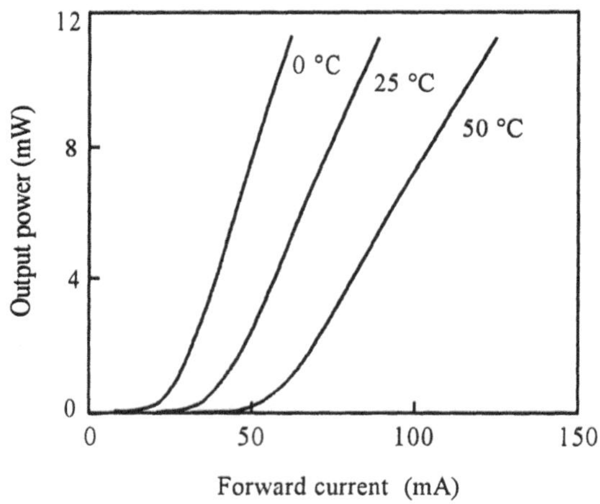

Figure 3.28 Typical P-I characteristic of a laser diode measured in CW operation at different temperatures.

An alternative approach is to acquire the P-I curve in pulsed regime by modulating the driving current at low duty cycle (less than 1%) with 100–500 ns pulses. Due to a rise in junction temperature, the CW lasing threshold is slightly higher than the pulsed one, while the slope efficiency is lower. Marked differences between pulsed and CW characteristics usually indicate poor die attachment and poor quality of the device (leaky junction).

Further information can be extracted by differentiating the P-I characteristic. The first derivative dP/dI (see Figure 3.29) gives the slope efficiency and hence the quantum efficiency versus the operating point [109]

$$\eta_q = \frac{q}{h\nu}\frac{dP}{dI} \tag{3.35}$$

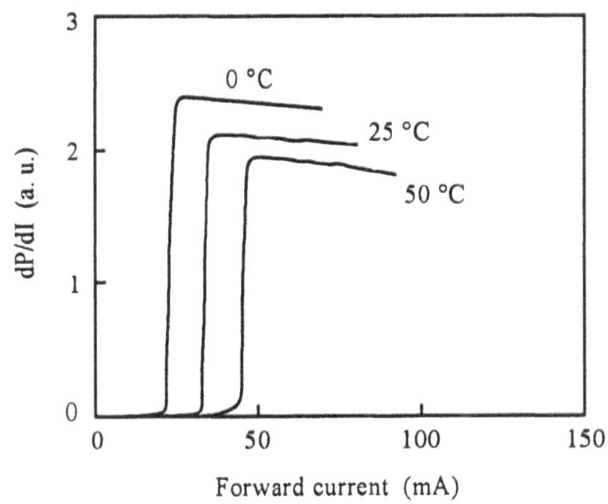

Figure 3.29 Typical plot of the first derivative dP/dI of the output power vs. the injection current for a laser diode.

For laser diodes, the second derivative d^2P/dI^2 (see Figure 3.30) exhibits a peak around the lasing threshold which can be assumed as the operative definition of I_{th}. Other accepted definitions of I_{th} are the intercept of the P-I curve linear fit above threshold with the I axis and the two segments intercept of the P-I curve linear fits below and above threshold. Many optical modules are provided with a built-in PIN photodiode for power monitoring purposes; this can be calibrated by comparing its output current with the previously described P-I curve. In this way a "tracking ratio" (mA/mW) can be fixed. This can provide a tool for direct measurement of the fiber coupling efficiency in pigtailed devices. Mechanical instabilities in the fiber mounting or changes in the near-field emission pattern of the laser can be detected from strong variations of the tracking factor with temperature.

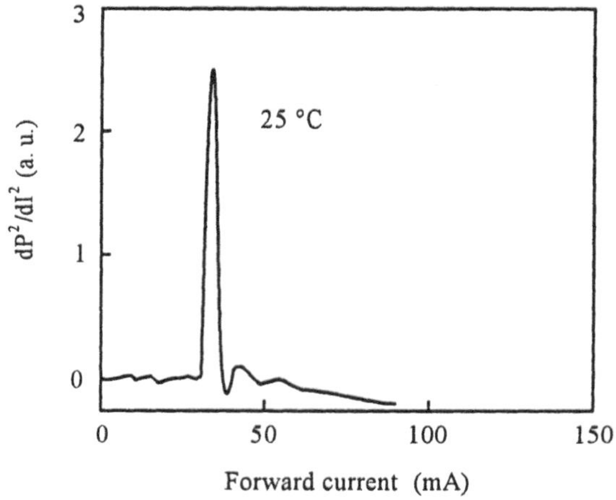

Figure 3.30 Typical plot of the second derivative d^2P/dI^2 of the output power vs. the injection current for a laser diode.

By using a voltmeter and an amperometer, the V-I characteristic of the diode can be recorded. This curve is similar to that of any other semiconductor diode and shows a weak dependence on the temperature. The typical voltage drop across the junction at useful operating current is 1.5 to 2 V.

The first derivative dV/dI represents the device dynamic resistance. The curve IdV/dI, along with suitable equivalent circuit modeling, provides the junction ideality factor, the contact resistance, and the leakage current; this allows laser diode manufacturers to have on line process control. Moreover from the IdV/dI characteristic the lasing threshold can be determined before mounting and packaging the device.

Linearity

The P-I characteristic of LEDs presents a significant nonlinearity. To measure the nonlinear behavior, the device is modulated with a high-purity sinusoidal signal superimposed to a dc bias. The amount of harmonic content in the emitted light is measured using a PIN or an *avalanche photodiode* (APD) connected to a wave analyzer [110]. A parameter of merit is given by the ratio of the power content of all the harmonics to the fundamental:

$$\mathrm{THD} = 20 \log_{10} \left| \frac{\sqrt{\sum_{n=2}^{\infty} V^2(nf_m)}}{V(f_m)} \right| \tag{3.36}$$

where f_m is the modulation frequency, $V(f_m)$ is the amplitude of the fundamental frequency electrical signal, and $V(nf_m)$ is the amplitude of the n-th harmonic component.

Taking care in avoiding any optical feedback, the linearity of laser diode response is far better than that of LEDs. However, for analog transmission purposes, such as TV signal broadcasting, it is important to evaluate this characteristic. To measure laser diode linearity, a sinusoidal modulation is applied to the device biased above threshold and the differential phase is obtained by subtracting from the photodiode response the phase delay obtained in linear conditions (low modulation depth). Care must be taken in order to avoid any saturation of the photodetector. The amplitude distortion can be obtained by using a spectrum analyzer connected to the photodetector [109].

3.2.2 Pulse Response

The time response of LEDs and laser diodes to fast driving pulses is a key parameter of these devices, especially for digital communication systems. However, the pulse response of laser diodes is far better than that of LEDs; thus, in the following we will refer to the techniques that can be used to evaluate this characteristic of laser diodes. To this purpose a very short current pulse (FWHM 100–200 ps) must be applied to the device biased at a proper dc level, typically 50–70% of the threshold current [110]. The light pulse duration can be measured by means of several techniques, both direct and indirect. Direct methods are simpler although they require costly equipment to achieve picosecond time resolution. Indirect methods, in turn, which will be described in Section 4.2.2, give even better time resolution with the drawback of greater complexity.

The current pulse can be generated, for instance, by using an oscillator driven at 100–300 MHz [111], followed by a step recovery diode [112, 113]. The current pulse is then amplified to the required level (50–100 mA typical) by an RF amplifier.

An easier method to measure the output light pulse relies on a fast detector, as PIN, APD or MSM diodes, attached to a sampling oscilloscope (see Figure 3.31). Since the time response of laser and detector are often comparable, the rise time of the detection chain must be known to perform the required deconvolution. Typically the rise time of a 50 Ω detection chain may be as short as 30 ps.

Even shorter optical pulses can be recorded with a streak camera [114–116]. This instrument can be used to resolve the ultrafast time variation of optical signals distributed over one spatial dimension. In conjunction with a spectrometer, a streak camera allows one to acquire a time resolved spectrum.

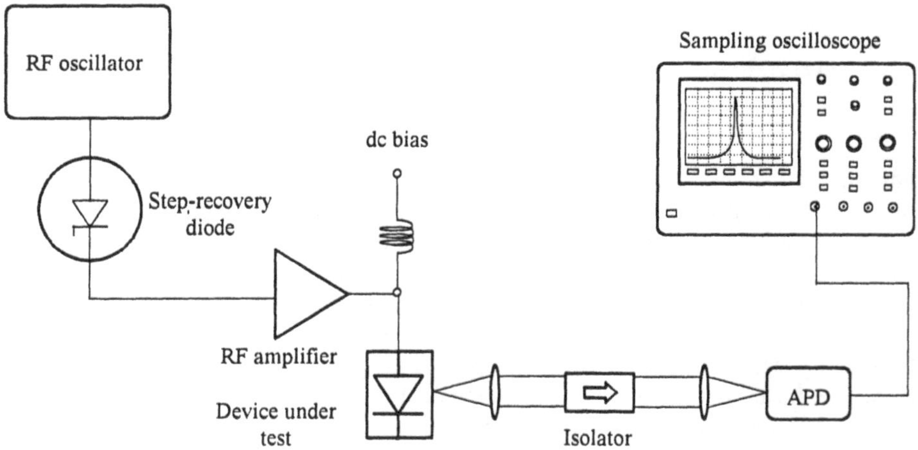

Figure 3.31 Block diagram of the experimental setup to evaluate the pulse response of a laser diode using a fast photodiode and a sampling oscilloscope.

The basic component of this instrument is the streak tube. This is a special kind of image intensifier mainly made of a photocathode, an accelerating grid, a pair of deflection plates, and a phosphor screen (see Figure 3.32). The light that hits the photocathode is converted into photoelectrons, which are accelerated by an electric field and focused by an electrostatic lens. The electrons are then swept in one direction by passing through the deflection plates. To this purpose a very fast high-voltage ramp is applied between these last electrodes, synchronously with the optical signal. Finally the photoelectrons hit the phosphor screen where the signal is back converted into light. Additionally a *multi-channel plate* (MCP) inside the streak tube before the phosphor screen can enhance the low signal level of photoelectrons, giving a significant improvement to sensitivity. A SIT camera or a cooled CCD camera can be used to record the light from the phosphor screen. The 2D image obtained in this way presents the spatial (or spectral) variation of the light intensity along one axis and its time variation along the other axis. The streak camera can be used in single shot mode or in synchroscan mode. In the latter case the signals from multiple scans of a repetitive phenomenon are averaged on the spatial detector (CCD or SIT). Streak cameras are commercially available that allow a direct measurement of optical pulses with time resolution better than 10 ps; however, their cost is rather high.

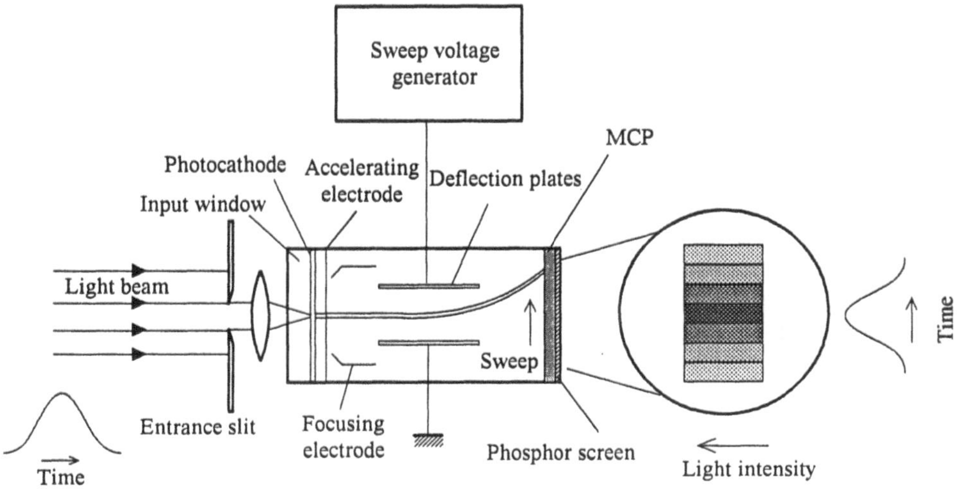

Figure 3.32 Schematic drawing of a streak camera tube.

3.2.3 Grating Spectrometer

The emission spectrum of broadband optical sources as LEDs and of MLM laser diodes can be measured using a grating spectrometer or monochromator. When needed, a higher resolution can be achieved by means of a Fabry-Perot interferometer. Since the measurement range (free spectral range) of a Fabry-Perot is limited to several tens of GHz, which corresponds to few angstroms, this instrument cannot perform an absolute wavelength measurement. Indeed the Fabry-Perot is usually employed to resolve the fine structure of a laser line in conjunction with a grating spectrometer. Refer to Section 2.2.4 for an overview of spectrometers and to Section 4.2.2 for a detailed treatment of the Fabry-Perot interferometer. In this section a description of the basic principles and operating conditions of the grating monochromator is given.

This process relies on a diffraction grating which can be realized either by mechanically ruling parallel grooves on a glass or metal substrate (ruled grating) or by recording the interference pattern from a highly coherent laser source onto a substrate with a photosensitive resist. Then the substrate is processed with photolithographic techniques (holographic grating).

Gratings can be realized either on plane or curved (concave) substrates and can operate both in transmission and reflection. However, grating monochromators typically mount a reflection-type device. The principle of operation of a grating consists in the dispersion of light impinging on its surface into several diffraction orders, at angles which depend on the wavelength, according to the formula

$$I(\alpha,\beta) = I_0 \frac{\sin^2\left[\frac{\pi a}{\lambda}(\sin\vartheta_i + \sin\vartheta_d)\right]}{\left[\frac{\pi a}{\lambda}(\sin\vartheta_i + \sin\vartheta_d)\right]^2} \frac{\sin^2\left[\frac{\pi w}{\lambda}(\sin\vartheta_i + \sin\vartheta_d)\right]}{\sin^2\left[\frac{\pi d}{\lambda}(\sin\vartheta_i + \sin\vartheta_d)\right]} \qquad (3.37)$$

where a is the groove width, d is the groove spacing, ϑ_i is the incidence angle, ϑ_d is the diffraction angle, w = Nd is the width of the grating being N the total groove number (see Figure 3.33) [117]. A plot of the light distribution resulting from (3.37) for a monochromatic source is shown in Figure 3.34.

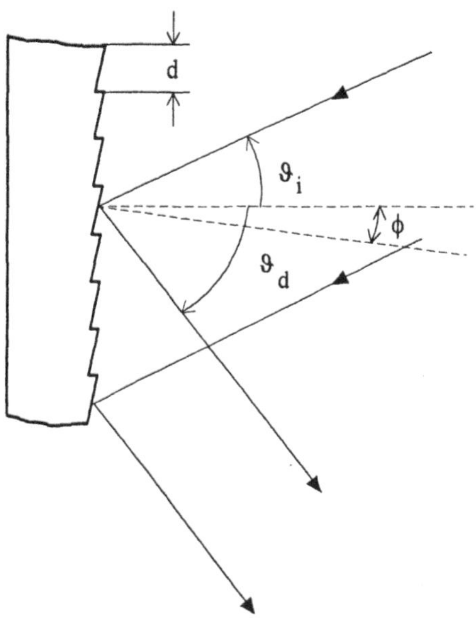

Figure 3.33 Reflection grating: ϑ_i is the incidence angle, ϑ_d is the diffraction angle, d is the groove spacing, and ϕ is the blaze angle.

The first factor in (3.37) corresponds to the diffraction pattern from a single slit of width a and gives the envelope of the interference peaks. The second factor originates from the interference of the light diffracted by each groove and consists of a sequence of narrow peaks corresponding to the different orders. The angular position of these peaks is individuated by the condition of constructive interference among the diffracted wavelengths

$$\sin\vartheta_i + \sin\vartheta_d = \frac{m\lambda}{d} \tag{3.38}$$

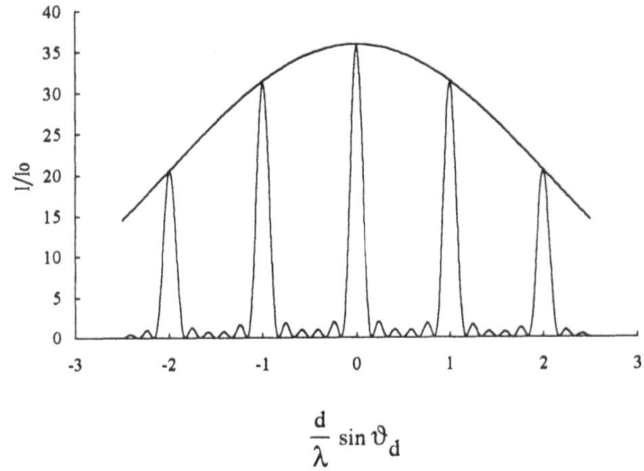

Figure 3.34 Fraunhofer diffraction pattern of a monochromatic source by a grating of six slits as a function of the sine of the diffraction angle ϑ_d. Normal incidence has been assumed (ϑ_i = 0). I_0 is the intensity which would be produced by a single slit.

The angular half-width of the peaks (i.e., from the maximum to zero) is given by

$$d\vartheta_d = \frac{\lambda}{Nd\cos\vartheta_d} = \frac{1}{w\cos\vartheta_d} \tag{3.39}$$

Besides each principal maximum there are several secondary maxima which are of negligible intensity when the number N of grooves is made sufficiently large. In fact their intensity falls as $1/N^2$.

Usually in a grating spectrometer the incidence angle ϑ_i is kept constant and the angular dispersion of a grating $d\vartheta_d/d$ gives the angular separation $d\vartheta_d$ obtainable for two wavelengths differing by $d\lambda$. This is obtained by differentiating equation (3.38) for a fixed incidence angle ϑ_i

$$\frac{d\vartheta_d}{d\lambda} = -\frac{m}{d\cos\vartheta_d} \tag{3.40}$$

As it can be argued from formula (3.38), the separation in terms of $\sin\vartheta_d$ between two adjacent peaks is proportional to the wavelength. Since all the wavelengths lead to a peak at the geometrical reflection angle ($\vartheta_i = -\vartheta_d$) which corresponds to the zeroth order, this is useless for spectral analysis purposes; hence, as the diffraction efficiency decreases with increasing orders, typically the first diffraction order is considered. In manufacturing of reflective ruled gratings, the grooves are shaped and tilted with respect to the grating surface by a suitable angle (the "blaze" angle) to gather more intensity inside the first order diffracted beam around the wavelength of interest; this allows one to maximize the reflection efficiency into this order (see Figure 3.33).

According to the Rayleigh criterion [118] the wavelength resolving power is given by the ratio of the angular dispersion power to the half width of the principal maximum

$$R_p = \frac{\lambda}{\Delta\lambda} = \frac{wm}{d} = Nm \tag{3.41}$$

3.2.4 Emission Spectrum Evaluation

A common arrangement used to measure the emission spectrum by means of a grating spectrometer is shown in Figure 3.35. The light emitted by the device under test is focused onto the entrance slit by the lens L_1, collimated by the mirror M_1 and steered toward the grating. The dispersed light is collected by the mirror M_2 and focused onto the exit slit. A photodiode measures the output intensity as long as the wavelength is scanned.

Beside the characteristic of the grating, the focal length f and the f-number f/# are two important parameters of a spectrometer. The f-number represents the ratio of the focal length of the collimating mirrors to its aperture, f/# = f/D. The f-number is also equal to

$$f/\# = \frac{1}{2\sin\vartheta} \tag{3.42}$$

where ϑ is the angular semi-aperture of the spectrometer (see Figure 3.35).

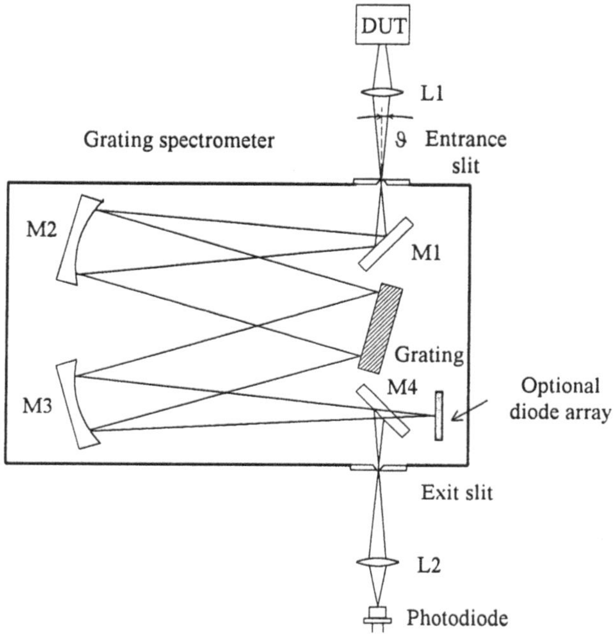

Figure 3.35 A common arrangement to measure the spectrum of an optical source using grating spectrometers.

For reaching maximum efficiency, the input beam should be focused onto the entrance slit with an optical system having approximately the same aperture as the spectrometer (in fact, using a higher aperture would lead to a waste of light); on the other hand, lower aperture optics would not exploit the full width of the grating, giving a resolving power lower than the nominal value. In the exit path (after the grating), the focal length transforms the angular dispersion into linear dispersion at the plane of the exit slit

$$dl = f\, d\vartheta \tag{3.43}$$

To preserve the nominal resolution of the spectrometer, the width Δl of the slits must be selected in such a way to fulfill the Rayleigh criterion. This implies that the slit transmits no more than half the width of the first order peak ($m = 1$) generated by a monochromatic source (see Equation (3.38)). By using the expressions (3.40) and (3.41), one gets the condition

$$\Delta l = \left(\frac{d\vartheta}{d\lambda}\right)\frac{\lambda}{R_p} f \tag{3.44}$$

However, for noncritical measurements, as in the case of spectral analysis of LEDs or other incoherent sources, wider slits can be acceptable; this leads to an easier alignment and a higher light-gathering power.

Before the measurement, the calibration of the instrument should be verified using suitable spectral lamps (Na, Cs, etc.) or reference laser sources as He-Ne or Ar lasers. Usually a stepping motor provides the angular scanning of the grating through a sine-bar mechanism which renders the rotation of the motor shaft proportional to the variation of the wavelength λ, according to the expression (3.38). A photodiode beyond the exit slit collects the transmitted light and gives a current signal which is amplified and recorded. Computed controlled setup can be arranged for automatic measurements using built-in standard interfaces.

A photodiode array in place of the exit slit can provide the whole spectrum at once. Yet in this case an additional limiting factor to the resolving power is given by the number of pixels of the array (N_p), typically 512 for 1/2" length. Assuming that the instrument should operate over a spectral range $\Delta\lambda$, the minimum line separation that can be resolved is $\Delta\lambda/N_p$. The choice of the spectrometer to couple to the photodiode array can be done according to the following guidelines.

The dispersing power of a grating is

$$\frac{d\phi}{d\lambda} = \frac{n}{\cos\phi} \tag{3.45}$$

The dispersed light overlaps the array over the range $\Delta\lambda$ if the following condition is fulfilled

$$f\Delta\phi = \frac{n}{\cos\phi} f\Delta\lambda = WN \tag{3.46}$$

where f is the focal length of the monochromator and W is the width of a pixel. Equation (3.46) gives a relationship between the monochromator parameters f, n and the array photodiode length WN. It results in

$$\frac{nf}{\cos\phi} = \frac{NW}{\Delta\lambda} \tag{3.47}$$

A further condition must be satisfied to optimize the instrument. The width of a line dispersed by the grating is

$$d\vartheta = \frac{n\lambda}{m\cos\phi} \tag{3.48}$$

where m is the total number of grooves of the grating. The maximum resolution provided by the photodiode array is preserved if the product $d\vartheta \cdot f$ is less than the pixel width W

$$\frac{nf}{\cos\phi}\frac{\lambda}{m} \leq W \tag{3.49}$$

From equation (3.49) and (3.47) the following condition on the resolving power R_p is obtained

$$R_p = m \geq \frac{\lambda}{\Delta\lambda} N \tag{3.50}$$

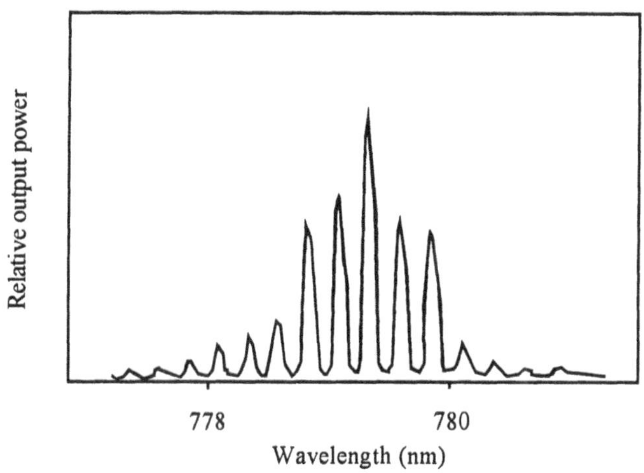

Figure 3.36 Example of the spectrum of a multi-longitudinal mode laser diode. Each peak corresponds to a cavity mode.

Off-the-shelf instruments are offered by a few companies operating in the electro-optic market; the photodiode array sensitivity can be greatly enhanced by coupling it to an image intensifier.

An LED emission spectrum is a bell-shaped broad band distribution, usually wider than 20 nm (FWHM).

The emission spectrum of a MLM laser consists of a series of peaks, corresponding to the longitudinal modes of the cavity. In the example shown in Figure 3.36 they are spaced by the about 0.25 nm, which corresponds to a cavity length of 350 μm. For visible-near infrared measurements, 600 to 300 grooves/mm grating having a 50×50 mm^2 area can be used into a 50–cm focal length monochromator. For this kind of instrument the wavelength resolution is typically of the order of 0.05 nm. Hence the mode spacing can be easily resolved by the instrument, whereas the width of the single peak (usually less than 10^{-3} nm) is beyond its resolving power.

3.2.5 Mode Partition Noise in Laser Diodes

In single-mode fiber communication systems operating under direct modulation, the chromatic dispersion of the optical fiber gives rise to the so-called mode partition noise when the optical power of the diode laser is randomly distributed among a number of longitudinal modes. In digital transmission schemes, during the current pulse, the mode competition leads to variation in the emitted wavelength. This results in pulse distortion and jitter with respect to the nominal time position at the end of the transmission channel [119]. The worst case occurs when a single-mode laser jumps between two or more longitudinal modes from pulse to pulse. Even laser diodes operating in single longitudinal mode, when driven at constant current, exhibit this behavior in modulated regime.

To measure the amount of mode partition noise, a mode partition coefficient k is to be defined [120]. The following assumptions are made:

1. In every pulse the total power of all longitudinal modes is constant. The amplitude a_i of each mode can be then normalized according to the expression

$$\sum_i a_i = 1 \tag{3.51}$$

2. Each longitudinal mode corresponds to stimulated emission, not to spontaneous emission.
3. The signal received after the distortion due to the medium can be decomposed into frequency components corresponding to each longitudinal mode according to the formula

$$s(t) = \sum_i f(\lambda_i,t)\ a_i \tag{3.52}$$

where λ_i is the i-th mode wavelength and $f(\lambda_i, t)$ is the i-th mode pulse waveform.

The fluctuations of the optical signal intensity are given by the second-order moment of s(t)

$$\sigma^2 = \overline{s^2(t)} - \left(\overline{s(t)}\right)^2$$

$$= \sum_i f^2(\lambda_i,t)\ \overline{a_i^{\,2}} + 2\sum_i \sum_{j>i} f(\lambda_i,t)\ f(\lambda_j,t)\ \overline{a_i a_j} \tag{3.53}$$

$$-\sum_i f^2(\lambda_i,t)\left(\overline{a_i}\right)^2 - 2\sum_i \sum_{j>i} f(\lambda_i,t)\ f(\lambda_j,t)\,\overline{a_i}\ \overline{a_j}$$

where the following definitions have been used

$$\overline{a_i} = \int a_i p(a_1,a_2,\ldots\ldots,a_n)da_1 da_2\ldots\ldots da_n$$

(3.54)

$$\overline{a_i a_j} = \int a_i a_j p(a_1,a_2,\ldots\ldots,a_n)da_1 da_2\ldots\ldots da_n$$

being $p(a_1, a_2,\ldots, a_n)$ the mode partition probability distribution.

After some algebra, the following expression is obtained

$$\sigma^2 = k^2\left[2\sum_i\sum_{j>i}\left(f(\lambda_i,t) - f(\lambda_j,t)\right)^2 \overline{a_i}\ \overline{a_j}\right] \qquad (3.55)$$

where

$$k^2 = \frac{\sum_i\sum_{j>i}\left(f(\lambda_i,t) - f(\lambda_j,t)\right)^2\left(\overline{a_i}\ \overline{a_j} - \overline{a_i a_j}\right)}{\sum_i\sum_{j>i}\left(f(\lambda_i,t) - f(\lambda_j,t)\right)^2 \overline{a_i}\ \overline{a_j}} \qquad (3.56)$$

Assuming stimulated emission, the condition $\overline{a_i a_j} \propto \overline{a_i}\ \overline{a_j}$ $(i \neq j)$ holds and the factor k only depends on the characteristics of the laser diode

$$k^2 = \frac{\sum_i\sum_{j>i}\left(\overline{a_i}\ \overline{a_j} - \overline{a_i a_j}\right)}{\sum_i\sum_{j>i}\overline{a_i}\ \overline{a_j}} = \frac{\overline{a_i}\ \overline{a_j} - \overline{a_i a_j}}{\overline{a_i}\ \overline{a_j}} \qquad (3.57)$$

using the relations $\sum \overline{a_i} = 1$ and $\overline{\left(\sum a_i\right)^2} = 1$ derived from assumption 1, k is finally expressed as

$$k^2 = \frac{\overline{a_i^2} - \left(\overline{a_i}\right)^2}{\overline{a_i} - \left(\overline{a_i}\right)^2} \qquad \text{for any } i \qquad (3.58)$$

From (3.58) it appears that k does not depend on the particular form of $f(\lambda_i, t)$ but is a characteristic of the laser diode. It can be measured by evaluating the values of $\overline{a_i}$ and of the square root of the variance $\Delta = \sqrt{\overline{a_i^2} - \left(\overline{a_i}\right)^2}$ for the i-th longitudinal mode [120].

A pulse amplitude modulated signal, having time interval T between pulses and pulse duration τ, is assumed. The amplitude $a_i(t)$ varies from pulse to pulse as shown in Figure 3.37. The experimental setup used in the "low pass filter method" (LPF) is shown in Figure 3.38. The time average of the amplitude $a_i(t)$ is ($\overline{a_i}\ \tau/T$), while the r.m.s. after a filter having cut-off frequency at half the baud rate (1/2T) is equal to ($\Delta\tau/T$). The i-th longitudinal mode is selected by means of a monochromator and the amplitude fluctuations are collected by a PIN photodiode followed by an ac coupled amplifier. Due to the ac coupling the expectation value $\overline{a_i}$ cannot be directly measured by means of a dc voltmeter. The signal is sent in parallel to two filters with cut-off at frequencies 1/2T and 1/T respectively, followed by two r.m.s. voltmeters.

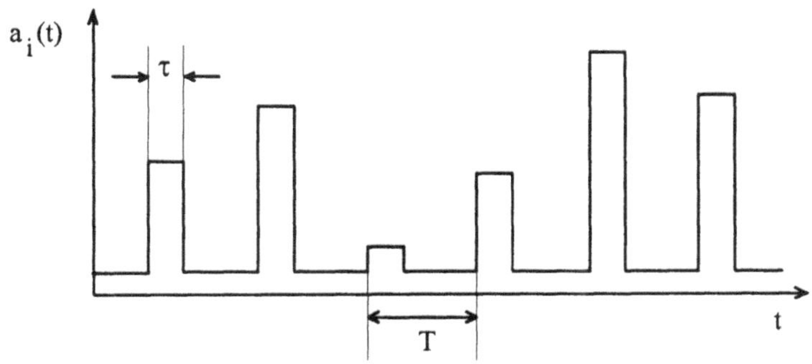

Figure 3.37 The amplitude $a_i(t)$ of the i-th longitudinal mode of a pulsed laser diode as a function of the time. The amplitude randomly varies from pulse to pulse.

The output voltage from the 1/2T filter gives the quantity $\Delta\tau/T$, while the output voltage from the 1/T filter is $\sqrt{2\left(\overline{a_i}\right)^2\tau^2/T^2 + 2\Delta^2\tau^2/T^2}$.

An alternative "sampling" method uses a sample-and-hold unit to achieve a 100 percent duty cycle pulse train so that a dc voltmeter and an r.m.s. voltmeter can be used to obtain $\overline{a_i}$ and $\sqrt{\overline{a_i^2} - \left(\overline{a_i}\right)^2}$. The sampling method measures k at a fixed time position within the pulse duration τ while the LPF method gives the average of k. The two

3.3.1 Dark Current and Responsivity Measurements

The electrical characterization of a photodetector involves several measurements that can be performed by reverse biasing the device in a suitable network. An example of a biasing tee is shown in Figure 3.39 [121]. The adjustable voltage +V is provided by a stabilized dc supply. The resistance R_1 and capacitance C_1 filter the residual noise that can be present in the dc supply. The choice of the value of resistance R_2 is critical since low values give rise to high thermal noise, whereas high values cause voltage drops during the measurement. The dc current is measured by means of a nanoammeter attached to the photodiode through a low-pass filter which suppresses any noise generated by the nanoammeter. The noise current is amplified through the transimpedance amplifier ac coupled to the photodetector.

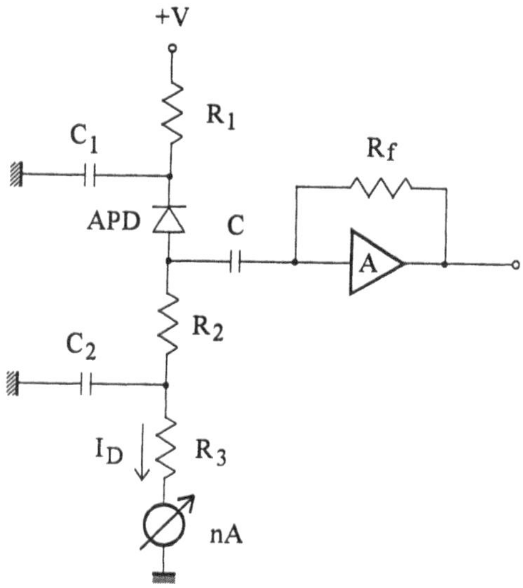

Figure 3.39 Biasing tee to measure the dark current and the responsivity of a photodetector (APD in the example). A is a transimpedance amplifier, nA is a nanoammeter (from [121]).

The first parameter which can be measured is the dark current; the photodiode is thermally stabilized and darkened, while special care must be devoted to prevent any stray light from impinging onto the device. The dark current slowly increases with the bias voltage until the breakdown condition is met when a sudden rise is observed, according to Figure 2.14. Care must be taken to not provoke damage to the photodiode by excessive reverse current.

The responsivity as a function of the bias voltage can be measured by illuminating the photodetector with a suitable source and measuring the current signal. This measurement is of fundamental importance for APD-type detectors. A suitable setup is shown in Figure 3.40. The light from a pigtailed LED is sent to a three-port fiber coupler. One of the output beams is monitored by a radiometer, the other is collected by an optical system (e.g., a microscope objective) and is focused onto the photodetector. The spot-size must be large enough to cover a significant portion of the sensitive area in order to minimize the influence of small scale inhomogeneities; on the other hand, vignetting effect must be avoided. The wavelength of the LED must correspond to the range of interest for the application of the photodiode and the power must be far from the saturation level, but well above the dark current level (e.g., 100 nW for a Si APD). For APDs the measured responsivity (A/W) is the product of the primary responsivity times the avalanche gain. By increasing the reverse bias voltage, the intrinsic layer undergoes to a progressive depletion until, for a certain value V_d, the complete depletion is achieved. Consequently the primary responsivity increases when the depletion region width increases until $V = V_d$, whereas the avalanche gain continues to grow also beyond the value V_d. Due to this fact the total responsivity curve exhibits a knee in correspondence with V_d.

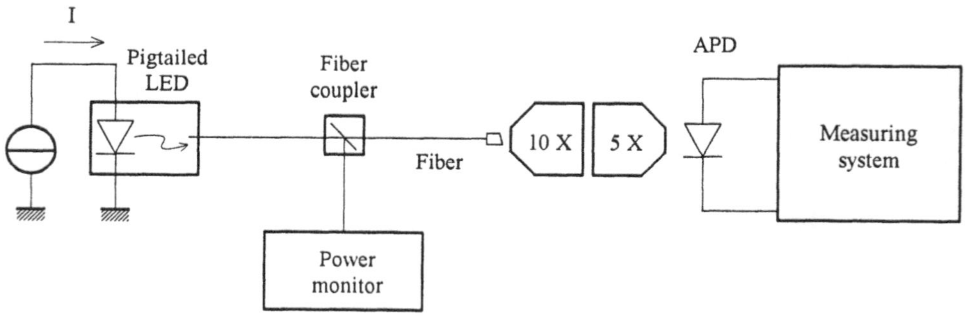

Figure 3.40 Experimental setup to measure the responsivity of a photodetector (APD in the example). The symbols 10X - 5X refer to microscope objective lenses (from [121]).

3.3.2 Baseband Frequency Response and Pulse Response

An important characteristic of a photodetector is its frequency response, which is related to the pulse response of the device. The last can be evaluated by sending a train of ultra-short laser pulses to the device. The pulses can be generated by directly modulating the driving current of a laser diode or by using a mode-locked laser. In the first case, a pulse duration of 100 ps can be achieved; in the second, pulse durations down to few ps can be generated. The output signal from the photodetector is recorded by means of a sampling oscilloscope or a fast digital oscilloscope. The Fourier transform of the output signal gives the frequency response of the detector measured in large signal regime. Equivalently, a wideband spectrum analyzer can be used to acquire the spectrum of the photocurrent.

A direct measurement of the frequency response of a fast photodetector can be obtained by illuminating the device with a high-frequency modulated optical signal. However modulation frequencies higher than 10 GHz cannot be easily achieved either by direct modulation of laser diodes or by using external modulators. To this aim the heterodyne scheme (see Section 4.2.3) has been proposed to produce a beat signal at the frequencies required for a wideband measurement [122]. The output beams from two identical SLM laser diodes are mixed by means of a fiber-optic coupler. The frequency of one of them is swept by varying the temperature of the device with a Peltier cooler. This technique produces fairly stable oscillating signals (drift on the order of 100 MHz/min) spanning a range of several tens of GHz, according to the formula [122]

$$I(t) \propto E_1^2 + E_2^2 + \cos\phi \; 2\, E_1 E_2 \cos(\omega_b t) \tag{3.59}$$

where E_1 and E_2 are the fields of the laser diodes, ω_b is the beat frequency and ϕ is the angle between the polarization directions of E_1 and E_2.

A frequency locking technique based on an external cavity can be used to provide active stabilization (see Figure 3.41) [123]. The outputs from two monolithic Nd:YAG lasers with emission at 1.06 μm are mixed on a beam combiner. The beat signal from one of the combiner's ouput ports is sent to a ring resonant cavity with 1 GHz free spectral range, while the output from the other port is used to test the device. Lasers are temperature tunable and are modulated at low frequency f_m. The output signal from the cavity is monitored by a low-frequency photodetector. On resonance the frequency $2f_m$ is measured while the frequency f disappears. By independently adjusting the temperature of both lasers, these can be locked to the same cavity mode. At this point a low-frequency heterodyne signal is obtained. Afterward the wavelength of one of the lasers is tuned until the matching with a different cavity mode is detected by the servo mechanism. In this way the beat signal increases by steps corresponding to the mode spacing of the cavity. The servo mechanism also provides the frequency stabilization.

Using a light beam having an intensity given by formula (3.59), the photocurrent from the detector under test is given by

$$i(t) \propto \frac{q\eta_o}{\hbar\omega} \left[E_1^2 + E_2^2 + H(\omega_b) \cos\phi \; 2 E_1 E_2 \cos(\omega_b t) \right] \tag{3.60}$$

where η_o is the quantum efficiency, ω is the frequency of the optical carrier, $\hbar$ is Planck's constant divided by 2π (that is, $\hbar = h/2\pi$), and $H(\omega_b)$ is the frequency response of the device. The maximum RF signal is achieved when ϕ is set equal to zero by properly aligning the principal axes of the coupler entrance fibers with the polarization vectors of both laser diodes. The fiber coupler must preserve the polarization. The photocurrent is monitored with a spectrum analyzer. The intermediate frequency (i.f.) of this instrument, set around few MHz, gives the resolution limit of the measurements.

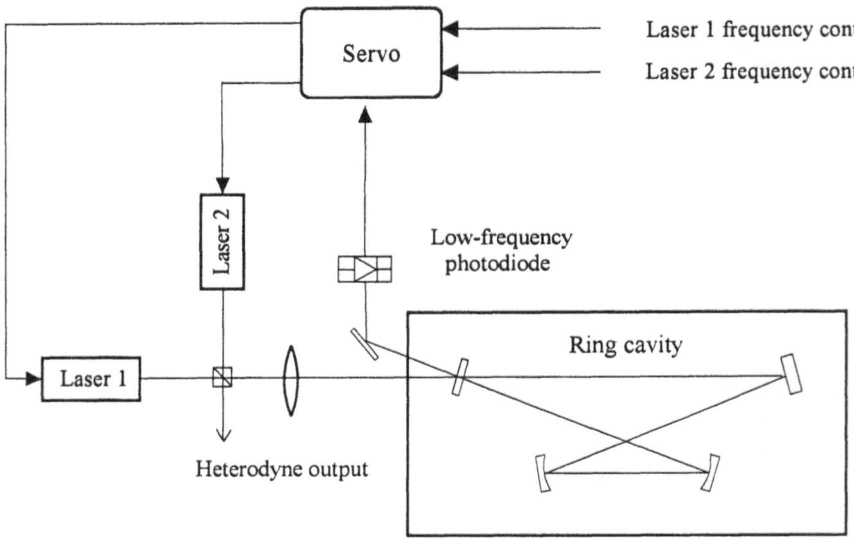

Figure 3.41 A frequency locking setup based on an external cavity usable to provide active stabilization of the heterodyne signal of two single longitudinal mode lasers (from [123]).

The frequency response of the device can be obtained by recording the peak of the spectrum while the frequency of the beat signal is scanned by varying the temperature of one of the two lasers (the center frequency of a GaAlAs laser diode drifts by approximately −12 GHz per degree Celsius). The linewidth of the heterodyne signal is given by the convolution of the line shapes of the two laser diodes. If these in turn are assumed to be Lorentzian, the heterodyne linewidth is the sum of those of the single laser diodes (see Section 4.2.3). The broadening of the beat signal results in a reduction of the resolution and the sensitivity of the measurements. Assuming a linewidth narrower than the i.f. of the spectrum analyzer, the peak of the power spectrum is given by

$$S_\omega = \frac{1}{2} R_L \left(\frac{2q\eta_o}{\hbar\omega} \cos\phi \, E_1 E_2 \right)^2 \tag{3.61}$$

where R_L is the load resistance of the spectrum analyzer. However if the linewidth of the beat signal is wider than the input spectral window, the peak of the spectrum is given by the previous formula (3.61) times an integration factor

$$P(\nu) \propto \frac{1}{B} \int_{\nu_0-B/2}^{\nu_0+B/2} \frac{\delta\nu_1 + \delta\nu_2}{(\nu - \nu_0)^2 + \left(\frac{\delta\nu_1 + \delta\nu_2}{2}\right)^2} d\nu \tag{3.62}$$

resulting from the ratio of the fraction of the photocurrent power spectrum falling within the bandwidth of the input filter of the spectrum analyzer to the total area of the spectrum. This leads to an amplitude reduction which lowers the sensitivity of the technique by several dB. However, linewidth stabilization methods, such as the one previously described, can be applied to overcome this limitation. Measurements of the frequency response of photodiodes up to 20 GHz can be performed with a resolution of a few MHz.

3.3.3 Photodiode Nonlinear Distortion

In an ideally linear system the response to an input signal i(t) can be written as r(t) = Gi(t), independently of the frequency. A practical system differs from the ideal case due to the limitation of the frequency response already addressed in the previous Section 3.3.2, and because of nonlinearities arising in the large signal regime. To take into account the first-order nonlinearities, the system response can be assumed to behave according to the following form

$$r(t) = G_1 i(t) + G_2 i^2(t) \tag{3.63}$$

Assuming that the input signal can be represented as a sinusoidal function with frequency, f

$$i(t) = A[1 + \cos(2\pi f t)] \tag{3.64}$$

the output signal from the system contains a component oscillating at the second harmonic frequency 2f, according to the formula

$$r(t) = AG_1 + A^2G_2 + (AG_1 + 2A^2G_2)\cos(2\pi ft) + A^2G_2\cos^2(2\pi ft)$$

$$= AG_1 + \frac{3}{2}A^2G_2 + (AG_1 + 2A^2G_2)\cos(2\pi ft) + \frac{1}{2}A^2G_2\cos(2\pi 2ft) \qquad (3.65)$$

In a similar fashion, when the input signal consists of a sum of two sinusoidal components at frequencies f_1 and f_2

$$i(t) = A[1 + \cos(2\pi f_1 t)] + B[1 + \cos(2\pi f_2 t)] \qquad (3.66)$$

the nonlinear transfer characteristic (3.63) gives rise to the sum frequency $(f_1 + f_2)$ and the difference frequency $(f_1 - f_2)$, under the assumption that $f_1 > f_2$

$$r(t) = \frac{3}{2}(A^2 + B^2)G_2 + 2ABG_2 + (A + B)G_1$$

$$+ A[G_1 + 2G_2(A + B)]\cos(2\pi f_1 t)$$

$$+ B[G_1 + 2G_2(A + B)]\cos(2\pi f_2 t)$$

$$+ ABG_2\cos[2\pi(f_1 + f_2)t] + ABG_2\cos[2\pi(f_1 - f_2)t]$$

$$+ \frac{1}{2}A^2G_2\cos(2\pi 2f_1 t) + \frac{1}{2}B^2G_2\cos(2\pi 2f_2 t) \qquad (3.67)$$

Higher order harmonics and frequency mixing arise when the parabolic assumption of the transfer characteristic fails and a polynomial form must be considered

$$r(t) = G_1 i(t) + G_2 i^2(t) + G_3 i^3(t) + G_4 i^4(t) + ... \qquad (3.68)$$

For the general case in which more than two frequency components are mixed together by a transfer characteristic given by the above formula, a series of terms at frequencies consisting of a linear combination of the frequency components f_1, f_2, f_3, ..., are generated by the nonlinear behavior of the system. In the simpler case in which just two frequencies are mixed by a system having a parabolic characteristic, a measure of the spectral power density at the fundamental frequencies f_1 and f_2 and at the second-order frequencies $f_1 + f_2$, $f_1 - f_2$, $2f_1$ and $2f_2$ allows to evaluate the nonlinear behavior of the system.

As an example the nonlinear response of a photodiode can be measured with the arrangement shown in Figure 3.42 [124].

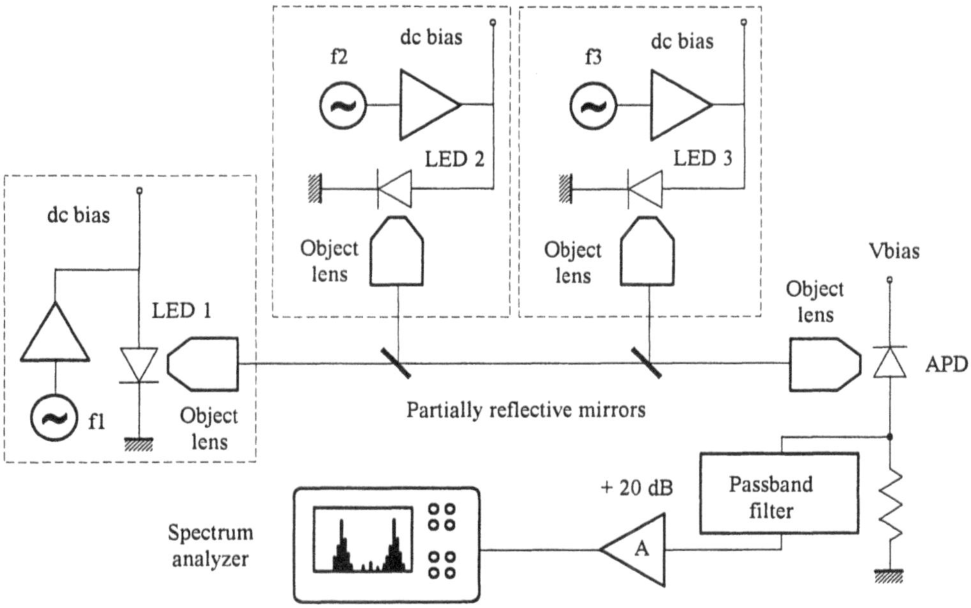

Figure 3.42 Arrangement that can be used to measure the nonlinear response of a photodiode.

Three GaAlAs LEDs are intensity modulated at frequencies f_1, f_2, f_3; their beams are collimated by suitable optics and mixed by means of partially reflective mirrors. The dc driving currents of the single diodes are carefully adjusted in order to equalize the intensities at the detector surface. The depth of the applied sinusoidal modulation is approximately 75% of the dc bias current, which is 100 mA. The photocurrent signal is passed through a bandpass filter and displayed on a spectrum analyzer. Since each LED is driven at a single frequency, the resulting optical signal contains only the fundamental frequencies f_1, f_2, f_3 and their harmonics produced by the nonlinearities of the LEDs. The center frequency and the bandpass of the filter are selected in such a way to analyze a portion of the spectrum containing only a second-order term $f_1 + f_2$ and a third order term

$f_3 - (f_1 + f_2)$ which are located between the second harmonic peaks $2f_1$ and $2f_2$. These two peaks provide two reference markers for the measurements. Note that nonlinear distortion components introduced by the measuring system, including LED nonlinearities of the type $(2mf_1 + 2nf_2 + 2pf_3)$ with m, n, and p integers, are positioned out of the filter band except for the already mentioned components $2f_1$ and $2f_2$. In [124], for example, the frequencies f_1, f_2, f_3 are 1498.2 kHz, 1503.2 kHz, and 6006.5 kHz, respectively. Figure 3.43 depicts the power of the fundamental and nonlinear components of the APD photocurrent as a function of the incident optical power. For Si-APD detectors with avalanche gain of about 20 dB, the magnitude of the second-order distortion is well approximated by a quadratic function of the incident power, whereas the third-order distortion behaves like a cubic function above the shot-noise level, which sets the limit of the experimental sensitivity.

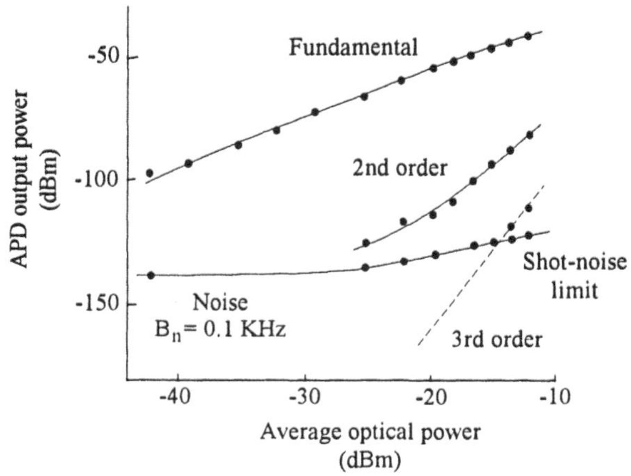

Figure 3.43 An example of the nonlinear distortion spectrum of an APD (from [124]).

A PIN photodiode, biased at relatively high reverse voltages, presents a nonlinear distortion below the shot-noise level.

3.3.4 Gain and Noise Figure Evaluation of APDs

The noise spectral density of an APD can be measured using a spectrum analyzer following a transimpedance amplifier (see Figure 3.44). The spectral density of the mean square value of the noise current after the amplification process is given by

$$\langle i_n^2 \rangle = \langle i_n^2 \rangle_{AMP} + 2q\langle i_{ph}^0 \rangle F(\langle M \rangle)\langle M \rangle^2 \tag{3.69}$$

where $\langle i_n^2 \rangle_{AMP}$ is the term due to the amplifier noise, q is the electron charge, $\langle i_{ph}^0 \rangle$ is the DC primary photocurrent, $\langle M \rangle$ is the average avalanche gain and $F(\langle M \rangle)$ is the excess noise factor (see Section 4.2.1) given by (2.94). It is worthwhile to remember that the coefficient k appearing in that expression is the ratio of holes to electrons ionization rates. For Si APDs, k is approximately equal to 0.02 [125]. In writing the equation (3.69) the noise terms associated to both the bulk and surface dark current have been neglected. This assumption is reasonable for a relatively high avalanche gain ($\langle M \rangle$ greater than 10).

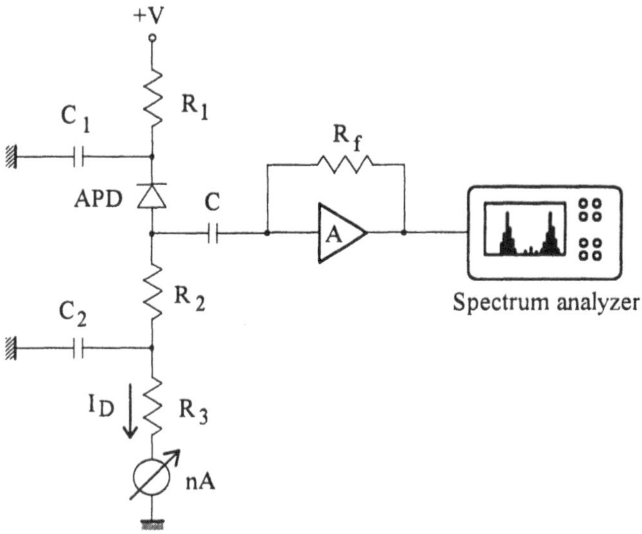

Figure 3.44 Biasing tee to measure the noise of a photodiode (APD in the example) using a spectrum analyzer.

The noise power spectral density after the transimpedance amplifier is given by

$$\sqrt{\langle v_n^2 \rangle} = R_f \sqrt{\langle i_n^2 \rangle} \tag{3.70}$$

If R_f is the input resistance of the spectrum analyzer, the voltage $\langle v_n^2 \rangle$ is related to the measured noise power P_n by

$$\langle v_n^2 \rangle = R_f P_n \tag{3.71}$$

Once the amplifier noise current contribution has been measured with LED emission switched off, this term can be subtracted from the overall noise power spectral density. The spectral power density of the APD noise current is

$$\sqrt{2q\langle i_{ph}{}^0\rangle F(\langle M\rangle)\langle M\rangle^2} = \sqrt{\langle i_n{}^2\rangle - \langle i_n{}^2\rangle_{AMP}} \tag{3.72}$$

From the measurement of the APD noise power spectral density

$$\langle i_n{}^2\rangle_{APD} = 2q\langle i_{ph}{}^0\rangle F(\langle M\rangle)\langle M\rangle^2 \tag{3.73}$$

it is possible to extract informations about the multiplication factor $\langle M\rangle$ and the excess noise factor $F(\langle M\rangle)$; however, a single point determination cannot lead to a complete evaluation of these factors.

Indeed, from expression (2.94), an approximate linear relationship between $F(\langle M\rangle)$ and $\langle M\rangle$ is immediately established if the assumption of relatively high gain values ($\langle M\rangle$ greater than 50) and of a low k value is made (which certainly holds for Si APDs). Under these conditions one gets

$$F(\langle M\rangle) \approx 2 + k\langle M\rangle \tag{3.74}$$

From equation (3.73) we obtain a second relationship connecting $F(\langle M\rangle)$ and $\langle M\rangle$

$$F(\langle M\rangle) = \frac{\langle i_n{}^2\rangle_{APD}}{2q\big(\langle i_{ph}{}^0\rangle\langle M\rangle\big)\langle M\rangle} = \frac{\langle i_n{}^2\rangle_{APD}}{2q\langle i_{ph}{}^0\rangle_M\langle M\rangle} \tag{3.75}$$

where the measurable quantity $\langle i_{ph}{}^0\rangle_M = \langle i_{ph}{}^0\rangle\,\langle M\rangle$ has been defined.

The primary responsivity is then given by

$$R = \frac{\langle i_{ph}{}^0\rangle_M}{P\langle M\rangle} \tag{3.76}$$

A series of measurements of the quantities $\langle i_{ph}{}^0\rangle_M$ and $\langle i_n{}^2\rangle_{APD}$ can be done for a fixed value of the optical power P by varying the applied reverse voltage above the knee value

V_d. In this condition the parameter R can be assumed constant. Two measurements of $\langle i_{ph}{}^0\rangle_M$ and $\langle i_n{}^2\rangle_{APD}$ at two different bias voltages can be used to extract the value of the primary responsivity R. This can be done by solving the system of six equations given by the relationships (3.74), (3.75), and (3.76) written twice for the six unknown quantities F_i, M_i ($i = 1, 2$), R and k; the subscript $i = 1, 2$ refers to the pairs of measured values of $\langle i_{ph}{}^0\rangle_M$ and $\langle i_n{}^2\rangle_{APD}$, which appear in the equations as known terms, e.g. the optical power P, which is the same for the two measurements. This procedure leads to an explicit expression for R

$$R = \frac{4q\langle i_{ph}{}^0\rangle^2{}_{M_1}(1 - x)/P}{\langle i_n{}^2\rangle_{APD_1} - x^3\langle i_n{}^2\rangle_{APD_2}} \tag{3.77}$$

where we defined x as the ratio $x = \langle i_{ph}{}^0\rangle_{M_1}/\langle i_{ph}{}^0\rangle_{M_2}$.

Indeed this procedure must be repeated over several pairs of measurements at different voltages, and the average value R must be taken in order to achieve higher precision and confidence over the range of bias voltages used in practice.

Once R is determined, the avalanche gain $\langle M\rangle$ versus the reverse voltage can be obtained from equation (3.76). The excess noise factor F versus $\langle M\rangle$ is given by (3.75) and (3.76) considered as a couple of parametric equations in the implicit parameter V, while the measured $\langle i_{ph}{}^0\rangle_M$ and $\langle i_n{}^2\rangle_{APD}$ values are referred to the corresponding V value. Finally, the ionization ratio k can be obtained either from the slope of the approximate straight line (3.74) under the above specified assumptions or by solving the rigorous equation (2.94), leading to the formula

$$k = \frac{\langle M\rangle}{(\langle M\rangle - 1)^2}\left[(F - 2) + \frac{1}{\langle M\rangle}\right] \tag{3.78}$$

The measurements described above can be performed by means of an automatic setup, which directly produces the diagrams for the function of interest (i.e., dark current versus V, responsivity R versus V, $\langle M\rangle$ versusV, $\langle i_n{}^2\rangle_{APD}$ versus V, F versus $\langle M\rangle$) [121].

REFERENCES

[1] Petermann, K., "Constraints for fundamental-mode spot size for broadband dispersion-compensated single-mode fibres", *Electron. Lett.*, Vol. 19, 1983, pp. 712–714.

[2] Di Vita, P., et al., "Optical fibre measurement standardization", *Proc. OFMC-NPL*, York, 1991, pp. 1–7.

[3] CCITT Blue Book, Recommendations G.601-G.654, "Transmission Media Characteristics", Geneva, 1989.
[4] International Standard IEC 793/1, "Optical Fibres – Part 1: Generic Specifications".
[5] International Standard IEC 793/2, "Optical Fibres – Part 2: Product Specifications".
[6] Namihira, Y., et al., "Optimum fibre parameters of low-loss single-mode optical fibres for use in 1.55 μm–wavelength region", *Electron. Lett.*, Vol. 23, 1987, pp. 963–964.
[7] Kane, W. T., "The first international fibre geometry round robin", *Proc. OFMC-NPL*, York, 1991, pp. 119–123.
[8] Young, M., "Fiber cladding diameter by contact micrometry", *Proc. OFMC-NPL*, York, 1991, pp. 123–126.
[9] Baines, J. G. N., et al., "Fiber diameter measurements and their calibration", *IEEE J. Lightwave Technol.*, Vol. LT-8, 1990, pp. 1259–1268.
[10] Colton, J. R., and A. G. Hallam, "Measurements on the uniformity of fibre cladding geometry", *Proc. OFMC-NPL*, York, 1991, pp. 127–130.
[11] Mechels, S., and M. Young, "Scanning confocal microscope for precise measurement of optical fiber diameter", *Proc. Soc. Photo-Optical Instrum. Engrs*, Vol. 1556.
[12] Collyer, A., et al., "Investigation into the use of a scanning confocal microscope as a primary measurement method for determining optical fibre diameter", *Proc. OFMC-NPL*, York, 1991, pp. 143–146.
[13] Inoue, A., et al., "Self-calibrating precise laser micrometer for optical fibers' diameter measurement", *Proc. OFMC-NPL*, York, 1991, pp. 139–142.
[14] Katagiri, T., et al., "Optical microscope observation method of a single-mode optical-fiber core for precise core-axis alignment", *IEEE J. Lightwave Technol.*, Vol. LT–2, 1984, pp. 277-283.
[15] Ide, A., et al., "Fiber misalignment measurement method by the reflected light from fibers", *IEEE J. Lightwave Technol.*, Vol. LT-8, 1990, pp. 1243–1249.
[16] Brinkman, R. J., et al., "Sideview: a method for measuring the geometrical parameters of optical-fibres", *Proc. OFMC-NPL*, York, 1991, pp. 151–154.
[17] Cancellieri, G., and A. Orfei, "Asymptotic effective cut-off condition in single-mode optical fibers", *IEEE J. Lightwave Technol.*, Vol. LT-5, 1987, pp. 199–205.
[18] Valotinen, T., et al, "The $\alpha 11_b(\lambda)$ method, a way to investigate the bending and bend sensitivity of single mode fibres", *Proc. OFMC-NPL*, York, 1991, pp. 20–23.
[19] Kuyt, G., "COST 217 interlaboratory comparison of uncabled and cabled fibre cut-off wavelength measurements", *Proc. OFMC-NPL*, York, 1991, pp. 12–15.
[20] Danielson, B. L., "Optical time-domain reflectometer specifications and performance testing", *Appl. Opt.*, Vol. 24, 1985, pp. 2313–2322.

[21] Barnoski, M. K., and S. M. Jensen, "Fiber waveguides: a novel technique for investigating attenuation characteristics", *Appl. Opt.*, Vol. 15, 1976, pp. 2112–2115.

[22] Personick, S. D., "Photon probe – An optical time domain reflectometer", *Bell Syst. Tech. J.*, Vol. 56, 1977, pp. 355–366.

[23] Le Boutet, A., "Analysis of backscattering: results applied to cable connection", *Proc. V ECOC*, Amsterdam, 1979, pp. 9.6/1–6.

[24] Costa, B., et al., "Splice loss evaluation by means of the backscattering technique", *Electron. Lett.*, Vol. 15, 1979, pp. 550–551.

[25] Rourke, M. D., "Measurement of the insertion loss of a single microbend", *Opt. Lett.*, Vol. 6, 1981, pp. 440–442.

[26] Hagemann, H. J., et al., "Optical Time-Domain Reflectometry (OTDR) of diameter modulations in single-mode fibers", *IEEE J. Lightwave Technol.*, Vol. LT-8, 1990, pp. 1641–1645.

[27] Barnoski, M. K., et al., "OTDR differential modal attenuation measurements", *Proc. IV ECOC*, Genova, 1978, pp. 133–139.

[28] Horiguchi, T., et al., "An acoustooptical directional coupler for an optical time-domain reflectometer", *IEEE J. Lightwave Technol.*, Vol. LT-2, 1984, pp. 108–115.

[29] Takeda, M., and T. Horiguchi, "Advances in optical time-domain reflectometry", *IEEE J. Lightwave Technol.*, Vol. LT-7, 1989, pp. 1217–1224.

[30] Bernard, J. J., at al., "1.3 μm portable reflectometer for the field test of single-mode fiber cables", *Tech. Dig. Symp. on Opt. Fiber Measurements*, Boulder, 1984, pp. 95–98.

[31] Bernard, J. J., and E. Depresles, "Submeter resolution OTDR for single-mode fiber LAN cabling tests", *Proc. EFOC-LAN 1988*, Amsterdam, pp. 127–130.

[32] Fontaine, J. J., et al., "Subpicosecond time-domain reflectometry", *Opt. Lett.*, Vol. 6, 1981, pp. 405–407.

[33] Youngquist, R. C., et al., "Optical coherence domain reflectometry. A new optical evaluation technique", *Opt. Lett.*, Vol. 12, 1987, pp. 158–160.

[34] Nakazawa, M., et al, "30–km–long fault location for single-mode optical fiber using 1.55 μm Q-Switched $Er^{(3+)}$:glass laser", *Opt. Lett.*, Vol. 9, 1984, pp. 312–314.

[35] Stensland, L., and G. Borak, "Raman time-domain reflectometry", *Proc. IOOC '81*, San Francisco, 1981, pp. 104–105.

[36] Wada, F., et al., "Variable wavelength optical time-domain reflectometry", *Proc. OFC '92*, San José, 1992, p. 150.

[37] Zoboli, M., and P. Bassi, "High spatial resolution OTDR attenuation measurements by a correlation technique", *Appl. Opt.*, Vol. 22, 1983, pp. 3680–3681.

[38] Suzuki, K., et al., "Optical time-domain reflectometer with a semiconductor laser amplifier", *Elecron. Lett.*, Vol. 20, 1984, pp. 714–716.

[39] Blank, L. C., and D. M. Spirit, "OTDR performances enhancement through erbium fibre amplification", *Electron. Lett.*, Vol. 25, 1989, pp. 1693–1694.

[40] Spirit, D. M., and P. J. Keeble, "High resolution optical time domain reflectometry enhancement using erbium-doped fibre amplifiers", *Proc. OFMC-NPL*, York, 1991, pp. 37–40.

[41] Healey, P., "Optical time-domain reflectometry by photon counting", *Electron. Lett.*, Vol. 16, 1980, pp. 631–633.

[42] Ripamonti, G., et al., "Correction of OTDR data for photodetector trapping phenomena", *Proc. OFMC-NPL*, York, 1991, pp. 49–52.

[43] Healey, P., and D. J. Malyon, "OTDR in single-mode fibre at 1.5 μm using heterodyne detection", *Electron. Lett.*, Vol. 18, 1982, pp. 862–863.

[44] Mac Donald, R. I., "Frequency-domain optical reflectometer", *Appl. Opt.*, Vol. 20, 1981, pp. 1840–1844.

[45] Kapron, F. P., et al., "Fiber-optic reflection measurements using OCWR and OTDR techniques", *IEEE J. Lightwave Technol.*, Vol. LT-7, 1989, pp. 1234–1241.

[46] Marcuse, D., *Principles of optical fiber measurements*, Academic Press, New York, 1981.

[47] Danielson, B. L., "Optical time-domain reflectometer specifications and performance testing", *Appl. Opt.*, Vol. 24, 1985, pp. 2313–2322.

[48] Reich, M., "Novel method for linearising the power response of optical time domain reflectometers", *Electron. Lett.*, Vol. 26, 1990, pp. 1029–1030.

[49] Maekawa, E., et al., "Measurement equipment for subscriber optical lines", *Elect. Comm. Lab.*, Vol. 34, 1986, pp. 669–675.

[50] Cox, J. D., et al., "First field demonstration of in-service fault location/supervisory using optical time domain reflectometry", *Electron. Lett.*, Vol. 26, 1990, pp. 110–112.

[51] So, V. C. Y., et al., "Automation of an optical time domain reflectometer to measure loss and return loss", *IEEE J. Lightwave Technol.*, vol. LT-8, 1990, pp. 1078–1083.

[52] Sankawa, I., et al., "Fault location technique for in-service branched optical fiber networks", *Phot. Tech. Lett.*, Vol. 2, 1990, pp. 766–768.

[53] Dawson, J. F., and T. C. Tozer, "Non-intrusive correlation optical time-domain reflectometry", *Proc. OFMC-NPL*, York, 1991, pp. 45–48.

[54] Walker, B., and D. J. Ives, "OTDR calibration techniques", *Proc. OFMC-NPL*, York, 1991, pp. 41–44.

[55] Foucault M., and J. L. Kemystetter, "OTDR calibration technique: splice simulator", *Proc. OFMC-NPL*, York, 1991, pp. 57–60.

[56] Ives, D. J., et al., "OTDR calibration device based on recirculating optical delay line", *Electron. Lett.*, Vol. 27, 1991, pp. 342–343.

[57] Safaai-Jazi, A., and L. J. Lu, "Accuracy of approximate methods for the evaluation of chromatic dispersion in dispersion-flattened fibers", *IEEE J. Lightwave Technol.*, Vol. LT-8, 1990, pp. 1145–1150.

[58] Cohen, L. G., "Comparison of single-mode fiber dispersion measurement techniques", *IEEE J. Lightwave Technol.*, Vol. LT-3, 1985, pp. 958–966.
[59] Cohen, L. G., and C. Lin, "A Universal Fiber-Optic (UFO) measurement system based on a near-IR Raman laser", *IEEE J. Quantum Electron.*, Vol. QE-14, 1978, pp. 855–859.
[60] Lin, C., et al., "Chromatic dispersion measurements in single-mode fibers using picosecond InGaAsP injection lasers in the 1.2 to 1.5 μm spectral region", *Bell Syst. Tech. J.*, Vol. 62, 1983, pp. 457–462.
[61] Barlow, A. J., et al., "Technique for direct measurement of single-mode fiber chromatic dispersion", *IEEE J. Lightwave Technol.*, Vol. LT-5, 1987, pp. 1207–1213.
[62] Costa, B., et al., "Phase-shift technique for the measurement of chromatic dispersion in single-mode optical fibres using LEDs", *Electron. Lett.*, Vol. 19, 1983, pp. 1074–1076.
[63] Sladen, F. M. E., et al., "Chromatic dispersion measurements on long fibre lengths using LEDs", *Electron. Lett.*, Vol. 22, 1986, pp. 841–842.
[64] Bomberger, W. D., et al., "Interferometric measurement of dispersion of a single mode optical fiber", *Electron. Lett.*, Vol. 17, 1981, pp. 495–496.
[65] Saunders, W. J., et al., "Precision interferometric measurement of dispersion in short single mode fibers", *Tech. Dig. Symposium on Optical Fiber Measurements*, NBS Special Publication No. 683, 1984.
[66] Sears, F. M., et al., "Interferometric measurements of dispersion spectra variations in a single mode fiber", *IEEE J. Lightwave Technol.*, Vol LT-2, 1984, pp. 181–184.
[67] Francois, P. L., et al., "Chromatic dispersion measurement from Fourier transform of white-light interference patterns", *Electron. Lett.*, Vol. 23, 1987, pp. 357–358.
[68] Cancellieri, G., *Single-mode optical fibres*, Pergamon Press, Oxford, 1991.
[69] Murata, H., *Handbook of optical fibers and cables*, Marcel Dekker, New York, 1988.
[70] CSELT Tech. Staff., "*Optical fiber communication*", CSELT, 1980.
[71] Mochida, K., et al., "Splicing of SM fibers by V-groove connector", Natl. Conv. 905, *IECE J*, 1977, pp. 4–209.
[72] Lischynsky, S., et al., "Plastic mechanical splice for fiber", *Proc. OFC '91*, San Diego, 1991, p. 140.
[73] Dean, D. L. Jr., "Extreme environmental testing of a new mechanical splice", *Proc. OFC '91*, San Diego, 1991, p. 141.
[74] Finzel, L., "The finger splice – A mechanical V-groove splicing technique for optical fibers", *Proc. VIII EFOC-LAN*, Munich, 1990, pp. 165–169.
[75] Shah, V., et al., "Cleaved oblique fiber end faces for high-performance mechanical splices", *Proc. OFC '91*, San Diego, 1991, p. 111.
[76] Ide, A., et al., "Fiber misalignment measurement method by the reflected light from fibers", *IEEE J. Lightwave Technol.*, Vol. LT-8, 1990, pp. 1243–1249.

[77] Kawata, O., et al., "Splicing and inspection technique for single-mode fibers using direct core monitoring", *IEEE J. Lightwave Technol.*, Vol. LT-2, 1984, pp. 185–191.

[78] Zheng, W., and O. Hulten, "Core alignment technique based on image processing of warm fibers", *Proc. OFC '92*, San José, 1992, p. 230.

[79] Yoshizawa, N., et al., "Strength improvement and fusion splicing for carbon-coated optical fiber", *IEEE J. Lightwave Technol.*, Vol. LT-9, 1991, pp. 417–421.

[80] Zheng, W., "Loss estimation for fusion splices of single-mode fibers", *Proc. Int. SPIE '91 Conference*, Boston, 1991.

[81] Suzuki, I., et al., "Splicing of 16 arranged fibers with arc fusion", *Proc. XV ECOC*, Paris, 1991, pp. 325–328.

[82] Kato, Y., et al., "Arc-fusion splicing of single-mode fibers: a highly efficient splicing technique", *Appl. Opt.*, Vol. 23, 1984, pp. 2654–2659.

[83] Costa, B., et al., "Splice loss evaluation by means of the back-scattering technique", *Electron. Lett.*, Vol. 15, 1979, pp. 550–551.

[84] Matthijsse, P., and C. M. De Blok, "Field measurement of splice loss applying the backscattering method", *Electron. Lett.*, Vol. 15, 1979, pp. 795–797.

[85] Gold, M. P., et al., "New approach to splice-loss monitoring using long-range OTDR", *Electron. Lett.*, Vol. 20, 1984, pp. 338–340.

[86] Marcuse, D., "Loss analysis of single-mode fibre splice", *Bell Syst. Tech. J.*, Vol. 56, 1977, pp. 703–718.

[87] Furukawa, S. I., and Y. Koyamada, "New simple one-end splice loss measurement method using an OTDR for single-mode optical fibre", *Electron. Lett.*, Vol. 22, 1986, pp. 535–537.

[88] Sugita, E., et al., "SC-type single-mode optical fiber connectors", *IEEE J. Lightwave Technol.*, Vol. LT-7, 1989, pp. 1689–1696.

[89] Satake, T., et al., "Very small single-mode ten-fiber connector", *IEEE J. Lightwave Technol.*, Vol. LT-6, 1988, pp. 269–272.

[90] Ashiya, F., et al., "Development of multifibre connector and their application", *Proc. IX EFOC-LAN*, London, 1991, pp. 304–308.

[91] Nagasawa, S., et al., "A single-mode multifiber push-on type connector with low insertion- and high return-losses", *Proc. XV ECOC*, Paris, 1991, pp. 49–52.

[92] Shah, V., et al., "Large fluctuations in transmitted power at fiber joints with polished endfaces", *Proc. OFC/IOOC '87*, Reno, 1987, paper TUF4.

[93] Kosiorska, H., et al., "Statistical method for predicting connector loss in the field using factory measurements", *Proc. OFC '91*, San Diego, 1991, p. 138.

[94] Das, S. K., et al., "Reflectance measurement in lightwave systems: a comparison of various techniques", *Proc. Tech. Dig. Symp. on Opt. Fiber Measurements*, Boulder, 1988, pp. 25–30.

[95] Das, S. K. et al., "Optical time-domain and CW techniques measure lightwave reflectance", *Laser Focus/Electro-Opt.*, 1988, pp. 129–134.

[96] Knecht, D. M., "Optical fiber connectors and splices", *Proc. OFC'92*, San Jose', 1992, p. 26.

[97] Richter, H., and H. Gruhl, "High dynamic optical far-field measurements to investigate the roughness of front faces of single-mode connectors", *Proc. OFMC-NPL*, York, 1991, pp. 65–68.

[98] Miller, S. E., and I. P. Kaminow, *Optical fiber telecommunications II*, Academic Press, San Diego, 1988.

[99] De Vries, W. J., et al., "Optical characterization and measurements of single-mode branching devices", *Proc. OFMC-NPL*, York, 1991, pp. 61–64.

[100] Poole, S. B., et al., "Characterization of special fibers and fiber devices", *IEEE J. Lightwave Technol.*, Vol. LT-7, 1989, pp. 1242–1255.

[101] Poole, S. B., et al., "Fabrication and characterization of low-loss optical fibers containing rare-earth ions", *IEEE J. Lightwave Technol.*, Vol. LT-4, 1986, pp. 870–876.

[102] Morkel, P., et al., "Spectral variation of excited-state absorption in neodymium-doped fiber laser", *Opt. Commun.*, Vol. 67, 1988, pp. 349–352.

[103] Evankow, J. D., and R. M. Jopson, "Nondestructive measurement of length dependence of gain characteristics in fiber amplifiers", *Phot. Tech. Lett.*, Vol. 3, 1991, pp. 993–995.

[104] Blank, L. C., "Optical time domain reflectometry on optical amplifier systems", *IEEE J. Lightwave Technol.*, Vol. LT-7, 1989, pp. 1549–1555.

[105] Spirit, D. M., et al., "Optical time domain reflectometry as a diagnostic for distributed optical fibre amplifiers", *Proc. OFMC-NPL*, York, 1991, pp. 33–36.

[106] Sato, Y., and K. Aoyama, "OTDR in optical transmission systems using Er-doped fiber amplifiers containing optical circulators", *Phot. Tech. Lett.*, Vol. 3, 1991, pp. 1001–1003.

[107] Mikami, O., Y. Noguchi, H. Yasaka, K. Magari, and S. Kondo, "Emission spectral width broadening for InGaAsP/InP superluminescent diodes", *IEE Proc. J*, Vol. 138, 1991, pp. 133–137.

[108] Hertsens, T., "Measuring diode laser characteristics", *Laser & Optron.*, Vol. 8, 1989, pp. 37-42.

[109] Puleo, M., and E. Vezzoni, "Caratterizzazione di componenti optoelettronici per sistemi di telecomunicazione in fibra ottica", *CSELT Technical Report*, No. 77, 1984.

[110] Straus, J., "The nonlinearity of high-radiance light-emitting diodes", *IEEE J. Quantum Electron.*, Vol. QE-14, 1978, pp. 813–819.

[111] Ito, H., H. Yokoama, S. Murata and H. Inaba, "Generation of picosecond optical pulses with highly RF modulated AlGaAs DH laser", *IEEE J. Quantum Electron.*, Vol. QE-17, 1981, pp. 663–670.

[112] Torphammar, P., R. Tell, H. Eklund, and A. R. Johnston, "Minimizing pattern effects in seminconductor lasers at high rate pulse modulation", *IEEE J. Quantum Electron.*, Vol. QE-15, 1979, pp.1271–1276.

measurements coincide for k values constant over the pulse. For high-speed single-mode fiber systems, k ranges between 0.4 and 0.7 and represents one of the performance limiting factors in intensity-modulation transmission systems.

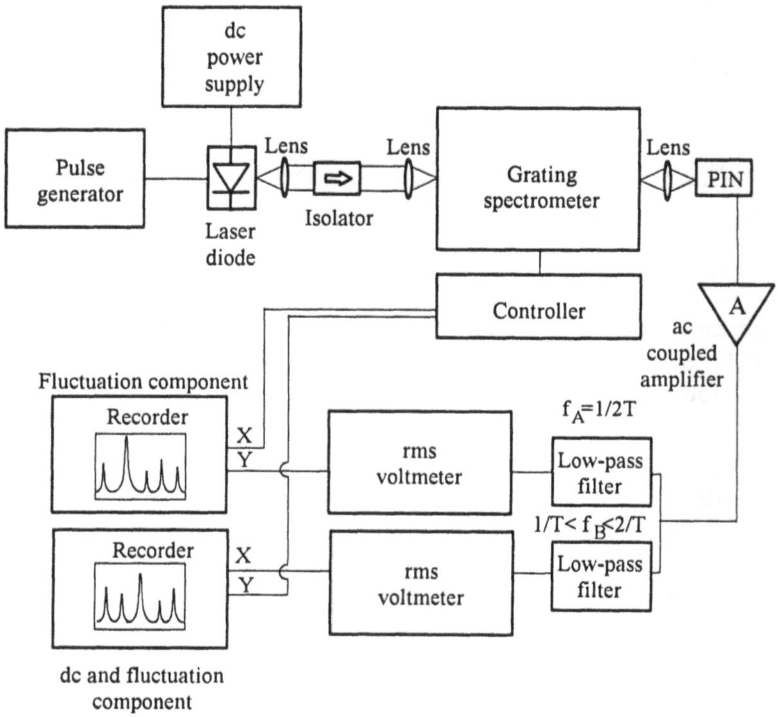

Figure 3.38 The experimental setup used to measure the mode partition noise of a laser diode using the low-pass filter method.

3.3 MEASUREMENTS OF PHOTODIODES

Photodetectors used with fiber optics are mainly PINs and APDs. A description of these devices is given in Section 2.3.2. In the following the measurement techniques to determine several parameters of these detectors are discussed. Most of the measurements, as that of the dark current, responsivity, frequency response, and nonlinear distortion, apply both to PINs and APDs. Section 3.3.4 addresses some specific measurements of APDs.

[113] Sze, S. M., *Physics of semiconductor devices*, second edition, Wiley, New York, 1981, p. 117.

[114] Lin, C., and J. E. Bowers, "Measurement of 1.3 and 1.55 μm gain-switched semiconductor laser pulses with a picosecond IR streak camera and a high-speed InGaAs PIN photodiode", *Electron. Lett.*, Vol. 21, 1985, pp. 1200–1202.

[115] Ito, H., H. Yokayama, S. Murata, and H. Inaba, "Picosecond optical pulse generation from an R.F. modulated AlGaAs D.H. diode laser", *Electron. Lett.*, Vol. 15, 1979, pp. 738–740.

[116] Tarucha, S., and K. Otsuka, "Response of semiconductor laser diode to deep sinusoidal injection current modulation", *IEEE J. Quantum Electron.*, Vol. QE-17, 1981, pp. 810–816.

[117] Jenkins, A. F., and H. E. White, *Fundamentals of optics*, fourth edition, McGraw-Hill, London, 1981, pp. 357–377.

[118] Born, M., and E. Wolf, *Principles of optics*, sixth edition, Pergamon Press, Oxford, 1980, pp. 334 et seq.

[119] Ogawa, K., "Analysis of mode partition noise in laser transmission systems", *IEEE J. Quantum Electron.*, Vol. QE-18, 1982, pp. 849–855.

[120] Ogawa, K., and R. S. Vodhanel, "Measurements of mode partition noise of laser diodes", *IEEE J. Quantum Electron.*, Vol. QE-18, 1982, pp. 1090–1093.

[121] Puleo, M., and E. Vezzoni, "An automatic testing system for avalanche photodetectors", *CSELT Technical reports*, Vol XI, 1983, pp. 163–170.

[122] Kawanishi, S., and M. Saruwatari, "Wide band frequency response measurement of photodetectors using optical heterodyne detection technique", *Electron. Lett.*, Vol. 22, 1986, pp. 337–338.

[123] Higgins, T. W., "Unique testbed evaluates high-speed photodetectors", *Laser Focus Word*, Vol. 28, 1992, pp. 27–28.

[124] Ozeki, T., and E. H. Hara, "Measurement of nonlinear distortion in photodiodes", *Electron. Lett.*, Vol. 12, 1976, pp. 80–81.

[125] Kanbe, H., T. Kimura, W. Mitsushima, and K. Kajiyama, "Silicon avalanche photodiodes with low multiplication noise and high speed response", *IEEE Trans. Electron. Devices*, Vol. ED-23, 1976, pp. 1337-1343.

Chapter 4

Measurements of Fiber Optics in Coherent Light

A. Galtarossa

E. Nava

G. Valentini

Among the coherent light measurements of optical fibers, we have taken only two examples in order to describe the main problems of such kind of test. These examples regard a particular kind of backscattering and polarization mode dispersion. Many other experimental setups could be discussed. Nevertheless they would appear oriented to particular problems, without giving any further insight into the aspects related to the use of coherent light.

On the other hand, in coherent light measurements on active optical devices, the generated or processed light is the object to be measured. Therefore a greater attention has been devoted to describe all the features regarding this light, such as:

1. emission spectrum,
2. intensity fluctuations,
3. 1/f noise.

These properties characterize the unmodulated light.

As for the modulated light, pulse and frequency response, and chirping appear as the most interesting parameters to be experimentally evaluated.

At the same time, the possibility of light direct amplification is investigated again, with particular attention to the selective properties of the optical gain and to the noise figure.

4.1 COHERENT LIGHT FIBER CHARACTERIZATION

In Section 3.1 we have described suitable methods for measuring attenuation loss on fibers, connectors, splices, and so on. All the methods were based on incoherent light sources, which are the most popular and diffused techniques. Such sources are generally less expensive and easier to be used either in factory or in the field. Moreover, coherent properties of the light can affect results in some of the measurements described earlier (this occurs, for example, in connector loss evaluation). On the other hand, in some situations coherent light use is necessary for particular purposes. This is the case of coherent backscattering and polarization mode dispersion measurements.

4.1.1 Coherent Backscattering

OTDR is presently the most common technique for evaluating loss characteristics of optical fiber links. As described in Section 3.1.5, this method is based on direct detection of the fraction of a probe pulse that is scattered back. However, the backscattered light has a very low level; consequently it is necessary to improve, in a significant manner, the signal-to-noise ratio. This can be done in various ways: enlarging the peak level of the probe pulse, enhancing detection sensitivity, or introducing a modulation technique in the transmitted signal. All these solutions employ an incoherent optical source intensity modulated.

Coherent backscattering is an alternative proposal in which the improvement of the S/N ratio of the backscattered light is obtained by means of coherent detection. Coherent detection is a linear process in which the receiver output current is proportional to the optical electric field incident upon the photodiode, as described in Section 2.3.4.

One of the disadvantages of coherent transmission systems is the necessity of employing an optical transmitter and a local oscillator, installed far apart from each other, with the same optical and electromagnetic characteristics. In coherent reflectometry, transmission and detection operations are necessarily both present in the same apparatus, and the same light source acts either as a transmitter or as a local oscillator. The backscattered electric field $E_{bs}e^{i\omega_o t}$ is combined with a high-level $E_{LO}e^{i(\omega+\omega_o)t}$ signal generated by a local oscillator. The two fields must have parallel polarizations, so that they can be vectorially superimposed. The total field detected by the photodiode is then

$$E_{TOT} = E_{bs}e^{i\omega_o t} + E_{LO}e^{i(\omega+\omega_o)t} \tag{4.1}$$

As the photodiode has a quadratic response, the output photocurrent is proportional to the square of the total field incident on the photodiode itself. Considering that the backscattered signal is very small compared to the electric field locally generated, the photocurrent exhibits two main components: a dc term due to the local oscillator and an ac current signal at the difference frequency ω whose amplitude is proportional to the field amplitude and not to the power amplitude. The mean square value of the photocurrent ac term is then

$$\langle i_{bs}^2 \rangle = 2P_{bs}P_{LO}\left(\frac{q\eta_o}{h\nu_o}\right)^2 \tag{4.2}$$

where P_{bs} and P_{LO}, respectively, are the backscattered power and the power generated by the local oscillator, q is the electron charge, η_o is the quantum efficiency of the photodiode, $h\nu_o = h\omega_o/2\pi$ is the photon energy.

The noise has three main sources: shot noise due to the local oscillator, shot noise due to the dark current of the photodiode, and circuit receiver noise. Under the assumption that the largest noise contribution is given by the local oscillator, the signal-to-noise ratio is expressed as [1]

$$\frac{S}{N} = \frac{\eta_o P_{bs}}{h\nu_o B} \tag{4.3}$$

where B is the baseband bandwidth.

The main difficulty for coherent backscattering is due to the possibility of having an optical coherent light source that is stable in time and has a very narrow linewidth. Up to now, some equipments were made using Fabry-Perot laser diodes with an external cavity, DFB laser diodes, or Raman lasers.

A first proposal, shown in Figure 4.1 [1], is based on a semiconductor laser diode coupled to a piece of single-mode fiber a few hundreds meters long acting as low reflectivity external resonator. Linewidth as narrow as 200 Hz has been obtained. Fluctuations in the fiber parameters determine frequency hopping inside the cavity several times per second. The range of frequencies on which the laser may lock is defined by the single-mode linewidth of the laser without external cavity.

The optical wave, generated by a CW laser diode, is divided, using a fiber directional coupler, into two beams. The first one, passes through a pulse modulator, and is then launched into the test fiber. The second beam acts as the local oscillator. Coherent detection improves sensitivity, compared to traditional OTDR. However, polarization fading produces an uncertainty into signal detection, which is of the same order of splice losses. Fading sources are the intrinsic birefringence of single-mode fibers and the random nature of Rayleigh backscattering. Their effects are strongly reduced by sampling a large number of independent Rayleigh and polarization states. In the equipment proposed in [1], the frequency hopping of the reference optical source allows signal fading reduction. In fact, when the source hops in frequency, the phases of the signal due to different scattering source changes, permitting an independent sampling of each contribution.

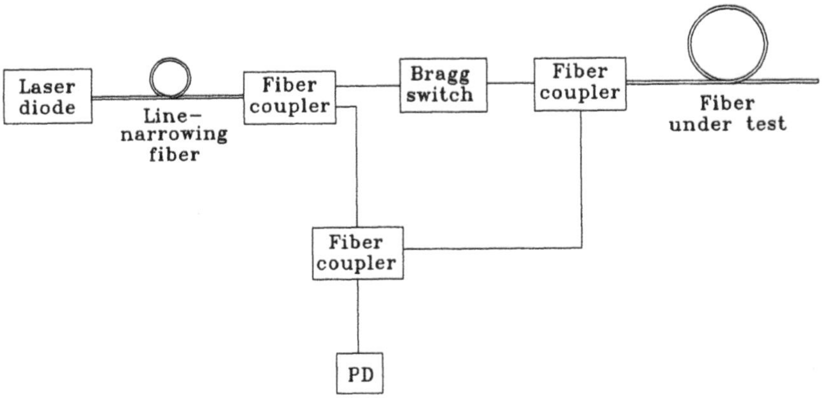

Figure 4.1 Experimental setup for coherent reflectometer (adapted from [1]).

The second proposal is a technique called coherent frequency modulated continuous wave reflectometry [2]. The experimental setup is shown in Figure 4.2. The optical source is a diode-pumped, single-frequency monolithic Nd:YAG ring laser, characterized by a very long coherence length and a low phase noise, oscillating in only one direction. The main advantages of this solution are: large dynamic range (due to coherent detection and low phase noise on the optical carrier); very narrow receiver bandwidth and, consequently, very small noise; high resolution; absence of dead-zone; and the possibility of characterizing active fibers without the risk of saturation. It is important to note that the long coherent length of the optical source enables one to measure fiber pieces several km long and reduces, at the same time, phase noise. Local oscillator signal is obtained from the power reflected by a reference arm. The signals originated by local oscillator and test fiber are then mixed, producing a beat frequency proportional to the differential time delay between them. Measuring the frequency spectrum of the detected current by means of an electrical spectrum analyzer, the attenuation behavior of the test fiber is derived. The proportionality factor between the position inside the fiber and the beat frequency is determined by a change (with time) of the optical frequency. In this solution, optical frequency chirping is obtained by tuning the temperature of the Nd:YAG crystal. This is reached, in turn, by means of a ramping signal generated by a temperature controller. This causes a frequency sweep rate of about 260 MHz/s, that is, relatively small. The main drawback of such technique is, however, that a Nd:YAG ring laser is not suitable for field use.

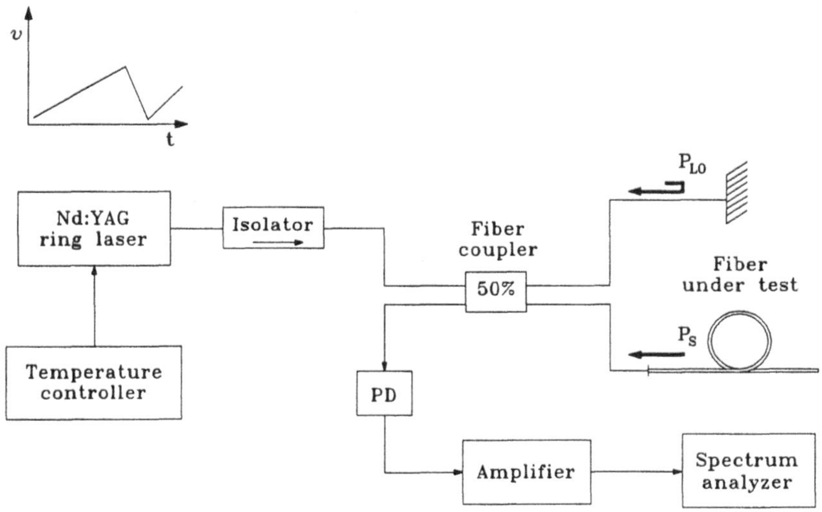

Figure 4.2 Experimental setup for coherent frequency modulated continuous wave reflectometry (adapted from [2]).

To overcome this problem, the employment of a DFB laser diode as the light source has been proposed [3], where frequency sweeping is done acting on the injection current. Nevertheless, DFB laser diodes have a shorter coherence length, compared to Nd:YAG ring laser. As a consequence, high-spatial resolution measurement are performed only over a specific fiber section, for example, two splices or two connectors close to each other. In order to obviate the short coherence length of the DFB laser diode, it is possible to use a delay line fiber inserted in the reference optical arm. Spatial resolution of about 0.5 m has been achieved for the backscattered signal from the far end of a 0.5–km–long test fiber. Higher resolution can be reached by using a more accurate frequency sweep.

An interesting application of coherent heterodyne detection [4] for optical time reflectometry of a long fiber link including semiconductor amplifiers has been proposed. A schematic diagram of such setup is reported in Figure 4.3. It is based on FSK modulation. A DFB laser diode is driven by a pulse generator. There are two signals: a reference signal concerning the local oscillator (at the shifted frequency f_r), and the backscattering signal (at the frequency f_b). The light generated by the DFB laser diode is divided into two beams by a fiber coupler. The upper beam is launched into the test fiber, while the other one acts as the local oscillator. Backscattering signal passes through fiber coupler I and is combined with the local oscillator signal by means of fiber coupler II. Such signals are detected by a PIN transimpedance receiver. Backscattering signal at the frequency f_r is converted down at the intermediate frequency $|f_r - f_b|$, while backscattered signal at the frequency f_b is converted at a zero frequency signal, that is out of the heterodyne receiver bandwidth. The detected signal is then demodulated and sent to an averaging unit.

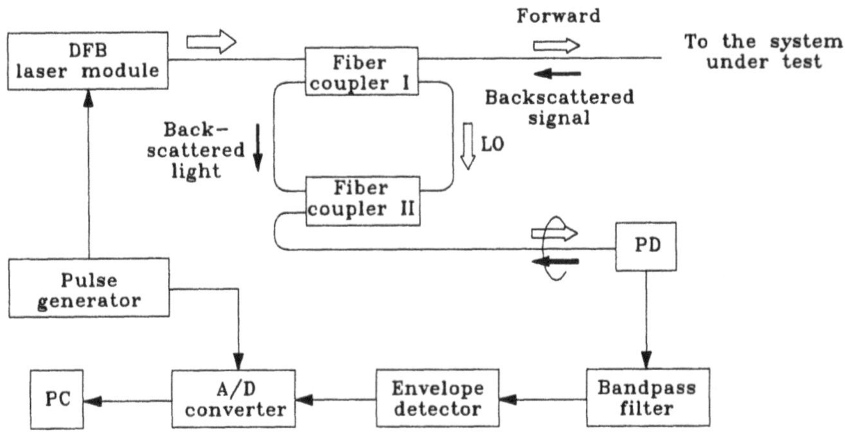

Figure 4.3 Experimental setup based on coherent heterodyne detection reflectometry (adapted from [4]).

Further improvements in coherent reflectometry performances have been recently obtained using active fiber amplifiers. For example, operating at $\lambda = 1.55$ μm, an equipment has been realized with a dynamic range of 33 dB, 1 μs probe pulse, and 1 MHz receiver bandwidth. Integration over 10^4 measurement samples leads to an improvement of about 13 dB with respect to traditional OTDR with the same light source [5]. Another recent proposal [6], still based on active fiber amplifiers, both as postamplifiers and in-line amplifiers, demonstrated a 40-dB dynamic range. Also in this application, the laser diode linewidth was accurately controlled using techniques similar to those described in [1]. The laser diode was coupled to a 1-km fiber piece to reduce source linewidth. In fact, microscopic refractive index fluctuations inside the fiber act as scattering centers and the backscattered signal is given by the sum of the individual scatters contributions. The amplitude of such a signal depends on the relative phases of the backscattered individual signals. As a result, the fiber operates as a low reflectivity external resonator and is able to reduce the intrinsic laser linewidth. Such fiber piece has been wound around a vibrating support for frequency source hopping enhancement.

4.1.2 Polarization Mode Dispersion: General Aspects

As demonstrated in Section 2.5.2, single-mode optical fibers always exhibit a more or less strong modal birefringence. This is due to undesired geometrical and material effects on the propagation constant of the two polarization states of the fundamental mode. Differences in the group velocities of the two polarization states induce pulse broadening. This effect is generally very small in comparison with chromatic dispersion contribution. However, polarization mode dispersion could become important in multigigabit systems, in which chromatic dispersion is strongly reduced by using a highly coherent light source whose wavelength is very close to the fiber zero chromatic dispersion wavelength. In direct detection systems the phenomenon is much less evident. In coherent systems, instead, it is also accompanied by depolarization.

A suitable method to analyze modal birefringence in uniform single-mode fiber has already been introduced. In this model the two orthogonal polarization eigenstates have their own propagation constants; the difference between the two propagation constants is proportional to the azimuthal asymmetry in the index of refraction. However, only fibers having short lengths and no external perturbations can be modeled in this way because mode coupling has, in them, negligible effects and the two polarization states propagate separately. Furthermore, splices and connectors rotate randomly fiber axes, so that polarization coupling is unavoidable in long optical links.

Polarization mode dispersion is well-characterized by means of the differential group delay between the two polarization states. For a fiber of short length L, the differential group delay [7] can be written as

$$\Delta\tau = -L\frac{\lambda^2}{2\pi c}\frac{d}{d\lambda}\Delta\beta \tag{4.4}$$

where $\Delta\beta$ is the difference between the propagation constants of the two polarizations. It is evident that polarization mode dispersion increases proportionally to the fiber length, when the fiber is a few meters long. On the other hand, a longer fiber, which presents power coupling between the two polarization states, has $\Delta\tau$ which evolves as the square root of the length [8].

As previously introduced, modal birefringence in single-mode fibers is due to both geometrical and stress contribution. The geometrical contribution produced by core ellipticity can be calculated using the following relation [7, 9]

$$\Delta\tau_g = \frac{ne^2[2\Delta]^2}{c}D_w(\lambda^*)\,L \tag{4.5}$$

where Δ is the equivalent step-index difference between core and cladding, $D_w(\lambda^*)$ is the normalized waveguide dispersion factor, n is the refractive index, which is almost the same for core and cladding, and e is the ellipticity. Observation of the far-field measurement [9] across the transverse orthogonal axis of the fiber core, permits to relate the core ovality to $\Delta\tau_g$. Although the prediction based on core ovality gives a value usually smaller than the measured one, a strong correlation between geometrical properties and the $\Delta\tau_g$ behavior of the fiber had been demonstrated.

Polarization coupling modifies polarization mode dispersion in long fiber cables [10, 11] according to the following properties:

1. the average dispersion is reduced, in comparison with the ideal case (without coupling);
2. fiber dispersion properties depend on environmental conditions, when a highly coherent source is employed;
3. polarization coupling is a random process and, as a consequence, the dispersion is analyzed from a statistical point of view.

Due to birefringence, a single-mode fiber becomes a two-mode waveguide because the two polarizations of the fundamental LP_{01} mode are no longer degenerate. If we consider the two phase constants β_j (j = x, y), we can expand them as a Taylor series around the central frequency ω_o as

$$\beta_j(\omega) = \beta_j(\omega_o) + \beta_j'(\omega_o)\Delta\omega + \frac{1}{2}\beta_j''(\omega_o)\Delta\omega^2 \qquad (4.6)$$

The term β_j' determines the group delay without signal distortion. When $\beta_j'' \approx 0$ the fiber is almost chromatic dispersion free, and the limiting factor to transmit the highest bit/rates is imposed by intersymbol interference due to polarization mode dispersion, which in turn is originated by $\beta_x' \neq \beta_y'$.

The principal states of polarization [12] constitute a basis set that completely describes polarization dispersion in a fiber in the presence of coupling. It has been observed that in any linear optical transmission medium with polarization-independent losses there are orthogonal output states of polarization, which correspond to orthogonal input states of polarization, wavelength-independent at first-order approximation. Principal states of polarization have the same role that polarization eigenmodes have in a fiber without coupling. The main difference between the two descriptions is that principal states of polarization depend on the birefringence distributed along the whole fiber. In the case of propagation through the fiber without mode coupling, the principal states of polarization coincide with its eigenmodes. Assuming that the light source has a linewidth narrow enough, the Jones matrix representing the fiber, which is frequency dependent, can be expanded around the central emission frequency of the light source.

If we consider the fiber as a linear medium described by the complex transfer matrix $T(\omega)$, the output electric field $\overline{E}_o$ is obtained from the input electric field $\overline{E}_i$ as follows

$$\bar{E}_o = T(\omega)\bar{E}_i \tag{4.7}$$

The transfer matrix T(ω), for a medium characterized by polarization-independent loss, is expressed as

$$T(\omega) = e^{\gamma(\omega)}U(\omega) \tag{4.8}$$

where γ(ω) is complex, while U(ω) is a unitary matrix. So we have

$$T(\omega) = e^{\gamma(\omega)}\begin{bmatrix} u_1(\omega) & u_2(\omega) \\ -u_2^*(\omega) & u_1^*(\omega) \end{bmatrix} \tag{4.9}$$

in which the matrix elements satisfy the constraint

$$|u_1|^2 + |u_2|^2 = 1 \tag{4.10}$$

When U(ω) is independent of z (uniform fiber), then U(ω) admits two eigenvectors, corresponding to the polarization eigenstates. The two polarization states are a basis for the representation of the electromagnetic field. When the fiber is no longer uniform, the two eigenstates are not suitable to represent the electromagnetic field. In this case, it is necessary to introduce a new basis. It can be done by means of the *principal states of polarization* (PSPs). The input PSPs are

$$\varepsilon_{i\pm} = \frac{1}{D_\pm}\begin{bmatrix} u_2' & -ik_\pm u_2 \\ -u_1' & ik_\pm u_1 \end{bmatrix} \tag{4.11}$$

where apostrophes indicate differentiation with respect to frequency, while

$$k_\pm = \pm k = \pm\sqrt{|u_1'|^2 + |u_2'|^2} \tag{4.12}$$

and

$$D_{\pm} = \sqrt{2k_{\pm}[k_{\pm} - \text{Im}(u_1{}^*u_1' + u_2{}^*u_2')]} \tag{4.13}$$

The differential group delay between the output PSPs is

$$\Delta\tau = k_+ - k_- \tag{4.14}$$

and the correspondent group delays $\tau_{\pm}$ then become

$$\tau_{\pm} = \beta' + k_{\pm} \tag{4.15}$$

where $\beta = \text{Im}(\gamma)$.

The orthogonality of the input and output principal states of polarization makes it possible to subdivide any input wave into two waves that, after propagation through the entire fiber length, remain separate. A pair of output signals are generated whose shape coincides with that of the input signal. The two output waves arrive at different times, determining pulse broadening that affects performances both for direct and coherent systems. In the frequency interval where PSP's definition is valid, no depolarization occurs. As shown earlier, principal states are not affected by depolarization up to the first order. It means that when the polarization state of an input signal coincides with one of the principal states of polarization, then the minimum system penalty due to polarization mode dispersion is achieved [12].

The transmission matrix of a long fiber link is then simply obtained from the product of the matrices corresponding to the single fiber pieces included in the cascade. The large number of parameters makes quite difficult the calculation of PSPs and their differential group delays. Operating some changes in the vector base [13], it is possible to simplify the analysis.

The general equation describing the evolution of the PSPs is expressed by [14]

$$\frac{\partial\bar{\Omega}(L, \omega)}{\partial L} = \frac{\partial\bar{W}(L, \omega)}{\partial\omega} + \bar{W}(L, \omega) \times \bar{\Omega}(L, \omega) \tag{4.16}$$

where $\bar{\Omega}$ is the Stokes vector representing the output positive PSPs, whose modulus, $|\bar{\Omega}|$, is equal to the differential group delay and $\bar{W}$ is the vector representing local modal birefringence and mode coupling.

Because of fiber longitudinal perturbation, $\bar{W}$ can be described as a random process and, as a consequence, PSPs and $\Delta\tau$ can be considered only in a statistical sense, by means of a probability function $P(\Delta\tau, L)$. Statistical description of PSPs is based on the definition of the correlation length L_c of the random mode coupling affecting the fiber. From a physical point of view, L_c represents the distance on which half the power launched on one fiber mode is coupled to the other mode. In all the cases where $L >> L_c$, the probability function is well represented by a Maxwell distribution [13]

$$P(\Delta\tau, L) = \frac{2\Delta\tau^2}{\sqrt{2\pi}\, q^3} e^{-\Delta\tau^2/(2q^2)} \tag{4.17}$$

where $q^2 = \tau_u{}^2 L L_c$ is its variance, having noted by τ_u the differential group delay per unit length of the unperturbed fiber. The differential group delay $\Delta\tau$ has mean value and standard deviation respectively given by

$$\langle\Delta\tau\rangle = \sqrt{\frac{8}{\pi}}\, \tau_u \sqrt{LL_c} \tag{4.18}$$

$$\sigma = \sqrt{\frac{3\pi - 8}{\pi}}\, \tau_u \sqrt{LL_c} \tag{4.19}$$

On the contrary, when $L << L_c$, the differential group delay is given by $\langle\Delta\tau\rangle = L\tau_u$, with $\sigma \approx 0$. From these relationships it is evident that the polarization dispersion increases linearly with the fiber length when the correlation length is much longer than the fiber length, while it evolves as the square root of the length in the opposite case. So, low birefringent fibers that are typically employed in telecommunication systems exhibit a polarization dispersion which increases as $\sqrt{L}$. A more complete analytical study on PMD can be found in [15].

4.1.3 Polarization Mode Dispersion: Measurement Techniques

The measurement of the parameter $\Delta\tau$ between the PSPs of a single-mode fiber can be performed by means of a Poincaré sphere. In fact, PSP evolution can be observed on a Poincaré sphere, as discussed in 2.1.2, and $\Delta\tau$ can be evaluated together with the bandwidth in which PSP theory is valid.

Starting from the PSP theory [12], the evolution of the state of polarization is described by [11, 16]

$$\frac{\partial \bar{S}(L, \omega)}{\partial \omega} = \bar{\Omega}(L, \omega) \times \bar{S}(L, \omega) \tag{4.20}$$

where $\bar{S}$ is the Stokes vector representing the output state of polarization and $\bar{\Omega}$ corresponds to the output positive PSP. According to [14], in the frequency interval of validity of PSP's approximation, $\bar{\Omega}$ is constant and a frequency change $\Delta\omega$ induces a rigid rotation of a Poincaré sphere around the vector $\bar{\Omega}$ by an angle $\phi = \Delta\tau\Delta\omega$ (see Figure 4.4). This property permits to obtain the differential group delay from the experimental evaluation of ϕ. In fact, varying the frequency of the input field, the point corresponding to the output state of polarization describes, on a Poincaré sphere, an arc of circle around Ω (see Figure 4.5). Some setups for polarization mode dispersion measurement based on this assumption have been proposed. They differ mainly in the choice of the light source. In any case this source must be tunable.

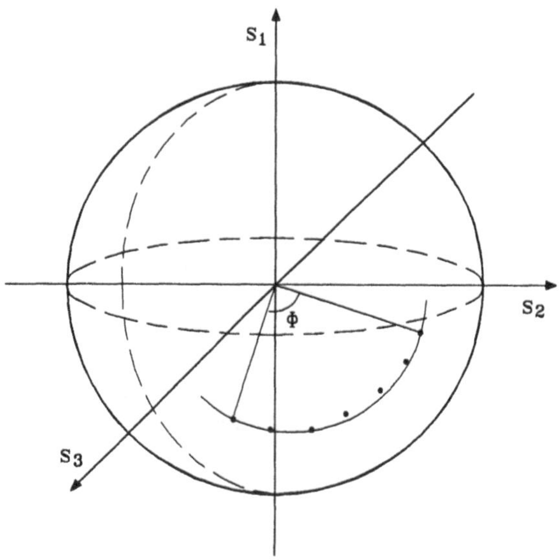

Figure 4.4 Polarization evolution versus wavelength on a Poincaré sphere.

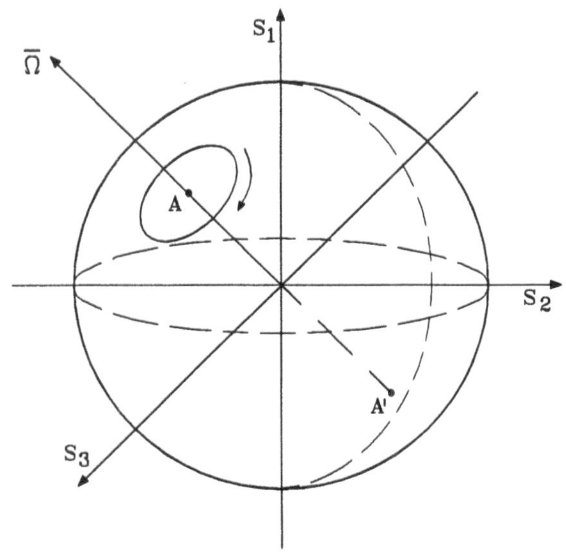

Figure 4.5 Rigid rotation of a Poincaré sphere around $\bar{\Omega}$.

A general setup for $\Delta\tau$ measurements is shown in Figure 4.6. The launching system is composed by a tunable light source and a polarization controller used for rotating the input state of polarization. The detection part is composed of a rotating linear polarizer, a photodiode, and a PC dedicated to data acquisition and processing. The polarization controller, made by means of a fiber loop, permits modification of the input state of polarization. It is needed, in particular, when the input state of polarization coincides with one of the PSP and no polarization dispersion would occur. Sometimes, a $\lambda/4$ plate and a $\lambda/2$ plate are included in the detection part. The former allows detection of the sense of polarization (right or left), while the latter enables us to rotate the output state of polarization if necessary. This is, in general, elliptical (circular and linear polarization can be considered as particular cases). The rotating linear polarizer, placed between the output fiber face and the photodiode, scans the elliptical output state of polarization. The current signal at the photodiode output exhibits, as a consequence, a sinusoidal shape.

The possibility of frequency tuning, which is required by the optical source, determines, at a fixed input state of polarization, changes in amplitude and phase of the elliptic output polarization state and, in turn, of the detected current. So, at any frequency with a fixed input polarization state, a different point on a Poincaré sphere exists, which represents the output state of polarization. All the points stand on an arc of circle as described earlier. From the knowledge of the tuning frequency range and from the measurement of the corresponding angle on an arc of a Poincaré sphere, it is possible to determine the differential group delay.

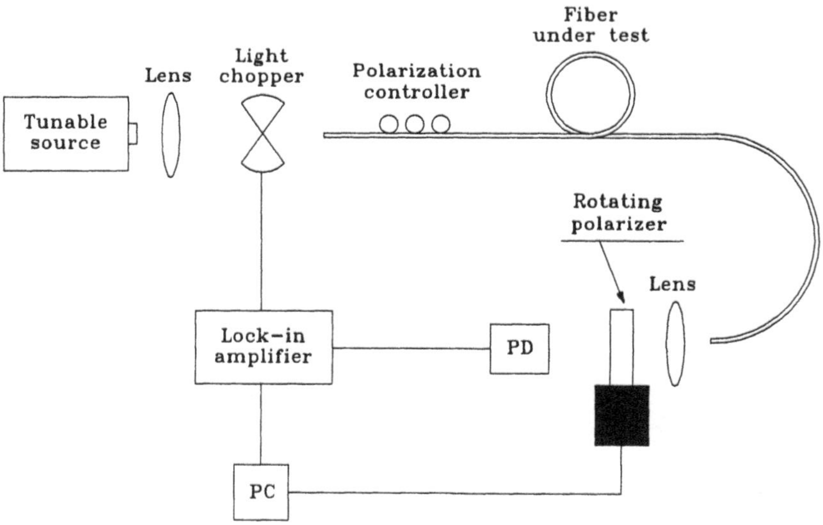

Figure 4.6 General setup for differential group delay measurement.

A few suitable tunable light sources have been proposed, although they appear more useful in laboratory tests than for field measurements. In [14] the source consists of a tunable external cavity laser with 45 nm of tunable range around 1550 nm. In [13, 17] the coherent source was chosen as a color center laser pumped by a neodymium laser. The frequency of the color center laser was varied in step of 31 GHz over two wavelength intervals set at 1552 and 1557 nm. An alternative proposal is based on an arrangement composed by a multi-mode laser diode and a monochromator.

All these solutions are dedicated to laboratory tests because of the complexity of the light source arrangement.

A more attractive solution for field tests is based on temperature tuned laser diode [18–21]. Considering that the wavelength tunability versus temperature is about 10–15 GHz/°C, it is possible to obtain a tunable source over a few hundreds GHz, typically 4 to 6, with a linewidth smaller than 1 GHz. The experimental setup is shown in Figure 4.7.

Measurements have been performed on standard single-mode fibers with the zero-dispersion wavelength at about 1300 nm, and on dispersion-shifted fibers at 1550 nm. Typical values obtained for differential group delay are in the range 0.1–1 ps/$\sqrt{\text{km}}$ [17, 18, 20].

When the light source is tunable over a large frequency range, as in the case of a color center laser or of an external cavity laser, it is possible to obtain the probability density function of the differential group delay repeating measurements of $\Delta\tau$ at different wavelengths [13]. The same measurement can be performed using a white lamp and a monochromator [22]. In this case the polarizer can be maintained as fixed and the statistics of the differential group delay can be obtained by tuning the source wavelength over a large spectral range.

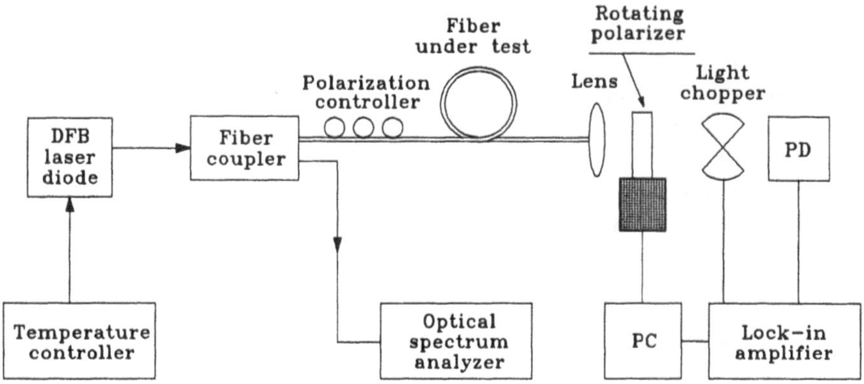

Figure 4.7 Experimental setup based on measurement performed in the frequency domain (adapted from [22]).

A DFB laser diode appears as a much more practical light source. However, its frequency tunability is quite small and, consequently, it is impossible to derive the statistics of differential group delay accurately. This limit can be overcome considering that the mean value of the differential group delay evaluated at different times coincides with that measured at different wavelengths [23] when the number of measurements is high enough.

PSPs are stable on a time interval of a few minutes in terrestrial buried cable [23]. Thermal variations represent the main source of PSP instability, in particular for fiber wound on drums. Buried cables, or cables installed inside a duct, are less sensitive to environmental temperature changes and good PSP stability over time intervals as long as 20 min. has been observed. Moreover, polarization mode dispersion in cabled fibers is much larger than in uncabled fibers [24], probably because mechanical stresses can be induced during cable manufacturing.

Field tests have shown that concatenation of polarization dispersion effects in the field can be predicted with a satisfactory degree of confidence on the basis of measurements performed in the factory on cables which are still wound on their drums [21]. It has been found that the differential group delay of a fiber cascade is well confined within theoretical bounds. This can be expressed as $\Delta\tau \pm \sigma$ and derived from the individual data by means of (4.18) and (4.19). Therefore, factory-based acceptance tests can be envisaged.

Polarization mode dispersion can also affect optical devices, like fiber couplers or isolators, usually included in active fiber amplifiers. In particular, it has been observed [25] a differential group delay of a few ps on some optical isolators, that is equivalent, regarding polarization mode dispersion, to some thousands of km of low birefringent optical fiber. Moreover, isolator behavior regarding polarization mode dispersion is sometimes dependent on temperature, both on a single isolator or on a cascade composed by various isolators. Furthermore, values of DGD of about 0.1–2 ps have been observed on pieces of erbium-doped fibers 40-m-long [26]. As a consequence, polarization mode dispersion may become the limiting factor in long-haul, high-bit-rate systems including optical fiber amplifiers.

An alternative way to measure polarization mode dispersion is based on an incoherent light source. An optical pulse is launched into the fiber and a time delay is measured because the two polarizations of the fundamental mode travel over different optical paths. Polarization coupling determines a power transfer between the two polarizations, described according to a statistical model. Using a broadband source, a spectrum averaging occurs and the dispersion measured corresponds to that obtained by means of coupled-power theory [24, 27–30]. The experimental setup proposed [28], based on an interferometric measurement, is of the type shown in Figure 4.8.

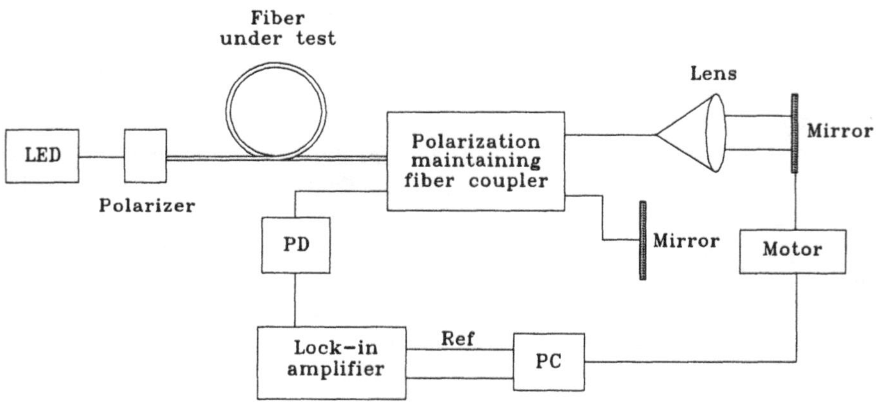

Figure 4.8 Experimental setup for measurement in the time domain (interferometric technique) (adapted from [28]).

Using this method, the r.m.s. width $\Delta\tau$ of the autocorrelation function of the light at the fiber output can be measured. Comparison between measurements based on the principal states of polarization and on autocorrelation function of the light at the fiber output has already been made, showing a good agreement between the results [31].

4.2 MEASUREMENTS OF COHERENT LIGHT OPTICAL SOURCES

Single-mode optical fibers are essential components in very high speed, long-distance communication links, because intermodal dispersion is eliminated. To fully exploit the potential of single-mode fibers, highly coherent lasers operating in single transverse and longitudinal mode are required as transmitters. Using single frequency, stabilized devices, both material and waveguide dispersions are reduced to a minimum, hence bandwidth is maximum. Several transmission schemes are employed for optical fiber communications. In *pulse code modulation* (PCM) the light intensity is switched on/off by using an external modulator or by directly varying the laser injection current. In this case the receiver sensitivity is limited by the detector noise. Recently, with the availability of narrowband single-mode semiconductor lasers, the interest for coherent transmission systems is growing. They rely on the modulation of a very stable optical carrier and on the homodyne or heterodyne detection. The most important coherent schemes are *amplitude shift keying* (ASK), *phase shift keying* (PSK), and *frequency shift keying* (FSK). According to the *frequency division multiplexing* (FDM) technique, a high number of transmission channels can be implemented in the low loss and low dispersion window of optical fibers around 1.55 μm by using a coherent transmission scheme and lasers having slightly different wavelengths.

Besides using external modulators, the coherent transmission can be accomplished by directly modulating the injection current of a laser diode. In this case, because of the change of the complex index of refraction as a function of the charge carrier density, both amplitude and phase modulation can be achieved. Hence the evaluation of the emission spectrum of laser diodes in CW regime and under modulation (FM response) is essential to design a coherent transmission system.

Single-mode laser diodes are also important for optical fiber sensors whenever an interferometer is employed.

Several types of highly coherent laser diodes are available; among them DFB, DBR, quantum well lasers, and external cavity lasers are the most important. They offer considerable brightness and narrow linewidth, and can be modulated at a high rate (up to several GHz) [32]. See Section 2.7 for a description of those devices. This section deals with the measurement of several fundamental laser diode characteristics: the pulse response, the emission bandwidth, and the frequency chirp. Afterwards the most common noises which affect the behavior of a semiconductor laser (AM noise, FM noise, and 1/f noise) are addressed [33].

4.2.1 Pulse Response and Frequency Response

Besides using direct methods for the measurement of the pulse response of a laser diode described in Section 3.2.2, even better time resolutions can be achieved by means of the optical autocorrelation technique [34]. The measurement system, shown in Figure 4.9, relies on *second harmonic generation* (SHG), which is a nonlinear phenomenon, quadratic in the optical intensity [35]. The optical pulse is divided by the beam-splitter plate of a Michelson interferometer (see Section 2.2.2) and is recombined on a $LiIO_3$ crystal. A high repetition rate pulse train generated by the device under test is sent to the interferometer while the length of one of the arms is continuously scanned within a range dependent on the pulse duration to be measured ($\Delta l > 3\,c\,\Delta t$ where Δt represents the FWHM). The time between successive pulses must be much shorter than the scan time. The two parts of the beam recombines into the nonlinear crystal with variable time delays; in this way all the possible reciprocal positions of two pulses are scanned. The second harmonic component is collected by a detector having a time response of the order of the scan time. Typically a high sensitivity *photomultiplier tube* (PMT) is employed since the signal level may be very low. The chopper is coupled to the lock-in amplifier and allows to subtract the uncorrelated noise.

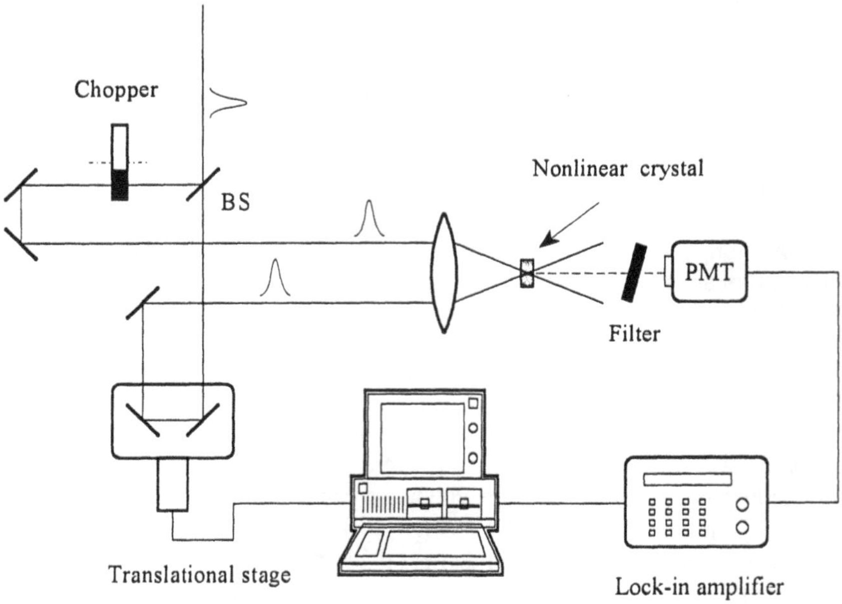

Figure 4.9 Optical autocorrelation technique based on second harmonic generation (SHG).

The photocurrent as a function of the delay τ is equal to the autocorrelation of the optical intensity i(t), given by

$$G(\tau) = \langle i^*(t)i(t+\tau)\rangle = \lim_{T\to\infty} \frac{1}{2T}\int_{-T}^{+T} i^*(t)i(t+\tau)dt \tag{4.21}$$

If the beams at the fundamental frequency ω impinge on the crystal being collinear, a dc offset appears in the autocorrelation curve due to the frequency conversion of the two distinct pulses. Indeed, the geometry shown in Figure 4.9 is usually adopted. The crystal is very thin (100 to 200 μm) and the beams are at an angle with respect to the crystal axis. Thereby, for the conservation of momentum, the signal converted to the second harmonic exits the crystal halfway only when both pulses are present. Consequently, a background free autocorrelation signal is obtained; moreover, the fundamental and SH components are angularly separated. A narrowband filter eliminates the residual radiation at circular frequency ω before the detector. The obvious drawback of this configuration is the low conversion efficiency, because the beams are superimposed only for a short path. An example of second harmonic autocorrelation is shown in Figure 4.10.

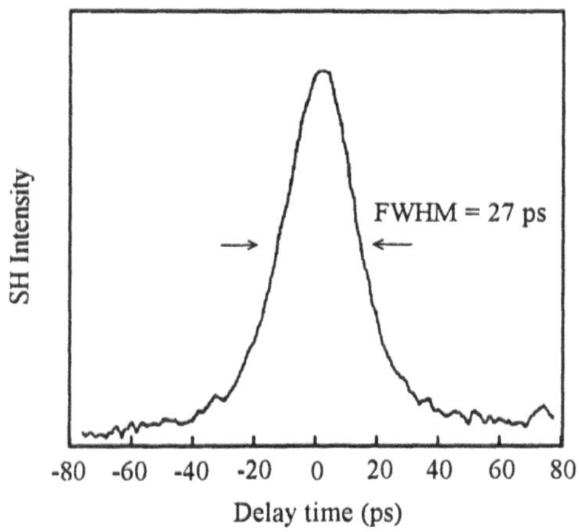

Figure 4.10 Example of optical autocorrelation signal obtained using background free second harmonic generation.

The maximum in Figure 4.10 corresponds to a condition of perfectly superimposed pulses, while the width of the curve depends on the temporal profile of the pulses that can be extracted by deconvolving the autocorrelation signal [36]. This in turn involves a reasonable guess over the pulse shape. For example the autocorrelation of a Gaussian pulse is again Gaussian with a FWHM 1.41 times the original one. A more refined deconvolution method does not require the knowledge of the pulse shape [37]. The Fourier transform of the pulse optical intensity i(t) can be expressed as

$$I(\omega) = |I(\omega)| \exp[i\phi(\omega)] \tag{4.22}$$

The amplitude of $I(\omega)$ can be immediately obtained from the Fourier transform $G(\omega)$ of the autocorrelation signal $G(\tau)$ as

$$|I(\omega)| = |G(\omega)|^{1/2} \tag{4.23}$$

The phase can be calculated from the Hilbert transform theory under suitable hypothesis. Thus we obtain

$$\phi(\omega) = \frac{2\omega}{\pi} \int_0^{\infty} \frac{\mathrm{Im}[G(\omega)]^{1/2}}{\xi^2 - \omega^2} d\xi \tag{4.24}$$

Hence, from the measurement of the optical intensity autocorrelation $G(\tau)$, the function $I(\omega)$ can be obtained and then $i(t)$ can be calculated by taking the inverse transform of $I(\omega)$. The laser emission is dominated by the stimulated emission and the light pulse rises after a build-up time which depends on the laser population inversion. Also the pulse width depends on the charge carried by the current pulse and the light pulse may be even shorter than the current pulse itself, whereas the shape of the current pulse is unimportant provided that it is short enough [38]. Assuming the current pulse to be an ideal Dirac $\delta(t)$ function, the measured light pulse $i(t)$ gives the pulse response $h(t)$ of the device and its Fourier transform $I(\omega)$ is equal to the transfer function $H(\omega)$.

The transfer function $H(\omega)$ obtained in this way may significantly differ from the small signal frequency response which refers to a sinusoidal modulation superimposed onto a dc bias. This response can be measured simply by detecting the emitted light with a fast photodiode (PIN or APD) connected to a sampling oscilloscope. The measured bandwidth depends on the bias current and on the modulation index. In this small signal regime, the frequency response of a single-mode laser (sometimes called modulation efficiency) is characterized by a resonance frequency ν_r in the GHz band, where the curve presents a sharp peak (see Figure 4.11). ν_r is given by [39, 40]

$$\nu_r = \frac{1}{2\pi}\left[\frac{1}{\tau\tau_p}\left(\frac{j}{j_{th}} - 1\right)\right]^{1/2} \tag{4.25}$$

where τ is the carrier recombination lifetime, τ_p is the photon lifetime for the lasing mode, j and j_{th} are the injection current density and the threshold current density respectively [41]. The resonance frequency increases with the injection current as a consequence of the rise in photon density and hence in stimulated emission; it can be also increased by reducing the cavity length which affects the photon lifetime τ_p. Laser diodes can be effectively modulated by an analog signal having a bandwidth significantly lower than the frequency ν_r, thereafter appreciable distortion and relaxation oscillation may occur.

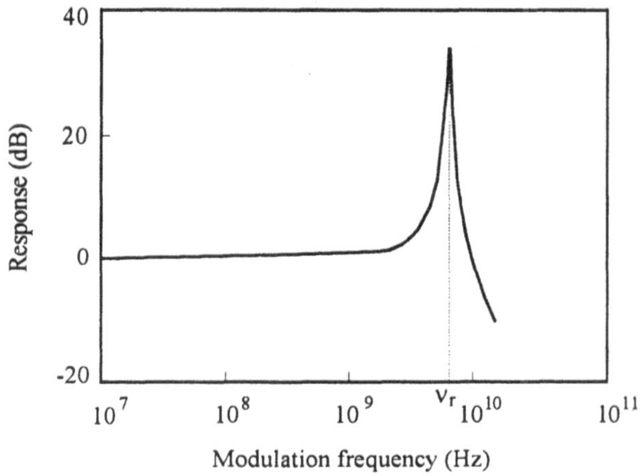

Figure 4.11 Typical frequency response of a laser diode to a sinusoidal modulation.

4.2.2 Emission Spectrum Evaluation

In Section 3.2.4 we described the emission spectrum measurement of an LED or an MLM laser using a grating spectrometer. When the linewidth to be measured is less than 0.1 nm (as for the SLM laser diodes) more refined techniques are required. A scanning Fabry-Perot interferometer in cascade to a grating spectrometer is by far the most common setup employed. If the measurement of the absolute wavelength is not required, the grating can be omitted.

Fabry-Perot Interferometer

A simple theoretical treatment of the *plane Fabry-Perot* (PFP) interferometer has been already given in Section 2.1.5 [42]. In this section we will describe the principles of operation of the *confocal Fabry-Perot* (CFP) spherical mirror interferometer which allows to achieve higher spectral resolution [43].

A CFP is made of two spherical mirrors having a reflectivity R, spaced exactly by the same distance as their radius of curvature r (see Figure 4.12). This arrangement constitutes a resonant cavity acting as a frequency filter with periodically spaced transmission peaks separated by $\Delta\nu = c/4r$, which is named the *free spectral range* (FSR) of the interferometer. By scanning the mirror separation over a length equal to $\lambda/4$ (for instance, by means of a piezo-actuator (PZT) attached to one mirror) it is possible to shift the transmission peaks by an entire FSR. By entering the spectrometer with a monochromatic beam, the transmitted power along the axis is significant only if the light frequency is close to a cavity resonance. By recording the transmitted light with a detector while scanning the FSR, the spectrum of the incoming light is acquired. The width $\delta\nu$ of the transmission peak pertaining to a monochromatic source depends on the Q factor of the cavity. This in turn is determined by several factors which contribute to the overall cavity losses: the mirror reflectivity, the mirror surface figure* and the diffraction losses. For a CFP it can be demonstrated that the mirror reflectivity R is by far the most weighty factor and that the FWHM of transmission peaks is given by

$$\delta\nu = c\,(1 - R^2)\,/\,4\pi r R \qquad (4.26)$$

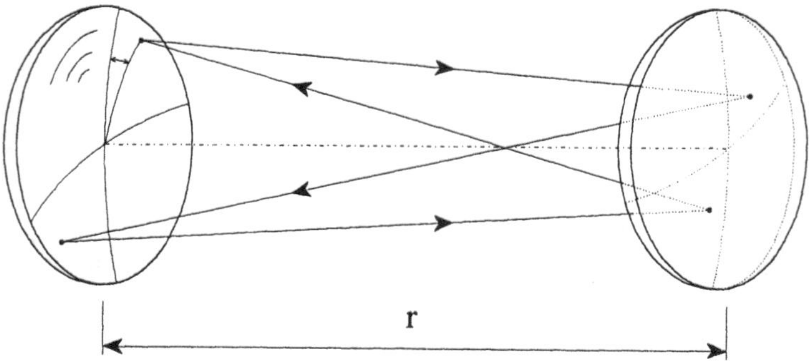

Figure 4.12 Confocal spherical mirror Fabry-Perot interferometer. The mirror separation r equals its curvature radius.

---

* The surface figure is a measure of the mirror roughness. It is usually defined with respect to the optical wavelength usually by λ/m. Typically m = 10, 20 with λ = 630 nm.

The ratio FSR/$\delta\nu$ is defined as *finesse* (F) of the interferometer. For R close to unity, as it is in practical cases, F is given by

$$F \approx \pi / [2(1 - R)] \qquad (4.27)$$

The quantity $\nu/\delta\nu$ is the spectral resolving power R_p:

$$R_p = 4rF / \lambda \qquad (4.28)$$

It has the same meaning as the resolving power of a grating spectrometer (see Section 3.2.3).

The capability to collect the off-axis light is given by the product of the solid angle Ω subtended at the limiting aperture of the system times the aperture area A (see Figure 4.14). This quantity is called the "étendue" U; high values of the étendue can be obtained using a large area detector. However the requirement of having high light-gathering power contrasts with the requirement of having high spectral-resolving power.

By analyzing the transverse mode properties of a general confocal resonator, it can be shown that, if the input beam transverse pattern matches the fundamental cavity mode, which has a Gaussian intensity profile, the FSR is doubled:

$$FSR_m = c / 2R \qquad (4.29)$$

It is remarkable that, whenever the confocal condition is precisely met, mode matching of the input field with the cavity is not required to obtain a clean spectrum. Multimode excitation simply leads to a reduction of the FSR.

It is worthwhile to compare the properties of the CFP to those of the PFP interferometer. While for the CFP the theoretical reflectivity-limited finesses can be easily achieved, this is not the case for the PFP. In fact the actual finesse is much lower than the theoretical one ($F = \pi\sqrt{R}/(1-R)$, see Section 2.1.5) due to alignment difficulties and to the greater influence of mirror surface figure and diffraction losses compared to the case of the CFP.

Moreover it can be demonstrated that for a given finesse (which depends on R) the ratio of the spectral resolving power to the étendue for the CFP is constant and given by

$$R_p / U \approx 0.7\ (2F / \pi\lambda)^2 \qquad (4.30)$$

the factor 0.7 being determined by a suitable choice of the detector area. This means that by increasing the mirror separation (and radius) both the resolving power, which explicity depends on r, and the étendue increase. Thereby, at high resolution corresponding to long radius and small FSR, the confocal Fabry-Perot is the best choice. However, for small mirror separation and consequently high FSR, it can be shown that the plane Fabry-Perot becomes competitive in terms of light gathering power and resolution since its drawbacks are far less pronounced at short cavity length (less than 4 cm for $F \approx 25$).

Obviously the major disadvantage of the CFP is the impossibility to change the FSR for a given couple of mirrors.

The basic experimental arrangement used with the plane Fabry-Perot is shown in Figure 4.13.

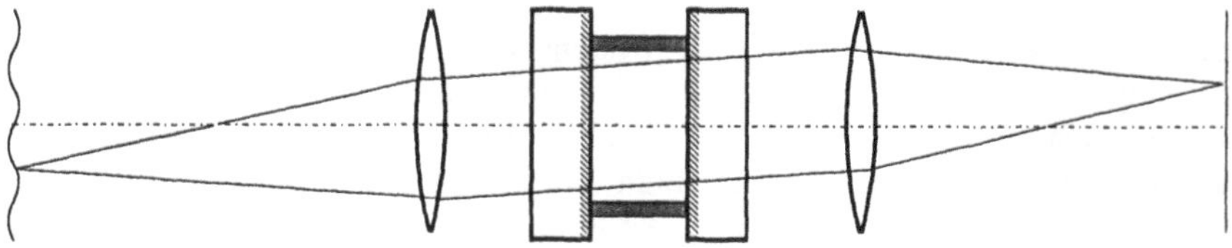

Figure 4.13 Experimental arrangement used with a plane Fabry-Perot interferometer. The light from an extended source is collected by the first lens, while the second lens forms a fringe pattern in its focal plane. Alternatively the interferometer can be used as spectral filter and the ligth from a point source (on axis) is transmitted or reflected depending on the wavelength.

The light source is placed at the focus of a collimating lens; the light exiting the interferometer is collected by a fringe-forming lens and observed at the lens focal plane. If the light source can be approximated by a point source, the illumination of the spot formed at the image plane depends on the interference condition; if the light source is an extended one, several bright rings will be observed anyway. By removing the collimating lens a system of annular fringes are formed even in the case of a point source. The interferometer acts as an angular filter. The practical function of the collimating lens is to increase the illumination in the fringe plane. Due to this fact, no special conditions on the optical quality of this component are imposed; on the contrary the fringe-forming lens must be of very high quality and the focusing condition must be precisely fulfilled.

A typical experimental setup used with a confocal interferometer is depicted in Figure 4.14. The light is collected from a lens and focused in the midplane of the interferometer; the detector is located just behind the rear mirror and collects the light which passes through a limiting aperture that can be the detector itself. If the beam incident on the lens is not collimated, the resolution of the instrument is slightly reduced; however, the amount of light received by the detector is significantly lower. If a collimated beam enters the interferometer from the rear side (being the detector removed), a fringe system forms in the focal plane of the front lens. A photodiode array or a camera placed in the image plane can be used to record the fringe pattern and to perform a nonscanning measurement.

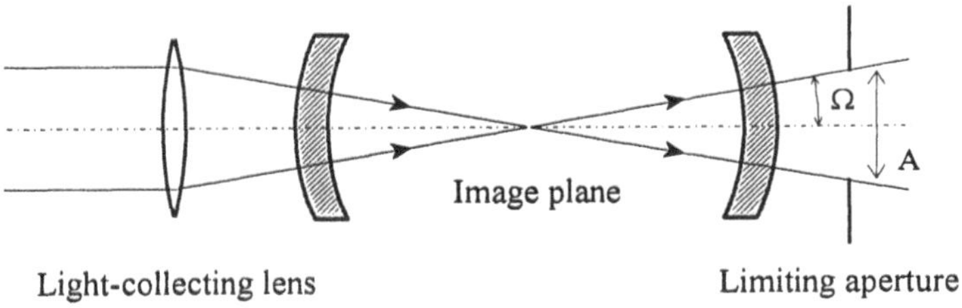

Figure 4.14 Experimental arrangement used with a confocal spherical mirror Fabry-Perot interferometer. The light exiting the device under test is collimated and focused in the middle plane of the interferometer.

A grating spectrometer can be coupled to a Fabry-Perot interferometer. A typical test bed is shown in Figure 4.15 [44]. The output of a thermally stabilized SLM laser diode is sent into a monochromator which measures the absolute wavelength of the device. The light exiting the monochromator is further analyzed by a scanning Fabry-Perot to resolve the laser linewidth.

Commercially available scanning Fabry-Perot's allow the achievement of spectral resolutions of about 1 MHz, (less than 300 kHz with patented design spherical, nonconfocal microcavities having finesse greater than 20,000).

Emission Linewidth and Frequency Jitter: Heterodyne Methods

With actively stabilized laser sources like DFB, DBR, or external cavity devices (see Section 2.7) linewidths below 1 MHz can be obtained [45]. To measure such narrow spectra, high resolution heterodyne techniques can be usefully employed.

Heterodyne is commonly used for coherent fiber optic systems [46]. The field of a modulated signal can be expressed as

$$E_S(t) = \mathcal{E}_S(t) \cos[\omega_S t + \phi_S(t)] \tag{4.31}$$

while the signal modulation may be included in the previous formula as

$$\mathcal{E}_S(t) = \mathcal{E}_{So}[1 + f(t)] \tag{4.32}$$

assuming amplitude modulation or

$$\phi_S(t) = \phi_{So}[1 + f(t)] \tag{4.33}$$

assuming phase modulation. A conceptually similar expression holds for frequency modulation, the frequency being the derivative of the phase.

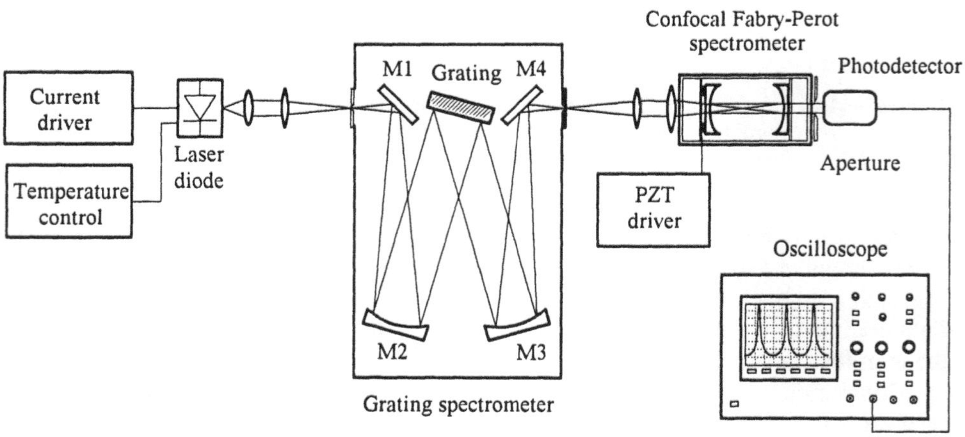

Figure 4.15 Experimental setup to measure the linewidth of a laser diode using a grating spectrometer followed by a Fabry-Perot interferometer.

The heterodyne technique relies on a *local oscillator* (LO) field, having a very stable frequency ω_{LO} close to that of the modulated signal; this reference field can be written as

$$E_{LO}(t) = \mathcal{E}_{LO}\cos(\omega_{LO}t) \tag{4.34}$$

The two fields (4.31) and (4.34) are superimposed onto a broadband detector taking care to preserve a good alignment. Since the detector responds to the intensity, the output current is proportional to

$$i(t) \propto [E_S(t) + E_{LO}(t)]^2 \tag{4.35}$$

Expanding the quadratic term, we obtain a dc component and several oscillating components at frequencies $2\omega_S$, $2\omega_{LO}$, $(\omega_S + \omega_{LO})$, and $(\omega_S - \omega_{LO})$. However, the photodiode only responds to the dc and difference frequency terms, while higher frequencies average to zero. Hence the photocurrent is

$$i(t) \propto \langle i \rangle + 2\mathcal{E}_S(t)\mathcal{E}_{LO}\cos[(\omega_S - \omega_{LO})t + \phi_S(t)] \tag{4.36}$$

where brackets $\langle\ \rangle$ represent time average as usual. It is then clear that the heterodyne signal preserves the amplitude or phase modulation contained in the original signal, but its frequency is down-converted from optical frequencies to the RF band.

This technique can be used to measure the linewidth of a laser. Under the quasi-monochromatic approximation, the optical field of a SLM laser diode employed as a local oscillator and biased well above threshold can be written as

$$E_L(t) = \mathcal{E}_L(t)\cos[\omega_L t + \phi(t)] \tag{4.37}$$

where ω_L is the average optical frequency. The time-dependent functions $\mathcal{E}_L(t)$ and $\phi(\tau)$ account for signal modulation and noise (amplitude and phase), both leading to line broadening; in this section we will address only the effects of the noise while the effects of the modulation will be considered in Section 4.2.5, which is devoted to chirp measurement. In this case the function $\phi(\tau)$ represents a stochastic phase fluctuation mechanism, whereas the contribution of the term $\mathcal{E}_L(t)$ (amplitude noise) to the laser spectrum is usually negligible; hence the field amplitude will be assumed constant in the following. To evaluate the power spectrum of the signal in formula (4.37), it is convenient to consider the autocorrelation function

$$G_E(\tau) = \left\langle E^*(t)E(t+\tau)\right\rangle = \left\langle \exp[i\Delta\phi(t,\tau)]\right\rangle \exp(i\omega_L\tau) \tag{4.38}$$

where $\Delta\phi(t,\tau)$ is the random phase change in time τ.

Assuming the phase jitter to be a zero-mean random process having a Gaussian probability distribution [47], its variance σ^2 is related to the instantaneous frequency fluctuation spectrum $S_\phi(\omega)$ by

$$\left\langle \Delta\phi^2(\tau) \right\rangle = \frac{\tau^2}{2\pi} \int_{-\infty}^{+\infty} \frac{\sin(\omega\tau/2)}{\omega\tau/2} S_\phi(\omega) d\omega \tag{4.39}$$

Assuming a flat spectrum $S_\phi(\omega)$ of amplitude γ [48], the above integral gives $2\gamma\pi/|\tau|$ so that the variance $\langle \Delta\phi^2(\tau) \rangle$ has a linear dependence on the time delay τ, according to

$$\left\langle \Delta\phi^2(\tau) \right\rangle = \gamma|\tau| \tag{4.40}$$

Using the relation

$$\left\langle \exp[i\Delta\phi(t,\tau)] \right\rangle = \exp\left[- \frac{\left\langle \Delta\phi^2(\tau) \right\rangle}{2} \right] \tag{4.41}$$

which holds for random phase jitter [48, 49], the autocorrelation function (4.38) can be written as

$$G_E(\tau) = \exp\left[- \frac{\left\langle \Delta\phi^2(\tau) \right\rangle}{2} \right] \exp(i\omega_L \tau) \tag{4.42}$$

The laser power spectrum can be obtained by Fourier-transforming the function $G_E(\tau)$. Using the expression (4.40) for $\langle \Delta\phi^2(\tau) \rangle$, the spectrum turns out to be a Lorentzian function:

$$S_E(\omega) = \frac{\Delta\omega/2\pi}{\Delta\omega^2/4 + (\omega_L - \omega)^2} \tag{4.43}$$

where $\Delta\omega = \gamma$ is the full width at half maximum (FWHM). The actual linewidth $\Delta\omega$ is given by the Schawlow-Townes formula corrected by the linewidth enhancement factor b [50] already introduced in Section 2.6.2 (see also Section 4.2.3):

$$\Delta\omega = \frac{v_g{}^2 h\nu g n_{sp} \alpha_{mir}}{4P_L}(1 + b^2) \tag{4.44}$$

where v_g is the group velocity of the light in the active medium, $h\nu$ is the photon energy, g is the laser gain constant, n_{sp} is the spontaneous emission factor [51] (which, for example, is approximately equal to 2.6 for GaAs lasers [50]), P_L is the single-ended output power, and α_{mir} is the mirror facet loss given by

$$\alpha_{mir} = -\frac{\log_e(r_{mir})}{L} \tag{4.45}$$

where r_{mir} is the reflectivity and L is the cavity length. The role of the b factor will be discussed in greater detail later on.

To measure the laser linewidth, a second laser having a frequency close to ω_L and negligible linewidth can be used as local oscillator. According to the heterodyne theory, the beat signal is centered at frequency $|\omega_L - \omega_{LO}|$ and has the same spectrum as the laser under test; a RF spectrum analyzer can then be used to evaluate the heterodyne signal.

However, a stable oscillator having a very narrow linewidth is often not available at the required wavelength.

To overcome this limitation, high-resolution self-heterodyne methods have been proposed to measure the output spectrum and the frequency fluctuations of a SLM laser diode [52–55].

In these methods the laser under test is also used as local oscillator. The laser field is divided in two, frequency-shifted in one branch of a Mach-Zehnder interferometer (see Section 2.2.3) and time-delayed of a time τ_o in the second branch; the beams are then recombined at the surface of a broadband detector. The total field is given by

$$E_T = E_L(t) + E_L(t + \tau_0)\exp(i\omega_m) \tag{4.46}$$

where ω_m is the frequency shift provided by an optical modulator; we have assumed for sake of simplicity that the fields in the two branches have the same amplitude. The interferometer converts the phase noise into intensity noise. Neglecting a shot-noise term, the autocorrelation function of the photocurrent $R_I(\tau)$ is a replica of the optical intensity correlation $G_{E_T}{}^2(\tau)$ [56]:

$$R_I(\tau) = s^2 G_{E_T}^2(\tau) = s^2 \left\langle E_T(t)E_T^*(t)E_T(t+\tau)E_T^*(t+\tau)\right\rangle \qquad (4.47)$$

where s is the detector sensitivity.

The product in brackets in the above expression gives rise to a sum of 16 terms; 10 of them average to zero due to the explicit presence of the oscillating phase factor $\exp(\pm i\omega_m t)$. After some algebra the following expression for $G_{E_T}{}^2(\tau)$ is obtained

$$G_{E_T}^2(\tau') = \begin{cases} 2E_L{}^4[2 + \cos(\omega'\tau')\exp(-\tau')] & \text{for } |\tau'| < \tau_0' \\ 2E_L{}^4[2 + \cos(\omega'\tau')\exp(-\tau_0')] & \text{for } |\tau'| > \tau_0' \end{cases} \qquad (4.48)$$

where the adimensional quantities $\omega' = \omega/\Delta\omega$, $\tau' = \tau\Delta\omega$, $\tau_0' = \tau_0'\Delta\omega$ have been used. The Fourier-transform of $R_I(\tau)$ gives the power spectrum $S_I(\omega)$ of the photocurrent as

$$S_I(\omega') = s^2E_L{}^4 \left\{ 4\delta(\omega') + \exp(-\tau_0')\delta(\omega - \omega_m') + \exp(-\tau_0')\frac{1/\pi}{1 + (\omega' - \omega_m')^2}\right.$$

$$\left. \cdot \left[\exp(\tau_0') - \frac{\sin(\omega - \omega_m')\tau_0'}{(\omega' - \omega_m')} - \cos(\omega - \omega_m')\tau'\right]\right\} \qquad (4.49)$$

This distribution is the physical quantity measured by the RF spectrum analyzer. It consists of a dc term, a monochromatic term at frequency ω_m, and a modified Lorentzian component centered at the same frequency. By increasing the delay time τ_0, the weight of the monochromatic component becomes lower and lower until a strictly Lorentzian shape is obtained for the power spectrum having a width which is twice that of the laser spectrum $\Delta\omega$.

Thereby a clear evaluation of the laser linewidth can be achieved providing delays much longer than the coherence time, which is related to the linewidth by the expression

$$\tau_{co} \geq \frac{2}{\Delta\omega} \qquad (4.50)$$

for a Lorentzian line shape.

An experimental arrangement to perform self-heterodyne linewidth measurements is shown in Figure 4.16. It consists of a modified version of a Mach-Zehnder interferometer where a single-mode optical fiber provides a suitable delay τ_0. The laser diode beam is split in two paths. The first path includes an acousto-optic modulator driven at a frequency (40 MHz) much higher than the spectral linewidth to be measured; the second path is a 1.5–km optical fiber. The outputs of the two Mach-Zehnder arms are recombined on an avalanche photodiode and the spectrum of the beat signal is analyzed and recorded.

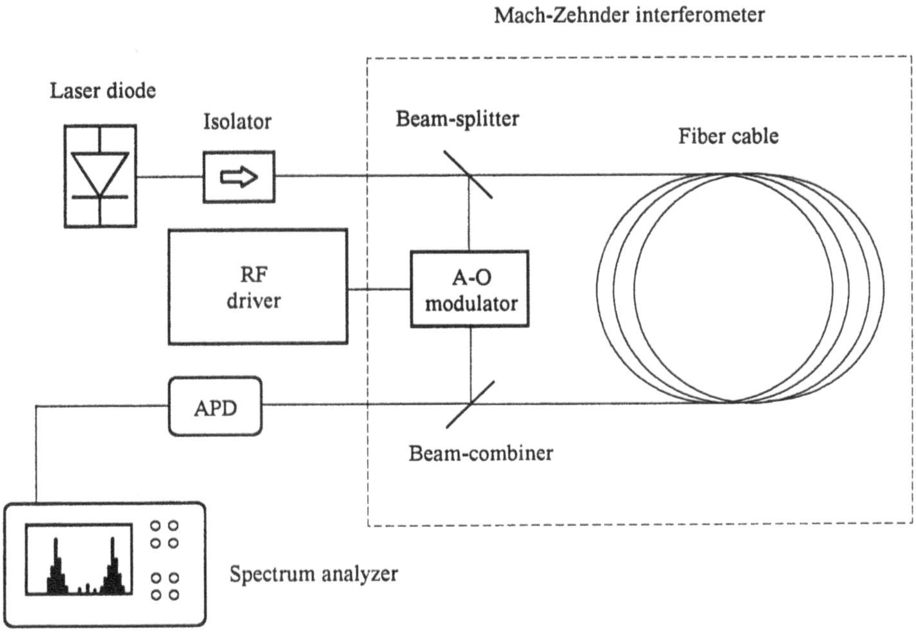

Figure 4.16 Experimental setup to measure the linewidth of laser diode by means of the self-heterodyne technique.

As shown in the previous theoretical treatment, the width of the measured spectrum is twice the actual value assuming a Lorentzian line shape. Assuming a Gaussian broadening, the width of the heterodyne spectrum turns out to be $\sqrt{2}$ times the actual width [57]. This arrangement relies on the condition that the two beams are uncorrelated, so that a resolution limit on the order of the inverse of the delay time can be achieved. An approximate estimate sets the delay to fulfill this requirement to about six times the coherence time [52]. For very narrow band SLM laser diodes, having linewidths of 10 kHz or less, this leads to the necessity of using several tens of km of optical fiber.

As it has been shown above (see Equation (4.49)) using delays significantly shorter than the laser coherence time, the observed self-heterodyne spectrum departs from the simple Lorentzian curve and some ripples appear, together with a sharp peak (delta function) centered at the modulator frequency. The theoretical prediction for $S_I(\omega)$, given in Equation (4.49) can be fitted to the experimental curve, providing the coherence time and hence the linewidth. This method has been suggested and experimentally verified in reference [52]. Figure 4.17 shows the self-heterodyne spectrum measured with delay greater (a) and lower (b) than the coherence time.

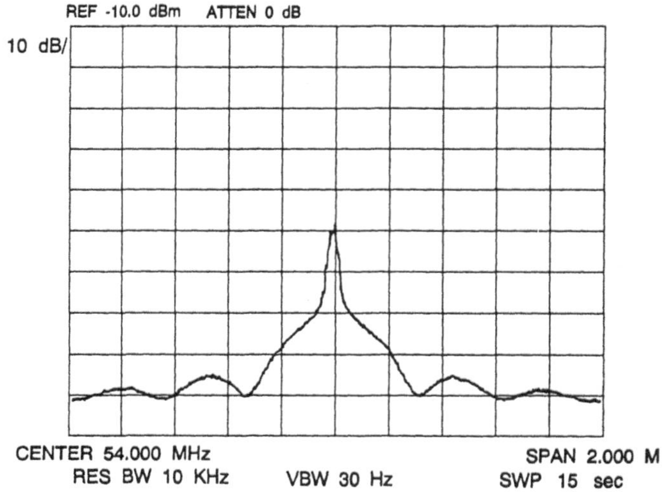

(a)

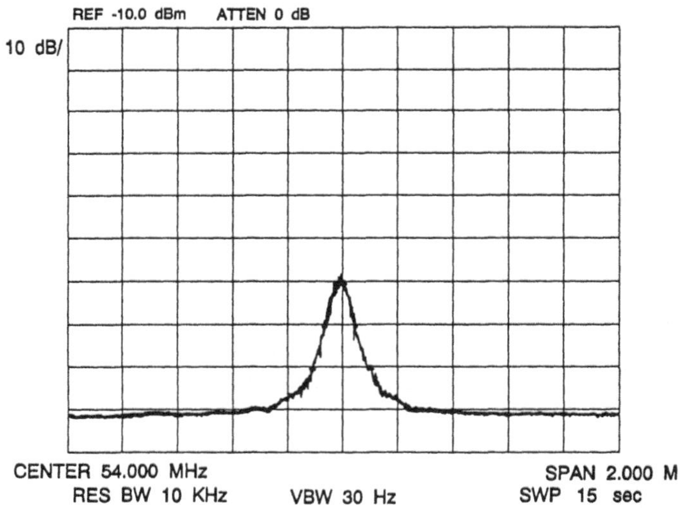

(b)

Figure 4.17 Self-heterodyne spectrum measured with delay (a) greater and (b) lower than the coherence time (from [52]).

One or more optical isolators must be used to prevent laser diode instabilities induced by optical feedback from light scattering in the fiber and reflections from connectors and coupling optics.

Other techniques have been proposed to measure narrow linewidth. Among them it is worth considering the fiber-optic ring resonator, which offers a measuring range from 100 kHz to 100 MHz with an accuracy within 10% [58]. The experimental setup is depicted in Figure 4.18. The resonator is made with a single-mode optical fiber (about 1 m long) and a low insertion loss directional coupler. A PZT cylinder with the fiber wrapped around, driven by a triangular wave modulator, acts as a phase modulator of the light in the resonator and scans a whole cavity free spectral range. Recording the output from the directional coupler with a photodiode, the finesse F (see Section 4.2.2) of the ring cavity can be measured. The value of F, obtained using a reference laser with a negligible linewidth, for instance a He-Ne, is compared to the finesse F' measured using the laser diode under test: F' depends both on the cavity finesse F itself and on the laser linewidth $\Delta\nu_{LD}$. Assuming Lorentzian profiles for the laser lineshape and the cavity transmission peak, it results

$$F' = \frac{cF}{FnL\Delta\nu_{LD} + c} \tag{4.51}$$

where L is the cavity length and n the fiber refractive index. The linewidth $\Delta\nu_{LD}$ can be extracted from the above formula.

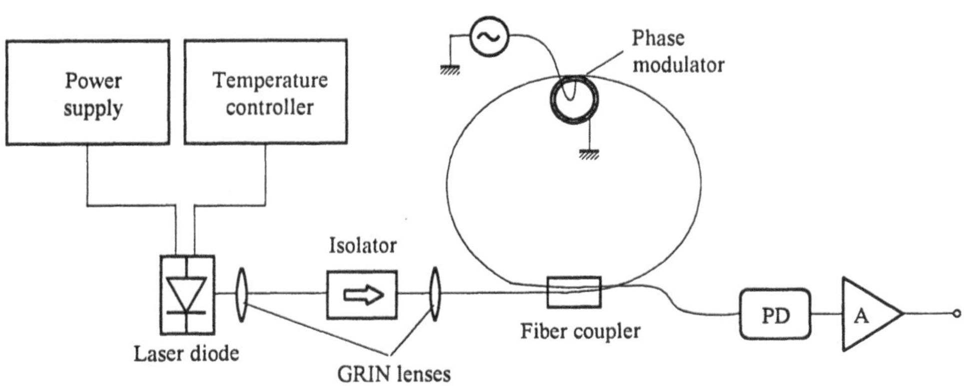

Figure 4.18 Laser linewidth measurement using a ring resonator. The finesse of the cavity is measured using the device under test and then compared to the one measured with a very stable reference laser.

4.2.3 FM Response Evaluation

In FSK transmission, the optical frequency of a SLM laser diode biased above threshold is modulated by varying the injected current. To this purpose, the modification of the optical frequency induced by amplitude modulation via the amplitude-phase coupling effect [59] can be usefully exploited. In addition, an *intensity modulation* (IM) is introduced, which leads to the insurgence of sidebands in the output spectrum.

The two main factors contributing to the optical frequency modulation are the carrier density variation, which induces a change in the refractive index, and the temperature variation due to heat generation at the junction. Both the effects lead to optical cavity length modulation and to a consequent frequency shift. The first effect is important at high modulation frequencies; the second one predominates at low frequencies. Other effects to be considered are the gain detuning due to the drift of the cavity resonances and the temperature dependence of the energy gap. As seen in Equation (4.44) the SLM diode laser linewidth contains a factor $(1 + b^2)$ which accounts for the broadening due to the phase-amplitude coupling.

A theoretical treatment of this phenomenon has been carried on using the semiclassical equation of motion for the cavity mode electric field [59]. Following a small signal approach, the field amplitude and the susceptibility (averaged over the cavity mode intensity distribution) can be written as

$$E_m(t) = E_0[1 + \eta(t)]\cos[\omega_m't + \phi(t)]$$

$$\chi_m(t) = \chi_{m0} + i\chi_{m1}(t) = \chi_{m0,r} + \chi_{m1,r}(t) + i[\chi_{m0,i} + \chi_{m1,i}(t)] \quad (4.52)$$

where small perturbations from the stationary values have been considered. ω_m' is slightly different from the cold cavity mode frequency due to the phase shift induced by the gain medium.

As a special case of Equation (4.52), the cavity mode electric field of a semiconductor laser modulated by a sinusoidal signal at frequency Ω can be written as

$$E_m(t) = E_0[1 + (m/2)\cos\Omega t]\cos[\omega_m't + \beta\cos\Omega t] \quad (4.53)$$

By substituting the expressions for $E_m(t)$ and $\chi_m(t)$ into the time evolution equation, assuming a rather large modulation frequency Ω, the intensity and phase modulation indexes, m and β, turns out to be linked by the relations

$$\beta = -\,1/2\; b_m(\Omega) m$$

$$b_m(\Omega) \equiv \chi_{m1,r} / \chi_{m1,i} \qquad (4.54)$$

The factor b_m represents the amplitude-phase coupling effect. If the electric susceptibility χ_m does not depend on the spatial coordinates and linearly depends on the carrier density n, the factor b_m turns out to be independent of Ω and coincides with the linewidth enhancement factor b already introduced in Section 4.2.2.

The spectrum of the optical field in Equation (4.53) contains several sidebands at frequencies multiple of the modulation frequency Ω, having intensities roughly proportional to

$$S(n\Omega) = J_n^2(\beta) \qquad (4.55)$$

where J_n is the Bessel function of order n. The power spectrum at the laser frequency ω' and at the first sideband at frequency $\omega_m + \Omega$ is

$$E_0^2[J_0^2(\beta) + m^2 J_1^2(\beta)]$$

$$E_0^2\{J_1^2(\beta) + [(m/2)(J_2(\beta) - J_0(\beta))]^2\} \qquad (4.56)$$

The intensity-modulation index m can be directly measured by using a photodiode, while the frequency modulation index β can be extracted from a power spectrum measurement. The relative sideband strength is measured and the value of β is obtained from Equations (4.56). The factor b can be found from formula (4.54) which gives

$$b_m = -\,2\beta / m \qquad (4.57)$$

The *frequency modulation* (FM) response heavily influences the coherent transmission system performances. It is obviously desirable to have flat FM response with modulation frequency.

It is possible to measure the FM response of laser diodes by direct observation of the spectrum with an optical spectrum analyzer [59–61]. However, due to the coupling of FM and AM modulation, it is more convenient to use indirect measurement techniques. One possibility is to convert the frequency modulation into amplitude modulation by using an optical interferometer [62]. An example of a suitable experimental setup is shown in Figure 4.19a; a plane mirror Fabry-Perot interferometer is used to this purpose. The transmission profile of the interferometer converts the frequency modulation into an intensity modulated signal (Figure 4.19b) detected with a fast photodiode connected to an electrical spectrum analyzer.

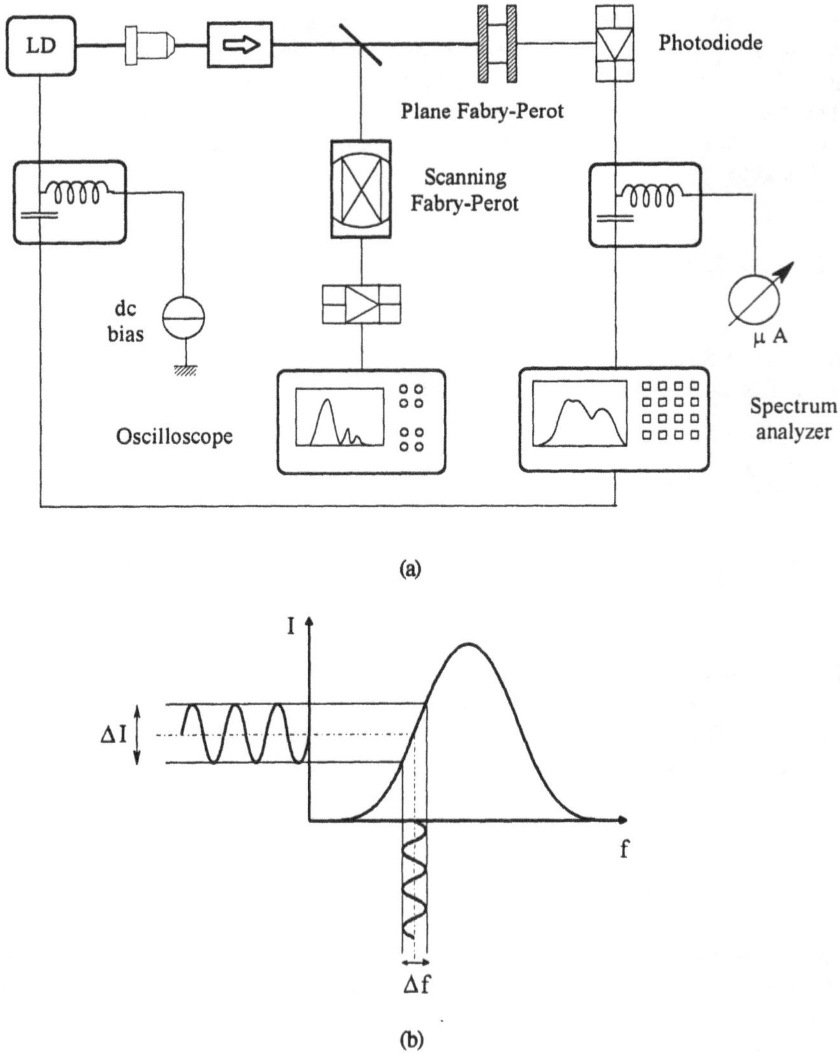

Figure 4.19 Measurement of the FM response of a SLM laser diode. (a) Experimental setup, (b) principle of operation of the frequency-amplitude conversion.

The effects of the nonlinear response of the FP can be kept to a negligible value by a proper choice of the working point of the device. A residual optical AM contribution to the signal can be removed by performing twice the measurement on both sides of the FP transmission peak at the same height (by adjusting the resonator length). The difference of the two measurements gives the FM response.

An alternative method to measure the FM response for a current-modulated laser diode relies on an approach similar to the self-heterodyne technique. The device under test is dc biased above threshold and RF modulated. The modulating carrier is alternatively switched on and off every period τ_o, equal to the differential time delay provided by an optical fiber Mach-Zehnder interferometer (see Figure 4.20) [63, 64]. The modulated and unmodulated laser fields are continuously recombined onto a fast photodetector.

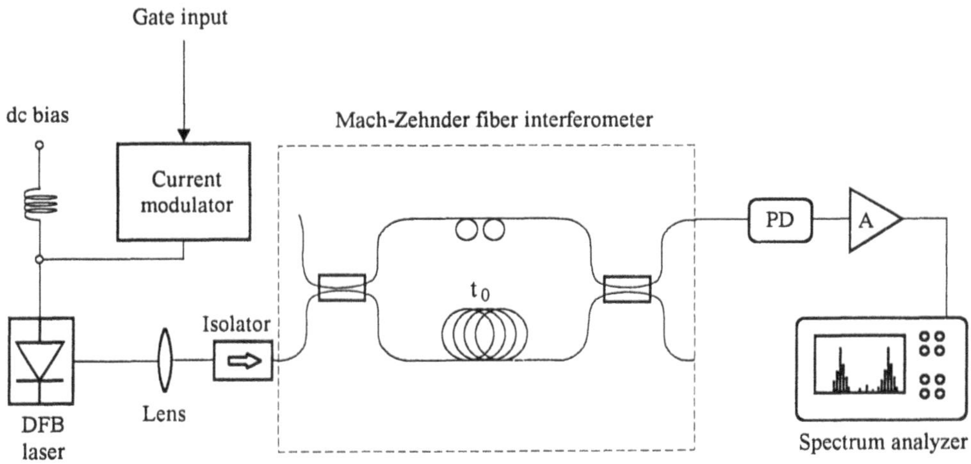

Figure 4.20 Self-heterodyne technique to measure the FM response of a laser diode. Experimental setup.

Typically a delay of few microseconds is chosen which is five times or more greater than the coherence time of a DFB or DBR laser. The current signal is sent to a microwave spectrum analyzer. Due to the gate applied to the modulation, the laser source to be analyzed in the heterodyne scheme is also alternatively used as the CW local oscillator, without need of an external AO modulator in the second arm of the Mach-Zehnder interferometer: for this reason the technique is called *gated delayed self-homodyne* (GDSH).

One important characteristic of this arrangement is that no thermal and current stabilization are required because it is selftracking within the time τ_o. The isolation from spurious optical feedback can be achieved by means of the usual techniques. A typical

GDSH photocurrent power spectrum includes several FM sidebands generated by the amplitude-phase coupling (see Figure 4.21). Under the hypothesis that the mixed fields are uncorrelated (i.e., $\Delta\omega\tau_0' \gg 1$) and that no appreciable wavelength shift is induced by the current modulation, the observed spectrum is given by

$$S_i(\omega) = S_D(\omega) + \frac{2\Delta\omega}{\Delta\omega^2 + \omega^2} \otimes \{S_m(\omega) + S_m(-\omega)\} \tag{4.58}$$

where symbol $\otimes$ denotes a spectral convolution.

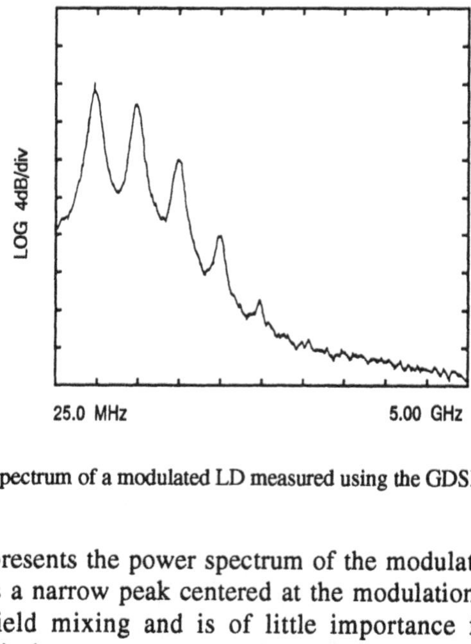

Figure 4.21 Power spectrum of a modulated LD measured using the GDSH technique (from [64]).

The term S_D represents the power spectrum of the modulated intensity and can be easily recognized as a narrow peak centered at the modulation frequency. It does not depend upon the field mixing and is of little importance for the laser spectrum measurement. $S_m(\omega)$ is the power spectrum of the optical modulated field and contains the information on the FM response of the device. In the photocurrent spectrum this term is convoluted with the static linewidth Lorentzian term. If $S_m(\omega)$ is an even function we can write

$$S_i(\omega) = \frac{4\Delta\omega}{\Delta\omega^2 + \omega^2} \otimes S_m(\omega) \tag{4.59}$$

The spectrum $S_m(\omega)$ contains several sidebands induced by the modulation. The amplitude of the sideband peak of order n is proportional to the Bessel function $J_n(\beta)$ of the same order calculated for the phase modulation index β considered in the former of Equations (4.54).

The static linewidth can be measured by applying a signal with a low modulation depth and considering the width of the first FM sideband. The phase modulation index β and, consequently, the linewidth enhancement factor b can be extracted by increasing the modulation depth until the first zero of the first-order Bessel function is achieved. This condition can be recognized by the lack of first sideband in the FM spectrum. This technique allows also to measure the laser spectrum induced by other transmission formats: for example a *pseudo-random binary sequence* (PRBS) can be applied to the device during the text. Frequency chirp up to $\pm$ 22 GHz has measured with a resolution approximately equal to the static linewidth $\Delta\nu$ of the laser [64]. This range corresponds to the typical frequency shifts obtainable from a DFB laser. Recently, a frequency-swept variant of the technique (S-GDSH) has been proposed [65] that allows the whole FM deviation curve to be recorded in a single measurement instead of collecting several single points measurements at fixed modulation frequencies.

4.2.4 Noise Figures Evaluation

In high-data-rate communication links, the performance of the system is limited by the noise. Depending on the detection scheme adopted (which can be direct or coherent) and on the modulation format, different noise terms become predominant.

A knowledge of statistical intensity fluctuations of the laser diode emission is important in high-speed direct detection systems; the intensity noise also influences the bit error rate in optical heterodyne detection systems.

In PCM the limiting phenomenon is usually the pattern effect experienced when long sequences of digital pulses are emitted by the diode. In this case the last pulses in the train may undergo distortion and amplitude reduction due to cumulative increments in electronic density caused by the preceding pulses [66]. This can lead to intersymbol interference which degrades the bit error rate. To measure the pattern effect, a high-rate logical sequence generator (up to 5 Gbit/s) is applied to the device and the output pulses are monitored by means of a direct measurement system.

A knowledge of the FM noise spectrum is relevant for FSK and PSK modulation schemes used in coherent transmission systems because this noise appears as an additive term in the noise of the final demodulated signal.

For long-distance fiber-optic communication systems operating at transmission rate in the range of Gbit/s under AM modulation, the ultimate limit to the bit rate is given by the dynamic line broadening or chirp rather than the static linewidth of the transmitter.

Intensity Fluctuations (IM Noise or AM Noise) in Highly Coherent Lasers

It has been shown that well above threshold the main contribution to the laser noise comes from random phase fluctuation (FM noise). In fact the nonlinear gain saturation mechanism effectively stabilizes the laser intensity so that at high-power oscillation regimes the intensity noise does not contribute to the laser linewidth (see Section 4.2.2). However, below and near the laser threshold the amplitude modulation noise is comparable to the phase noise. A theoretical treatment of the laser noise is given in [67–69].

The intensity noise can be measured as electrical noise by using a photodetector and a spectrum analyzer. The noise power spectrum is given by two main contributions [70, 71]

$$\Delta P(\omega) = |b(\omega)|^2[\eta_o\langle I(t)\rangle F + \eta_o{}^2\Delta P_I(\omega)] \tag{4.60}$$

where η_o is the detector quantum efficiency supposed to be known, $b(\omega)$ is the frequency response function of the complete system (i.e., the Fourier transform of the intensity-pulse response function of detector and attached electronic chain). $\langle I(t)\rangle$ is the time averaged light intensity (in photons per second), F is a factor accounting for the shot-noise increase due to the multiplication process inside the photodetector, and $\Delta P_I(\omega)$ is the power spectrum of the laser intensity fluctuations. Using an APD detector the system response $|b(\omega)|^2$ can be written in terms of operative parameters as

$$|b(\omega)|^2 = 2q^2R_LG(\omega)\langle M\rangle^2 \tag{4.61}$$

where q is the electron charge, $G(\omega)$ is the transfer function of the photodiode-amplifier combination, R_L is the load resistance, and $\langle M\rangle$ is the APD multiplication factor.

F is the APD excess noise factor, which basically accounts for the gain process in the avalanche photodiode, already given in Section 2.3.2 (Equation (2.94)). We recall that the expression for F can be put in the form

$$F(\langle M\rangle) = \left[k\langle M\rangle + \left(2 - \frac{1}{\langle M\rangle}\right)(1 - k)\right] \tag{4.62}$$

where k is the holes to electron ionization ratio coefficient (see Section 3.3.4) (approximately 0.02 for Si-APD) [71].

The first term in parenthesis in Equation (4.60) represents shot noise multiplied by the factor F, whereas the second term is usually called excess noise. Due to the integration over the finite bandwidth Δf of the spectrum analyzer, the actual measurable quantity is the r.m.s. of $\Delta P(\omega)$ given by

$$\text{r.m.s.}(f) = (2\Delta f)^{1/2} |b(2\pi f)| \{\eta_o \langle I(t)\rangle + \eta_o^2 \Delta P_I(2\pi f)\}^{1/2} \tag{4.63}$$

The average intensity $\langle I(t)\rangle$is related to the dc photocurrent $\langle J(t)\rangle$ by the relation

$$\langle J(t)\rangle = |b(0)|\eta_o\langle I(t)\rangle \tag{4.64}$$

A measurement of r.m.s.(f) cannot lead to a direct evaluation of $\Delta P_I(\omega)$ due to the presence of the shot noise term and to the presence of the instrumental transfer function $b(\omega)$. It is then convenient to express the AM noise spectrum $\Delta P_I(\omega)$ with respect to the shot noise. For this purpose a calibration procedure can be followed by measuring the r.m.s.(f) signal of an incoherent broadband light source. For instance, an LED source or a filtered halogen lamp present the required characteristics of large linewidth $\Delta\nu$ of the emission spectrum in comparison with the frequency f. For this source, under easily verified conditions [68] the r.m.s.(f) only consists of white spectrum shot noise and is given by

$$\text{r.m.s.}(f)_{LED} = (2\Delta f)^{1/2} |b(2\pi f)| \{\eta_o \langle I(t)\rangle\}^{1/2} \tag{4.65}$$

For each frequency f, the power of the incoherent source must be varied until the r.m.s.(f) signal equals the one acquired with the laser under test. In this condition the corresponding average currents $\langle J(t)\rangle_{LED}$ and $\langle J(t)\rangle_{Laser}$ are measured using a dc current meter. With the aid of relations (4.63), (4.64), and (4.65), one gets the equation

$$\Delta P_I(\omega) = \frac{[\langle J(t)\rangle_{LED} - \langle J(t)\rangle_{Laser}]}{\eta_o^2 |b(0)|} \tag{4.66}$$

In the above formula the $\Delta P_I(\omega)$ is expressed as difference between two shot-noise terms (proportional to $\langle J(t)\rangle$). This leads to the excess noise term due to the fully shot-noise characteristic of the incoherent source emission.

A significant measure of the intensity noise of a laser diode is given by the ratio of the electrical noise $\Delta P(\omega)$ (including the shot noise term) to the average optical output power. This figure is called *relative intensity noise* (RIN):

$$\mathrm{RIN} = \frac{\Delta P(\omega)}{\langle I(t) \rangle} \tag{4.67}$$

Assuming a unit load resistance and a unit bandwidth of the spectrum analyzer [71] the noise power spectral density is given by the noise current terms so that

$$\mathrm{RIN} = \frac{\langle i_n^2 \rangle_{AM} + 2q\langle i_{ph}^0 \rangle}{\langle i_{ph}^0 \rangle^2} =$$

$$= \frac{\Delta P(\omega)}{2\Delta f\, R_L G(\omega) \eta_o^2 \langle M \rangle^2 \langle i_{ph}^0 \rangle^2} + \frac{2q}{\langle i_{ph}^0 \rangle}\left[1 - \frac{F}{\eta_o}\right] \tag{4.68}$$

where $\langle i_n^2 \rangle_{AM}$ is the spectral density of the noise current and $\langle i_{ph}^0 \rangle$ is the dc photodetector current assuming $\eta_o = 1$. To measure the RIN, the transfer function $G(\omega)$ can be evaluated by using an incoherent source (LED) as described above while the other factors (R_L, $\langle M \rangle$, F and Δf) are assumed to be known.

The intensity noise power spectrum can be measured using the experimental setup shown in Figure 4.22 [70]. Due to thermal noise level in low-impedance receivers, including an APD photodetector and a broadband amplifier, the intensity noise cannot be directly measured by means of a spectrum analyzer; in fact, the thermal noise at typical photocurrent level is higher than the optical noise of the laser [70, 72]. To overcome this limitation, the light source after a proper optical isolation stage must be modulated by means of a low-frequency chopper. The optical component of the noise which is correlated with the chopper frequency is measured by a spectrum analyzer followed by a lock-in amplifier, while the uncorrelated photodetector noise (i.e., the electronic amplifier thermal noise and the APD dark current shot noise [71]) component is rejected. The noise signal is acquired as a function of the frequency with the spectrum analyzer span set at zero. The second lock-in amplifier shown in Figure 4.22 measures the average photocurrent through an integration network with suitable time constant. The output of this instrument is related to the average optical intensity. The computer controls the acquisition process and calculates the ratio between the term $\Delta P(\omega)$ and the dc component.

FM Noise – Chirp and Power Spectrum of a Modulated Laser

The FM quantum phase noise in semiconductor lasers is generally higher than the intensity noise. It does not appear in direct detection systems but it is converted into intensity fluctuations in coherent systems. Hence it is significant in heterodyne detection adopted for PSK or FSK modulation schemes.

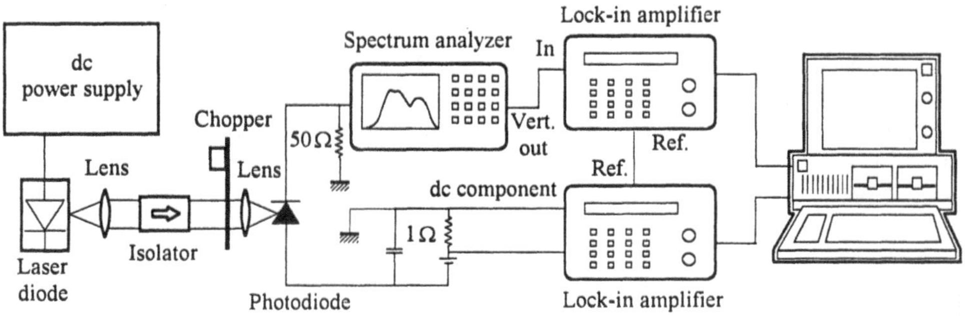

Figure 4.22 Experimental setup to measure the intensity noise of a laser diode.

The linewidth of a semiconductor laser is determined by the phase noise. In Section 4.2.2 we have considered the static linewidth of a laser diode; however, the dynamic linewidth induced by the modulation (chirp) is typically much greater.

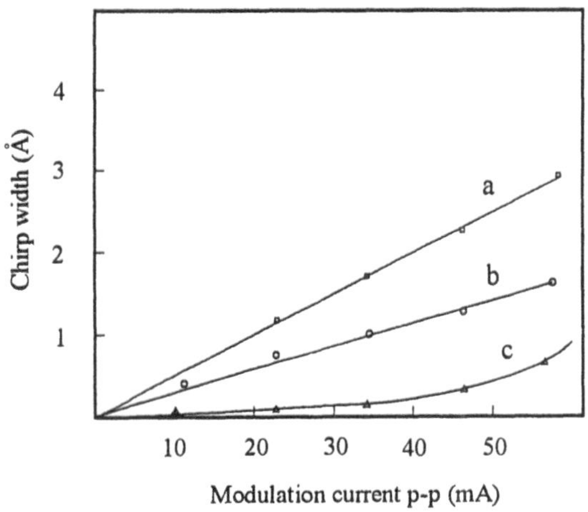

Figure 4.23 Chirp width as a function of the modulation current for: (a) a buried crescent C^3 laser, (b) a DCPBH C^3 laser, and (c) a GRIN-rod external cavity ridge waveguide laser (from [73]).

While a modulation is applied to the device biased above threshold, the chirp can be measured using a high-resolution monochromator [73]. Depending on the current modulation depth and frequency (100 to 800 MHz) the device experiences a wavelength shift during every time period: a photodiode array at the monochromator output can measure the chirp width as a broadening of the laser emission line averaged over several modulation cycles. A typical plot of line broadening due to the chirp for different modulation depths is shown in Figure 4.23.

The major influence on the emission frequency at high modulation rate comes from the variation of the injected carrier density at the junction; to observe this effect, time-resolved chirp measurements can be performed using a streak camera (see Section 3.2.2) with the experimental setup shown in Figure 4.24 [74].

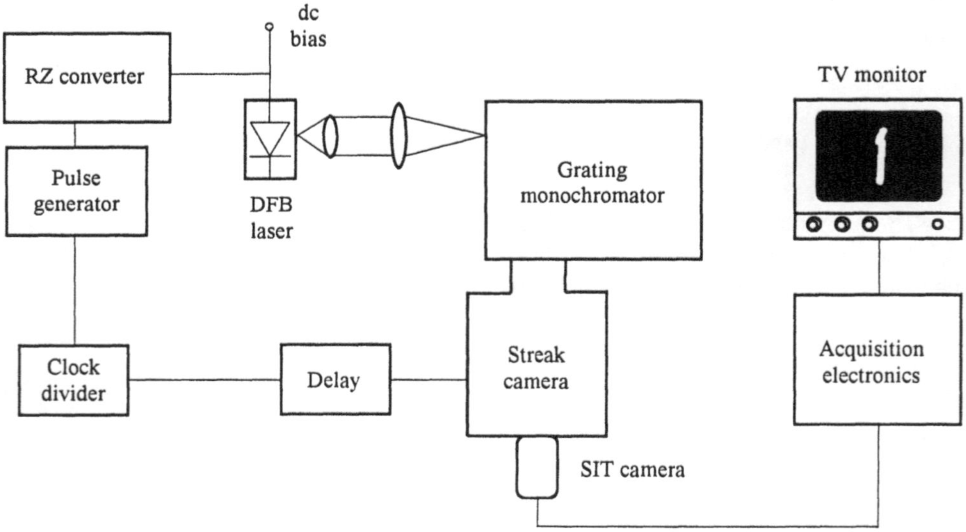

Figure 4.24 Chirp measurement setup based on a streak camera.

The light from the laser diode is wavelength dispersed by a grating spectrometer coupled to a synchroscan streak camera. The SIT camera is a high-sensitivity, high-resolution TV camera which converts the time resolved signal into a standard TV signal. This allows to record the time behavior of the wavelength shift and to correlate the direction of the frequency deviation with the phase of the driving pulse. The wavelength chirp is of the order of 3–5 Å. Wavelength resolution as low as 0.1–0.2 Å and time resolutions of few ps can be achieved. Other methods to measure the chirp rely on the self-heterodyne scheme (see Section 4.2.3).

1/f Noise

Measurements of the linewidth of laser diodes as a function of the emitted power revealed a departure from the modified Schawlow-Townes formula (4.44), which accounts for the spontaneous emission and gives a P^{-1} dependence of the linewidth [75]. This power independent noise sets the lower limit to the linewidth of semiconductor lasers. Several different explanations have been proposed for this line-broadening effect [76], including statistical fluctuations of the number of electrons in the active layer which produce variations in the refractive index, thermal fluctuations of the occupancies in the conduction and valence bands, mode competition, and the 1/f noise in the FM noise spectrum [77, 78].

The 1/f noise is present also in the AM noise spectrum, however, it has been shown that the correlation between the low-frequency 1/f AM and FM noise spectra is rather weak [78] in contrast with the strong correlation occurring between the power dependent AM and FM noises. This last correlation gives rise to the b enhancement factor (see Section 4.2.3). The FM noise can be measured by converting it to intensity noise by means of a Fabry-Perot interferometer [78] (see Section 4.2.3). The laser driving current must be carefully stabilized. To avoid any low-frequency noise component in the driving current, a lead battery can be used to bias the device under test. The laser frequency is temperature tuned to the maximum slope of the interferometer transmission curve with a temperature stabilization within 0.01 K. The resulting AM noise is detected by a photodiode attached to a spectrum analyzer.

In a different approach the 1/f noise can be measured by considering the marked dependence of the linewidth on the measurement time [77] due to the low-frequency components of this noise. The linewidth can be evaluated with a delayed self-heterodyne technique with two different lengths of the fiber delay line [79]. As an example, in the experimental work reported in [79], two fiber lengths of 1.5 and 27 km have been used. The FM noise spectrum can be assumed equal to

$$S(f) = C + \frac{k}{f} \tag{4.69}$$

The term C accounts for the white noise and inversely depends on the output power, whereas k is a constant. The laser power spectrum measured with the delayed self-heterodyne method considering a delay time τ_d turns out to be

$$I(f) = \mathbf{F}\left\{ \exp\left[\int_0^\infty S(f') \frac{\sin^2 \pi f' t}{f'^2} (1 - \cos 2\pi f' \tau_d) df' \right] \right\} \tag{4.70}$$

Let us consider a frequency much lower than the inverse of the delay time τ_d; the phase difference between the reference and signal beam is small and this component does not contribute to the FM spectrum detected by the self-heterodyne method. Hence the noise components at frequencies below $1/\tau_d$ are not detected by this technique. The presence of a low-frequency component in the FM spectrum can be revealed because the linewidth measured using the longer delay is wider.

For instance, the linewidth of a DFB laser biased six times above the threshold measured with a fiber length of 1.5 km is 14 MHz, while that measured with the 27-km fiber is 21 MHz. From a power spectrum measurement the constant k in Equation (4.69) has been estimated as $k = 1.44\ 10^{12}\ Hz^2$. Assuming $C = 0$ far above threshold, Equation (4.69) predicts a linewidth of 21 MHz using the 27-km fiber and of 17 MHz using the 1.5-km fiber. These values agree with the experimental results and lead the authors to conclude that the low-frequency 1/f noise component is the predominant source of the power independent line broadening in DFB lasers.

Terminal Electrical Noise

The measurement of electrical noise at the *laser diode terminal* (TEN) allows one to determine several important parameters of SLM lasers, such as the threshold current, the emission linewidth, and the mode-hopping behavior [80]. The experimental setup to measure the TEN as well as the optical intensity noise (see Section 4.2.3) is shown in Figure 4.25. The temperature of the laser is carefully controlled using a Peltier module. A tilted attenuator, or a Faraday isolator, is inserted in the optical path to reduce the optical feedback. The electrical noise is amplified by a low-noise amplifier and evaluated by a spectrum analyzer. For very low noise levels, a lock-in amplifier must be connected to the spectrum analyzer. In this case the laser diode injection current must be modulated with a square wave from zero to the selected bias. A simultaneous detection of the optical intensity noise shows that the two noises exhibit similar behavior versus frequency.

4.2.5 Radiation Pattern Evaluation: Near-Field, Far-Field, Astigmatic Distance, and Polarization

The near-field radiation pattern of a laser diode is the intensity distribution of the laser light immediately beyond its output facet. The angular spectrum of the laser is represented by the far-field pattern which is the intensity distribution at a distance much greater than the Rayleigh range, defined as

$$z_R = \frac{\pi w_0^2}{\lambda} \tag{4.71}$$

where w_0 is the beam waist* radius.

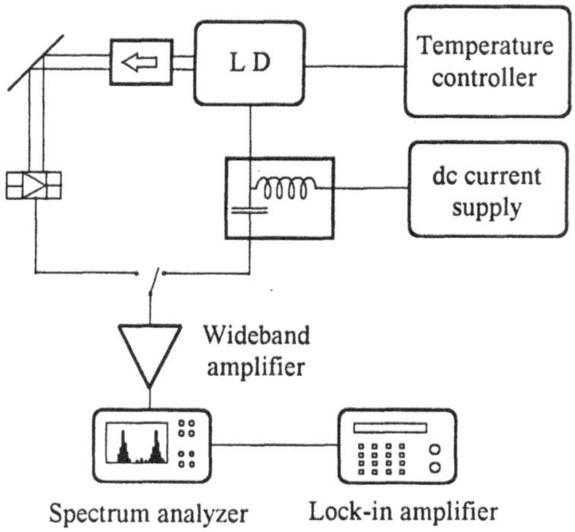

Figure 4.25 Experimental apparatus to measure the electrical noise of a laser diode. The intensity noise is measured in the meantime.

Since the laser diode emission is strongly astigmatic, the waist has two different radii along the directions parallel and normal to the junction. The divergence in the normal plane is higher than in the parallel plane, even though in the normal plane the beam is almost diffraction limited because of the smaller dimension of the emitting facet. A knowledge of these characteristics is important to optimize the design of collimators and fiber-coupling optics.

In practice the near-field distribution is obtained by imaging the output facet with a suitable optical system onto a TV camera [81].

A proper setup can be arranged to simultaneously acquire both the near- and the far-field. A possible scheme is shown in Figure 4.26. The device under test is mounted on a three-axis translation stage to permit fine alignment and focusing. O_1 is a high-quality microscope objective that images the output facet of the laser device onto the CCD matrix of camera C_1. To achieve a suitable magnification, O_1 should be a 40X-50X objective which has a focal length of 4–5 mm. The distance D_1 is set equal to 2 m in order to have a magnification factor around 300–400. The beam-splitter BS divides the beam in two portions; the lens L_2 allows to acquire the far-field pattern of the laser by making an image of the second focal plan of O_1 onto the CCD matrix camera C_2. Special care must be taken to perform a precise alignment of the device with the optical axis of the system. Moreover,

* The waist of a laser beam is the position of minimum transverse dimension.

the output facet of the laser must be carefully positioned by means of the micrometric translator until the focusing condition is met.

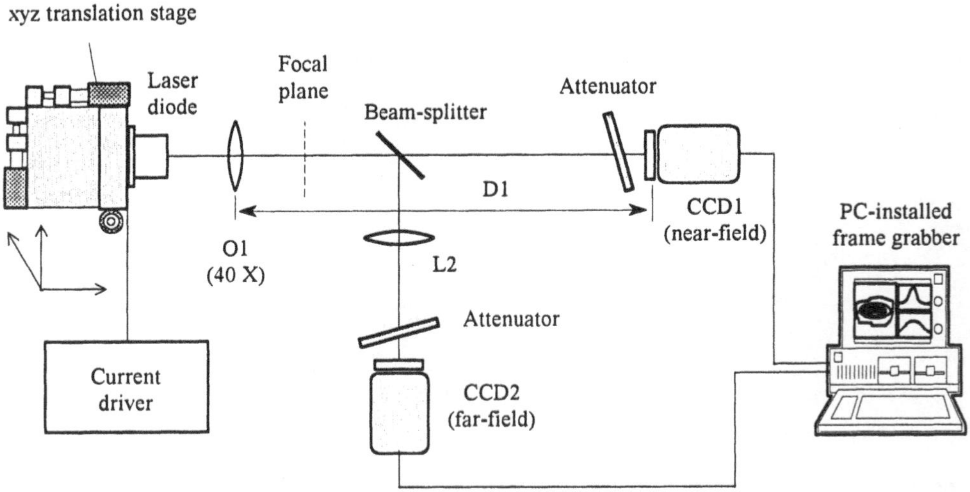

Figure 4.26 Experimental setup to contemporaneously acquire both the near- and the far-field profiles of a laser diode.

To evaluate the magnification factor of the near-field, the diode can be laterally displaced by few μm and the consequent displacement of the image at the camera surface can be measured. An absolute measurement requires the knowledge of the CCD pixel dimensions given by the manufacturer. The far-field is obtained in the focal plane of O_1 which makes the Fourier transform of the near-field of the laser. The magnification factor M_2 of L_1 is around 1. The scaling factor between the angular divergence and the linear dimension at the CCD surface is given by

$$\Delta\alpha = \frac{\Delta l}{f M_2} \tag{4.72}$$

where f is the focal length of O_1. The images from C_1 and C_2 can be recorded using a frame-grabber installed into a PC. Commercially available software packages provide 2-D/3-D display of the near and far-field patterns, with possibility of Gaussian fittings, sections along selected axes, and calculations of power/energy contained within user defined apertures.

A simpler method for measuring the beam profiles relies on a slit facing a large-area photodiode. For the near-field case, the slit/photodiode assembly, mounted on a fine translation stage, must be positioned in place of camera C_1. A transverse scan of the slit position gives the near-field profile along the direction orthogonal to the slit orientation.

For the far-field angular profile due to the high divergence of a laser diode, a different setup is suggested (see Figure 4.27). The device is mounted at the center of a rotational stage whereas the slit is fixed and positioned some centimeters apart from the laser facet. After a careful centering of the diode facet on the rotation axis, the stage is turned and the output from the photodiode is recorded and correlated to the goniometer position.

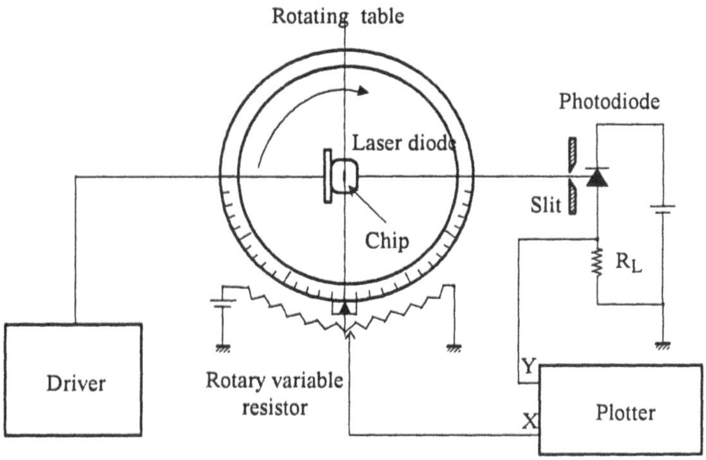

Figure 4.27 Experimental arrangement to measure the far-field angular profile of a laser diode.

Typical far-field profiles of a laser diode are shown in Figures 4.28(a) and 4.28(b). A knowledge of the near- and far-field intensity profiles gives information about the properties of spatial coherence of the laser beam. Actually, assuming a perfect spatial coherence, the near- and far-field profiles (amplitude and phase) are related by the Fourier transform operation. In an experimental evaluation the amplitude profiles (both near- and far-field) can be obtained as the square root of the measured intensity profiles, while the perfect coherence condition translates into the assumption of a uniform phase distribution over the near- and far-field cross sections of the beam. Under this hypothesis the magnitude of the numerically calculated 2D Fourier transform of the near-field amplitude profile should be equal, except for a scale factor dependent on the far-field lens focal length, to the far-field amplitude profile [82]. A departure from this behavior indicates the lack of a perfect spatial coherence in the laser beam, which can be due to multiple transverse mode oscillation and to phase aberration.

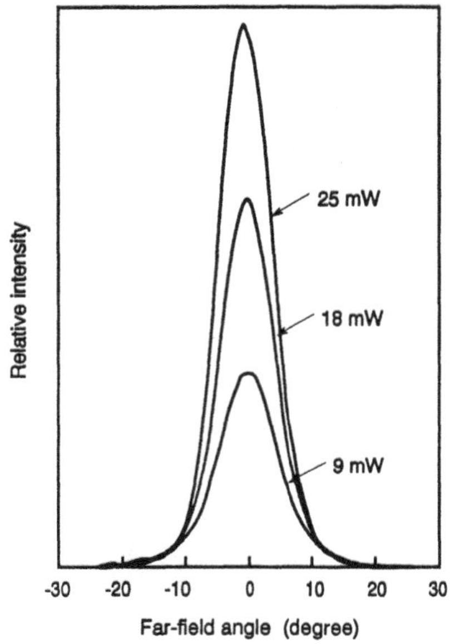

(a)

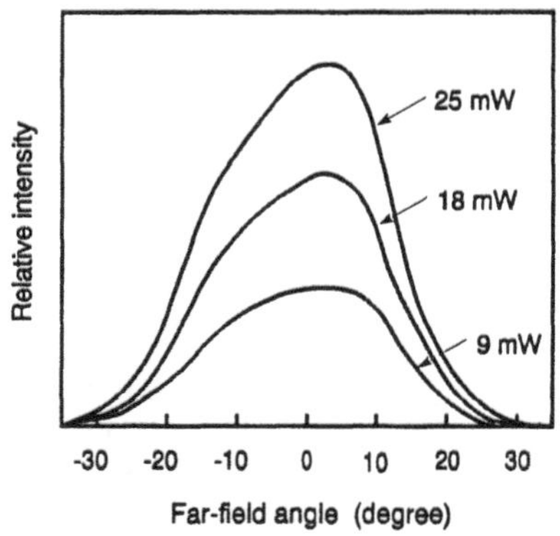

(b)

Figure 4.28 Typical far-field profiles of a laser diode (a) parallel and (b) orthogonal to the junction.

A typical feature of the emission of a laser diode is its intrinsic astigmatism. The beam exhibits two different beam waists along the direction parallel and orthogonal to the junction. A relatively simple method to measure both beam waist positions and radii is the knife-edge technique, already introduced in Section 3.1.1 for the spot-size measurement and shown in Figure 4.29. The microscope objective O focuses the beam while the knife edge cuts the beam at a fixed plane. The beam power beyond the knife edge is collected by a large area PIN detector while the knife edge is moved across the beam. The photocurrent intensity versus the knife-edge position is recorded; the beam diameter along one direction can be obtained as the FWHM of the derivative of the measured curve. Multiple measurements are performed by varying the distance z between the diode and the microscope objective in order to obtain a curve of the beam diameter versus z. This must be repeated for the two orthogonal directions; the astigmatism is evaluated by measuring the distance between the two focal planes (astigmatic distance).

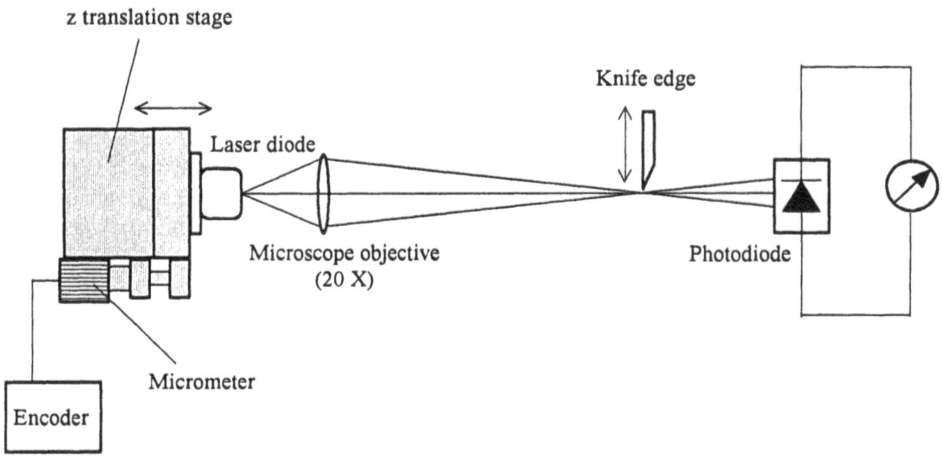

Figure 4.29 Knife-edge method to measure the beam waist positions and radii along the orthogonal directions.

Another measurement concerning the optical beam properties is the determination of the polarization ratio. To this purpose the light from a laser diode is collimated with suitable optics and analyzed using a high extinction ratio polarizer (e.g., a Glan-Thompson prism) mounted on a rotating stage. The transmitted light is measured as a function of the polarization angle by a photodiode or a radiometer.

4.3 MEASUREMENTS OF OPTICAL FIBER AMPLIFIERS

Optical amplifiers are becoming key components in high-performance fiber communication links. In the near future they are predicted to replace digital regenerators, giving substantial improvements in terms of transparency, bandwidth, and reliability. Optical amplifiers are also suitable for *wavelength division multiplexing* (WDM) applications in coherent transmission systems. To this purpose a key factor is the transparency of amplifiers to future upgrading of the link capacity which is limited by amplification noise and bandwidth. Optical amplifiers can be usefully employed in optical broadcasting networks (CATV). Also direct detection systems can take advantage of the use of optical preamplification stages.

Three basic optical amplifiers have been proposed: *semiconductor laser amplifiers* (SLA), fiber Raman amplifiers, and rare-earth-doped fiber amplifiers. Of these, the *erbium-doped fiber amplifier* (EDFA), already introduced in Section 3.1.12 looking at its properties that do not require a particular light degree of temporal coherence, is undoubtedly the most developed and important device. Even though, in principle, SLAs offer the smallest size and the highest efficiency, they are polarization sensitive devices and involve high losses in fiber connections. On the other hand, fiber amplifiers can be directly coupled to the optical link. Fiber Raman amplifiers are polarization sensitive and require higher pump levels in comparison to EDFA. The latter is then the most promising device [83].

EDFA consists of Er^{3+} doped single-mode fiber typically (but not exclusively) pumped with 980 nm or 1480 nm radiation from a laser diode. Different pump schemes are shown in Figure 4.30 [84]. In configuration (a) the input signal and the pump enter the amplifier from the same side. The band-pass filter separates the amplified signal and the residual pump power, the optical isolator prevent from undesired feedback reaching both the amplifier and the signal source. In configuration (b) the signal and the pump are counter-propagating, whereas in configuration (c) bidirectional pumping is employed. The fiber length affects the overall behavior of the device and an optimal length must be selected as a trade-off when noise figures in addition to gain are considered.

In the following we will address the evaluation of amplifier performances, involving measurements of the gain, the noise characteristic, and the bandwidth.

4.3.1 Amplifier Gain

At room temperature, Er^{3+} ions in a silica glass matrix can be modeled as a three-level laser system [85]. To treat the fiber amplifier gain behavior, it is better to introduce the normalized signal and pump intensities I_s', I_p':

$$I_p' = \frac{I_p}{I_{th}}$$
$$I_s' = \frac{1}{\varepsilon}\frac{\sigma_s \nu_p}{\sigma_p \nu_s}\frac{I_s}{I_{th}} \tag{4.73}$$

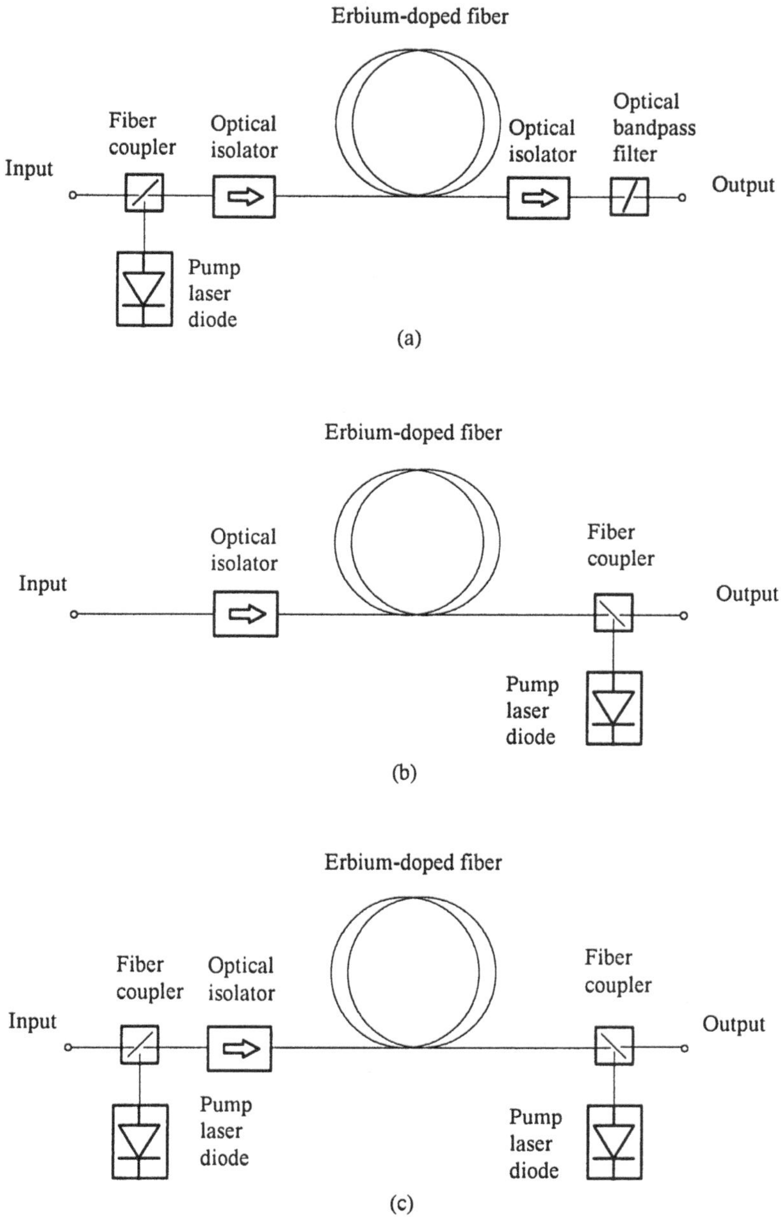

Figure 4.30 Different schemes used to pump an erbium-doped fiber amplifier: (a) forward-pumped EDFA, (b) backward-pumped EDFA, and (c) bidirectional-pumped EDFA (from [84]).

where $\varepsilon = \tau_{31}/(\tau_{32} + \tau_{31})$ is the branching ratio, τ_{ij} is the relaxation time between energy levels i and j, σ_s and σ_p are the transition cross section at the signal wavelength and the pump absorption cross section, respectively; I_{th} is the pump threshold intensity given by

$$I_{th} = \frac{h\nu_p}{\varepsilon\sigma_p\tau_{21}} \tag{4.74}$$

The equations describing the pump and signal intensity along the fiber amplifier in small signal regime (assuming monochromatic pump and signal radiation) are [84, 86]

$$\frac{dI_p}{dz} = -\frac{1 + I_s}{1 + 2I_s + I_p}\alpha_p I_p \tag{4.75}$$

$$\frac{dI_s}{dz} = -\frac{I_p - 1}{1 + 2I_s + I_p}\alpha_s I_s \tag{4.76}$$

where, z is the propagation coordinate, α_s and α_p are the absorption coefficient at signal and pump wavelength. From the above Equations (4.75) and (4.76), the gain $G = I^s_{out}/I^s_{in}$ for small input signal can be expressed as an implicit function of the normalized pump power:

$$\log_e\left(\frac{P_p}{P_{th}} - \frac{1}{2}\alpha_p L - \frac{1}{2}\frac{\alpha_p}{\alpha_s}\log_e G\right) + \frac{1}{2}\alpha_p L - \frac{1}{2}\frac{\alpha_p}{\alpha_s}\log_e G = \log_e\left(\frac{P_p}{P_{th}}\right) \tag{4.77}$$

where $P_{th} = I_{th} A_{eff}$ (A_{eff} being the fiber effective core area) is the pump power necessary to build up the population inversion of the amplifier modeled as a three-level system. It can be expressed as

$$P_{th} = \frac{A_{eff}\, h\nu_p}{\varepsilon\sigma_p\, \tau_{21}} \tag{4.78}$$

Once α_p is known, by adjusting the pump power until the output signal power equals the input ($G = 1$) and substituting the measured pump power P_0 in the above equation, one gets an indirect measurement of P_{th} according to the formula

$$\log_e\left(\frac{P_0}{P_{th}} - \frac{1}{2}\alpha_p L\right) + \frac{1}{2}\alpha_p L = \log_e\left(\frac{P_0}{P_{th}}\right) \tag{4.79}$$

The characterization of an optical amplifier can be carried out by measuring the values of α_s, α_p and P_{th}. The absorption coefficients can be measured by using the cut-back method or by considering the outputs from two fiber trunks having different lengths successively coupled to the same system, as already discussed in Section 3.1.12. Equation (4.77) can be used to obtain the gain as a function of the normalized pump power and the length of the fiber amplifier. The fiber length corresponding to maximum gain can be selected. By rearranging Equations (4.75) and (4.76) a differential equation in I_s' I_p' is obtained. This can be solved assuming that the pump power vanishes at the end of the fiber amplifier. This leads to the following expression for the maximum output signal as a function of the input signal and of the pump power at the amplifier entrance

$$\log_e\left(\frac{\gamma}{\varepsilon}\frac{P_s^{out}}{P_{th}}\right) - \log_e\left(\frac{\gamma}{\varepsilon}\frac{P_s^{in}}{P_{th}}\right) + \left(\frac{\gamma}{\varepsilon}\frac{P_s^{out}}{P_{th}} - \frac{\gamma}{\varepsilon}\frac{P_s^{in}}{P_{th}}\right) = \frac{\alpha_s}{\alpha_p}\left[\frac{P_p^{in}}{P_{th}} - 1 - \log_e\left(\frac{P_p^{in}}{P_{th}}\right)\right] \tag{4.80}$$

where $\gamma = \sigma_s \nu_p / \sigma_p \nu_s$.

A direct measurement of the gain can be performed by means of the setup shown in Figure 4.31. To pump the amplifier at 980 nm a GaAlAs laser diode or a Ti:Sapphire laser can be used. Alternatively a F-center laser or a InGaAsP laser diode can be employed to pump the amplifier at 1.48 μm. The optical isolators are polarization independent [87] and prevent from the onset of laser oscillation; the isolation factor should be as high as 60 dB. The pump radiation and signal are sent to the amplifier using a WDM fiber coupler. One of the coupler output port is used to monitor the power of the input signal; the other is connected to the amplifier. The output signal is measured by means of an optical spectrum analyzer and gives the dependence of the amplifier gain upon the input wavelength. Many laser diodes having different emission wavelengths can be used as signal sources to cover the amplifier bandwidth (ranging from 1.5 μm to 1.6 μm). Alternatively a broadly tunable external cavity laser diode can be used. The pump power must be kept at relatively low levels to avoid the depletion of the inversion by amplified spontaneous emission.

Gains higher than 45 dB can be achieved with fiber length of a few tens of meters. EDFA having 3 dB bandwidth as wide as 35 nm have been demonstrated using Er^{3+} doped fibers with a SiO_2-Al_2O_3-P_2O_5 core [88].

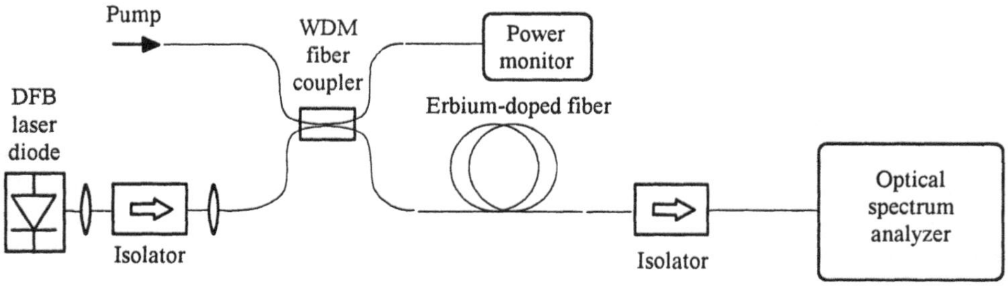

Figure 4.31 Experimental setup to measure the gain of an optical amplifier.

4.3.2 Amplifier Noise

The main source of noise in an optical amplifier is given by *amplified spontaneous emission* (ASE). For an EDFA operating in the nonsaturated portion of the characteristic, the two main contributions to noise come from signal-spontaneous (s-sp) beat noise and spontaneous-spontaneous (sp-sp) beat noise. However, since sp-sp beat noise is broadband, it can be easily eliminated using a spectral filter. Let us assume the optical output power n_{out} in photons per second given by

$$n_{out} = Gn_{in} + 2(G - 1)n_{sp}B \tag{4.81}$$

where n_{in} is the signal input power (photons per second), G is the gain, B is the amplifier bandwidth, and n_{sp} is the spontaneous emission factor defined as [89]

$$n_{sp} = \frac{N_2}{N_2 - \sigma_{12}/\sigma_{21}N_1} \tag{4.82}$$

where N_2 and N_1 are the populations of the upper and lower laser levels, respectively, σ_{12} and σ_{21} are the absorption and stimulated emission cross sections. The second term in (4.81) represents the amplified spontaneous emission. If we assume the noise to be a random process following Poisson statistics its mean square power is

$$\sigma_n^2 = Gn_{in} + 2(G - 1)n_{sp}B + 2n_{in}G(G - 1)n_{sp} + 2(G - 1)^2 n_{sp}^2 B \tag{4.83}$$

The first term in the above equation is the shot noise of the signal, the second is the shot noise of the amplified spontaneous emission, the third is the signal-spontaneous beat, and the last is the spontaneous-spontaneous beat. For high gain and strong input signal the only significant term in Equation (4.81) is the amplified signal, while the noise reduces to the s-sp beat.

The signal-to-noise ratio at the amplifier input is

$$\left(\frac{S}{N}\right)_{in} = \frac{(n_{in})^2}{n_{in}} = n_{in} \tag{4.84}$$

The signal-to-noise ratio at the amplifier output is

$$\left(\frac{S}{N}\right)_{out} = \frac{(n_{in}G)^2}{2n_{in}G(G-1)n_{sp}} \approx \frac{n_{in}}{2n_{sp}} \tag{4.85}$$

Under the previous assumptions the noise figure of the amplifier becomes

$$F = \frac{\left(\frac{S}{N}\right)_{in}}{\left(\frac{S}{N}\right)_{out}} \approx 2n_{sp} \tag{4.86}$$

For an ideal amplifier, completely inverted, the factor n_{sp} is equal to 1 and the noise figure is $F \approx 3$ dB. Hence a simple characterization of the amplifier noise reduces to a measure of the factor n_{sp}. For this purpose both electrical (photocurrent) and optical measurement techniques can be used.

Figure 4.32 shows an example of a suitable setup [90]. The pump radiation at 980 nm or 1.48 μm is mixed with the signal using a polarizing beam-combiner and sent to a fiber coupler having one port butt-coupled to the fiber amplifier. The other port is used to monitor the input light power. The signal can be obtained from an external cavity laser diode or from a diode-pumped Er^{3+} oscillator. The amplified output is filtered by a 2 to 3 nm passband filter centered at the signal frequency. The signal is detected by means of a InGaAs PIN photodiode followed by a wideband amplifier and a spectrum analyzer. The noises, both sp-sp and s-sp, depend on the pump power because this parameter affects the gain and the amplified spontaneous emission. At high pump power and, hence, high gain, the main contribution comes from the s-sp beat noise, as already stated. A measurement of the mean square photocurrent of the main noise sources as a function of the pump power is shown in Figure 4.33 for a fixed signal level. The dependence of the s-sp noise photocurrent on the input signal is shown in Figure 4.34 for pump power high enough so that this term dominates.

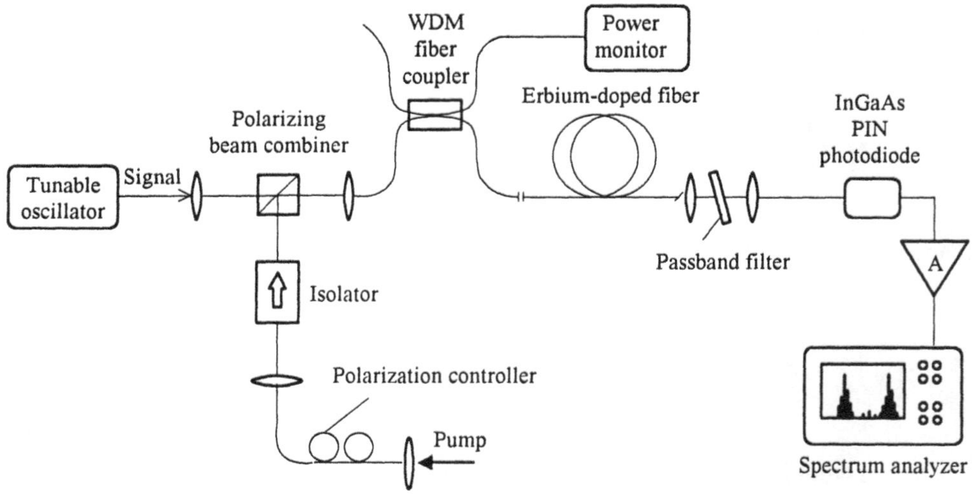

Figure 4.32 Experimental setup to measure the noise figure of an optical amplifier.

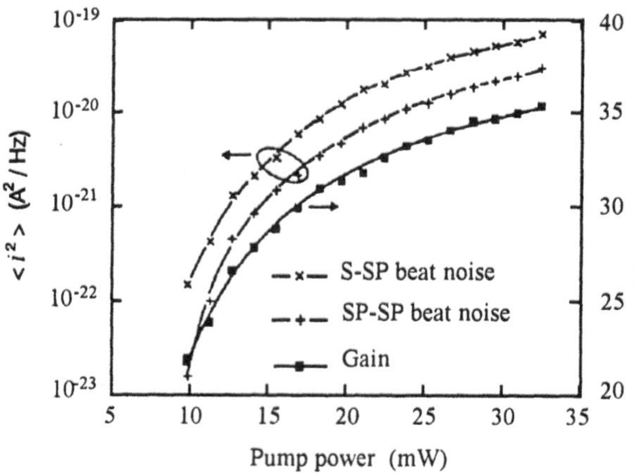

Figure 4.33 Mean square photocurrent of the main noise sources as a function of the pump power in an optical amplifier (from [90]).

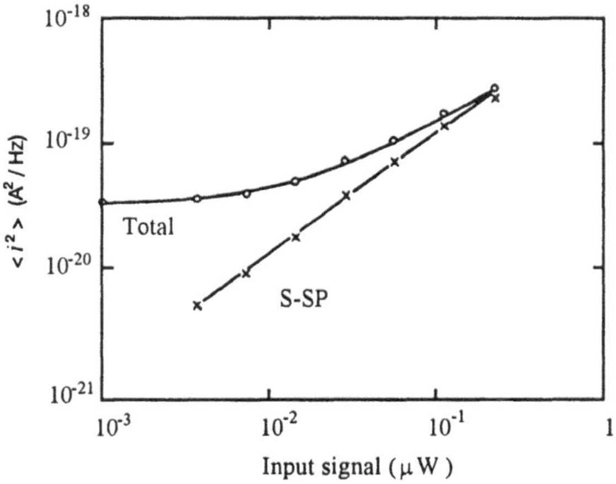

Figure 4.34 Dependence of the s-sp noise photocurrent on the input signal for pump power high enough so that this term dominates (from [90]).

The noise figure F is evaluated as the ratio of the mean square photocurrent noise $\langle i^2_{s\text{-}sp}\rangle$ to the noise predicted for an ideal amplifier:

$$F = \frac{\langle i^2_{s\text{-}sp}\rangle}{2\eta_o q(G-1)I_s} \tag{4.87}$$

where I_s is the mean signal photocurrent and η_o is the detector quantum efficiency.

From data of Figure 4.34 (for a gain of 37.4 dB $\lambda_s = 1531$ nm and $\lambda_p = 1493$ nm) a value of F = 5.2 dB is calculated.

Equivalently, the noise factor can be obtained from a measurement of the optical power of the amplified spontaneous emission within the passband filter:

$$P_{sp} = 2(G-1)n_{sp}Bh\nu \tag{4.88}$$

$$F = \frac{P_{sp}}{(G-1)Bh\nu} \tag{4.89}$$

A value of 4.8 dB has been calculated [90] from this measurement. It is remarkable that those values agree fairly well and are close to the theoretical limit.

For pumping at 980 nm a value of F from 3.2 dB to 3.42 dB has been estimated [85].

REFERENCES

[1] King, J. P., et al., "Development of a coherent OTDR instrument", *IEEE J. Lightwave Technol.*, Vol. LT-5, 1987, pp. 616–624.

[2] Sorin, W. V., et al., "Coherent FMCW reflectometry using a temperature tuned Nd:YAG ring laser", *IEEE Photon. Technol. Lett.*, Vol. 2, 1990, pp. 902–904.

[3] Shimizu, K., et al., "Measurements of Rayleigh backscattering in single-mode fibers based on coherent OFDR employing a DFB laser diode", *IEEE Photon. Technol. Lett.*, Vol. 3, 1991, pp. 1039–1041.

[4] Horiuchi, Y., et al., "Novel coherent heterodyne optical time domain reflectometry for fault localization of optical amplifier submarine cable systems", *IEEE Photon. Technol. Lett.*, Vol. 2, 1990, pp. 291–293.

[5] Koyamada, Y., and H. Nakamoto, "High performances single mode OTDR using coherent detection and fibre amplifiers", *Elettron. Lett.*, Vol. 26, 1990, pp. 573–575.

[6] Izumita, H., et al., "Coherent optical time-domain reflectometer with a 40 dB dynamic range enhanced with erbium-doped fiber amplifiers", *Proc. OFC'92*, San José, 1992, p. 147.

[7] Quoi, K. W., et al, "Screening dispersion-shifted fibers for polarization-mode dispersion due to core ellipticity", *IEEE J. Lightwave Technol.*, LT-8, 1990, pp. 1684–1687.

[8] Poole, C. D., "Polarization properties of single-mode fibers: intrinsic limits and uses", *Tech. Dig. OFC '88*, New Orleans, 1988, paper WA1.

[9] Love, J. D., et al., "Birefringence in elliptically deformed optical fibers", *Electron. Lett.*, Vol. 15, 1979, pp. 615–616.

[10] Poole, C. D., "Polarization-mode dispersion measurements in conventional single-mode fiber", *Proc. OFMC-NPL*, York, 1991, pp. 77–80.

[11] Matera, F., "Evolution of polarization mode dispersion in single-mode fibres", *Proc. X EFOC-LAN*, Paris, 1992, pp. 16–21.

[12] Poole, C. D., and R. E. Wagner, "Phenomenological approach to polarization dispersion in long single-mode fibres", *Electron. Lett.*, Vol. 22, 1986, pp.1029–1030.

[13] Curti, F., et al., "Statistical treatment of the evolution of the principal states of polarization in single-mode fibers", *IEEE J. Lightwave Technol.*, Vol. LT-8, 1990, pp. 1162–1166.

[14] Poole, C. D., et al., "Dynamical equation for polarization dispersion", *Opt. Lett.*, Vol. 16, 1991, pp. 372–374.

[15] Matera, F., and C. G. Someda, "Random birefringence and polarization dispersion in long single-mode optical fibers", in *Anisotropic and nonlinear optical waveguides*, G. I. Stegeman and C. G. Someda, eds., Elsevier, Amsterdam, 1992.

[16] Simon, A., and R. Ulrich, "Evolution of polarization along a single-mode fiber", *Appl. Phys. Lett.*, Vol. 31, 1977, pp. 517–520.

[17] Curti, F., et al., "Concatenation of polarization dispersion in single-mode fibres", *Electron. Lett.*, Vol. 25, 1989, pp. 290–292.

[18] Andresciani, A., et al., "Measurement of group-delay difference between the principal states of polarization on a low-birefringence terrestrial fiber cable", *Opt. Lett.*, Vol. 12, 1987, pp. 844–846.

[19] Poole, C. D., et al., "Polarization dispersion and principal states in a 147–km undersea lightwave cable", *IEEE J. Lightwave Technol.*, LT-6, 1988, pp. 1185–1190.

[20] Galtarossa, A., et al., "Concatenation of polarization dispersion: relationship between measurements before and after the installation of a long terrestrial cable", *Proc. OFC '91*, San Diego, p. 87.

[21] Galtarossa, A., et al., "Relationship between polarization dispersion measurements before and after installation of optical cables", *Electron. Lett.*, Vol. 27, 1991, pp. 595–596.

[22] Poole, C. D., "Measurements of polarization-mode dispersion in single-mode fibers with random mode coupling", *Opt. Lett.*, Vol. 14, 1989, pp. 523–525.

[23] De Angelis, C., et al., "Time evolution of polarization mode dispersion in long terrestrial links", *IEEE J. Lightwave Technol.*, Vol. LT-10, 1992, pp. 552–555.

[24] Gisin, N., "Polarization mode dispersion measurements in optical fibers, cables, installed terrestrial cables and fiber ribbons", *Proc. OFMC-NPL*, York, 1991, pp. 93–96.

[25] Bernard, J. J., et al., "Measurements of polarization dispersion in the components of in-line erbium doped fiber amplified links", *Proc. Opt. Amplifiers and Their Appl. Top. Meeting*, Colorado, 1991, pp. 272–274.

[26] Galtarossa, A., and M. Schiano, "Complete characterization of polarization mode dispersion on erbium doped optical amplifiers", to be published.

[27] Thèvenaz, L., et al., "Overall polarization dispersion after propagation through different fibres", *Proc. IX EFOC-LAN*, London, 1991, pp. 156–158.

[28] Gisin, N., et al., "Polarization mode dispersion of short and long single-mode fibers", *IEEE J. Lightwave Technol.*, Vol. LT-9, 1991, pp. 821–827.

[29] Passy, R., "Simultaneous measurements of beat length and polarization mode dispersion with the interferometric technique", *Proc. OFMC-NPL*, York, 1991, pp. 85–88.

[30] Namihira, Y., and H. Wakabayashi, "Fiber length dependence of polarization mode dispersion measurements in long-length optical fibers and installed optical submarine cables", *J. Opt. Commun.*, Vol. 12, 1991, pp. 1–9.

[31] Gisin, N., et al., "Comparison of measurements of polarization-mode dispersion in single-mode fibers", *Proc. OFC '92*, San Jose', 1992, p. 146.

[32] Meland, E., R. Holmstrom, J. Schlafer, R. B. Laucr and W. Powazinik, "Extremely high frequency (24 GHz) InGaAsP diode lasers with excellent modulation efficiency", *Electron. Lett.*, Vol. 26, 1990, pp. 1827–1829.

[33] Miller, S. E., and I. P. Kaminov, *Optical fiber telecommunications II*, Academic Press, New York, 1988.

[34] Shapiro, S., *Ultrashort light pulses – Picosecond techniques and applications*, Springer Verlag, Berlin, 1977, pp. 84–122.

[35] Yariv, A., *Quantum electronics*, Wiley, New York, 1975, pp. 421–428.

[36] Ito, H., H. Yokayama, S. Murata, and H. Inaba, "Picosecond optical pulse generation from an R.F. modulated AlGaAs D.H. diode laser", *Electron. Lett.*, Vol. 15, 1979, pp. 738–740.

[37] Andersson, T., and S. T. Eng, "Determination of the pulse response from intensity autocorrelation measurements of ultrashort laser pulses", *Opt. Commun.*, Vol. 47, 1983, pp. 288–290.

[38] Downey, P. M., J. E. Bowers, R. S. Tuker, and E. Agyekum, "Picosecond dynamics of a gain-switched InGaAsP laser", *IEEE J. Quantum Electron.*, Vol. QE-23, 1987, pp. 1039–1047.

[39] Lau, K. Y., C. Harder, and A. Yariv, "Ultimate frequency response of GAAs injection lasers", *Opt. Commun.*, Vol. 36, 1981, pp. 472–474.

[40] Paoli, T. L., and J. E. Ripper, "Direct modulation of semiconductor lasers", *Proc. IEEE*, Vol. 58, 1970, pp. 1457–1465.

[41] Kressel, H., and J. K. Butler, *Semiconductor Lasers and Heterojunction LEDs*, Academic Press, New York, 1977, p. 561.

[42] Vaughan, J. M., *The Fabry-Perot interferometer – History, theory, practise and applications*, Adam Hilger, Bristol, 1989.

[43] Hercher, M., "The spherical mirror Fabry-Perot interferometer", *Appl. Opt.*, Vol. 7, 1968, pp. 951–966.

[44] Takakura, T., K. Iga, and T. Tako, "Linewidth measurement of a single longitudinal mode AlGaAs laser with a Fabry-Perot interferometer", *Japan. J. Appl. Phys.*, Vol. 19, 1980, pp. 725–727.

[45] Ohtsu, M., *Highly coherent semiconductor lasers*, Boston, 1992, p. 40.

[46] Gowar, J., *Optical communication systems*, Prentice Hall International, London, 1984.

[47] Armstrong, J. A., "Theory of interferometric analysis of laser phase noise", *J Opt. Soc. Am.*, Vol. 56, 1966, pp. 1024–1031.

[48] Yariv, A., *Quantum electronics*, Wiley, New York, 1975, pp. 311–313.

[49] Rowe, H. E., *Signals and noise in communications systems*, D. Van Nostrand Company, Princeton, 1965.

[50] Henry, C. H., "Theory of the linewidth of semiconductor lasers", *IEEE J. Quantum Electron.*, Vol. QE-18, 1982, pp. 259–264.

[51] Fleming, M. W., and A. Mooradian, "Fundamental line broadening of single-mode (GaAl)As diode lasers", *Appl. Phys. Lett.*, Vol. 38, 1981, pp. 511–513.
[52] Richter, L. E., H. I. Mandelberg, M. S. Kruger, and P. A. McGrath, "Linewidth determination from self-heterodyne measurements with subcoherence delay times", *IEEE J. Quantum Electron.*, Vol. QE-22, 1986, pp. 2070–2074.
[53] Okoshi, T., K. Kikuchi, and A. Nakayama, "Novel method for high resolution measurement of laser output spectrum", *Electron. Lett.*, Vol. 16, 1980, pp. 630–631.
[54] Ries, R., F. Favre, and D. Le Guen, "Influence of extrinsic FM noise on laser diode linewidth measured by self-heterodyning", *Opt. Commun.*, Vol. 57, 1986, pp. 269–273.
[55] Ishida, O., "Delayed self-heterodyne measurement of laser frequency fluctuations", *IEEE J. Lightwave Technol.*, Vol. LT-9, 1991, pp. 1628–1533.
[56] Gallion, P. B., and G. Debarge, "Quantum phase noise and field correlation in single frequency semiconductor laser systems", *IEEE J. Quantum Electron.*, Vol. QE-20, 1984, pp. 343–349.
[57] Nazarathy, M., W. V. Sorin, D. M. Baney, and S. A. Newton, "Spectral analysis of optical mixing measurements", *IEEE J. Lightwave Technol.*, Vol. LT-7, 1989, pp. 1063–1096.
[58] Tai, S., K. Kyuma, and T. Nakayama, "Novel measuring method for spectral linewidth of laser diodes using fibre-optic ring resonator", *Electron Lett.*, Vol. 21, 1985, pp. 91–93.
[59] Harder, C., K. Vahala, and A. Yariv, "Measurement of the linewidth enhancement factor of semiconductor lasers", *Appl. Phys. Lett.*, Vol. 42, 1983, pp. 328–330.
[60] Welford, D., and S. B. Alexander, "Magnitude and phase characteristics of frequency modulation in directly modulated GaAlAs semiconductor diode lasers", *IEEE J. Lightwave Technol.*, Vol. LT-3, 1985, pp. 1092–1099.
[61] Vpdhanel, R. S., and S. Tsuji, "12 GHz FM bandwidth for a 1530 nm DFB laser", *Electron Lett.*, Vol. 24, 1988, pp. 1359–1361.
[62] Gambini, P., N. P. Caponio, M. Puleo, and E. Vezzoni, "Accurate measurement of the FM response in magnitude and phase of different DFB laser structures", *CSELT Technical reports*, Vol XVIII, 1990, pp. 123–130.
[63] Baney, D. M., and W. V. Sorin, "Measurement of a modulated DFB laser spectrum using gated delayed self-homodyne technique", *Electron. Lett.*, Vol. 24, 1988, pp. 669–670.
[64] Baney, D. M., and P. B. Gallion, "Power spectrum measurement of a modulated semiconductor laser using an interferometric self-homodyne technique: influence of quantum phase noise and field correlation", *IEEE J. Quantum Electron.*, Vol. QE-25, 1989, pp. 2106–2112.
[65] Baney, D. M., "Extending measurement domains with interferometry", *Laser&Optron.*, Vol. 10, 1991, pp. 49–52.

[66] Torphammar, P., R. Tell, H. Eklund, and A. R. Johnston, "Minimizing pattern effects in seminconductor lasers at high rate pulse modulation", *IEEE J. Quantum Electron.*, Vol. QE-15, 1979, pp. 1271–1276.
[67] Armstrong, J. A., and A. W. Smith, *Progress in optics*, E. Wolf, ed., Wiley, New York, 1964, Vol. VI, pp. 211–257.
[68] McCumber, D. E., "Intensity fluctuations in the output of cw laser oscillator. I", *Phys. Rev.*, Vol. 141, 1966, pp. 306–322.
[69] Yamamoto, Y., "AM and FM quantum noise in semiconductor lasers–Part I: theoretical analysis", *IEEE J. Quantum Electron.*, Vol. QE-19, 1983, pp. 34–46.
[70] Paoli, T. L., "Noise characteristics of stripe-geometry double-heterostructure junction lasers operating continuously–I. Intensity noise at room temperature", *IEEE J. Quantum Electron.*, Vol. QE-1, 1975, pp. 276–283.
[71] Yamamoto, Y., "AM and FM quantum noise in semiconductor lasers–Part II: comparison of theoretical and experimental results for AlGaAs lasers", *IEEE J. Quantum Electron.*, Vol. QE-19, 1983, pp. 47–58.
[72] Puleo, M., "Ultra-high sensitivity technique for characterization of laser diode intensity noise in the GHz range", *CSELT Technical reports*, Vol XVII, 1989, pp. 199–200.
[73] Olsson, N. A., N. K. Dutta, and K. Y. Liou, "Dynamic linewidth of amplitude modulated single-longitudinal mode semiconductor lasers operating at 1.5 m wavelength", *Electron Lett.*, Vol. 20, 1984, pp. 121–122.
[74] Kuwazuru, M., S. Yamamoto, K. Mochizuki, and M. Fujise, "Observation of chirp in a 1.5 m DFB-LD using a streak camera", *Electron Lett.*, Vol. 22, 1986, pp. 108–109.
[75] Welford, D., and A. Mooradian, "Observation of linewidth broadening in (GaAl)As diode lasers due to electron number fluctuations", *Appl. Phys. Lett.*, Vol. 40, 1982, pp. 560–562.
[76] Osinski, M., and J. Buus, "Linewidth broadening factor in semiconductor lasers – An Overview", *IEEE J. Quantum Electron.*, Vol. QE-23, 1987, pp. 9–29.
[77] O'Mahony, M. J., and I. D. Henning, "Semiconductor laser linewidth broadening due to 1/f carrier noise", *Electron. Lett.*, Vol. 19, 1983, pp. 1000–1001.
[78] Kikuchi, K., and T. Okoshi, "Measurement of spectra and correlation between FM and AM noises in GaAlAs lasers", *Electron Lett.*, Vol. 19, 1983, pp. 812–813.
[79] Kikuchi, K., and T. Okoshi, "Dependence of semiconductor laser linewidth on measurement time: evidence of predominance of 1/f noise", *Electron Lett.*, Vol. 21, 1985, pp. 1011–1012.
[80] Andrekson, P. A., P. Andersson, A. Alping, and S. T. Eng, "In situ characterization of laser diodes from wide-band electrical noise measurements", *IEEE J. Lightwave Technol.*, Vol. LT-4, 1986, pp. 804–812.
[81] Peled, S., "Near- and far-field characterization of diode lasers", *Appl. Opt.*, Vol. 19, 1980, pp. 324–328.

[82] Duncan, M. D., and R. Mahon, "Beam quality measurements using digitized laser beam images", *Appl. Opt.*, Vol. 28, 1989, pp. 4569–4575.
[83] Walker, G. R., N. G. Walker, R. C. Steele, M. J. Creaner, and M. C. Brain, "Erbium-doped fiber amplifier cascade for multichannel coherent optical transmission", *IEEE J. Lightwave Technol.*, Vol. 9, 1991, pp. 182–192.
[84] Nakagawa, K., S. Nishi, K. Aida, and E. Yoneda, "Trunk and distribution network application of erbium-doped fiber amplifier", *IEEE J. Lightwave Technol.*, Vol. 9, 1991, pp. 198–208.
[85] Koechner, W., *Solid-state laser engineering*, Second edition, Springer Verlag, Berlin, 1988, p. 64.
[86] Desurvire, E., J. R. Simpson, and P. C. Becker, "High-gain erbium-doped traveling-wave fiber amplifier", *Opt. Lett.*, Vol. 12, 1987, pp. 888–890.
[87] Masafumi, K., and T. Matsumoto, "High-gain polarization-insensitive optical amplifier consisting of two serial semiconductor laser amplifiers", *IEEE J. Lightwave Technol.*, Vol. 9, 1991, pp. 284–290.
[88] Yamada, M., M. Shimizu, M. Horiguchi, M. Okayasu, and E. Sugita, "Gain characteristics of an Er^{3+}-doped multicomponent glass single-mode optical fiber", *IEEE Photon. Technol. Lett.*, Vol. 2, 1990, pp. 656–658.
[89] Yariv, A., *Quantum electronics*, Wiley, New York, 1975, pp. 293–300.
[90] Giles, C. R., E. Desurvire, J. L. Zyskind, and J. R. Simpson, " Noise performance od erbium-doped fiber amplifier pumped at 1.49 mm, and application to signal preamplification at 1.8 Gbits/s", *IEEE Photon. Technol. Lett.*, Vol. 1, 1989, pp. 367–369.

Chapter 5

Single-Mode Fiber-Optic Sensors

V. Annovazzi-Lodi

S. Merlo

5.1 INTRODUCTION

The main application of optical fibers is presently in the area of telecommunications, and this is expected to continue for many years to come. Over two decades of research and development have led to a firmly established technology which exploits the great capacity of this medium, its very low attenuation, and its reduced cost and lower size with respect to traditional cables.

On the other side, the benefits of optical fibers are also being exploited in the field of sensing, control, and instrumentation. In these applications, one can even take advantage of some of the technical problems that have been found in telecommunications, i.e., generally speaking, of fiber sensitivity to environment.

The field of sensors has initially expanded utilizing fibers and components that had been developed for telecommunications, such as low-loss fibers for different transmission windows, couplers, wavelength multiplexers, high-quality LEDs, and semiconductor lasers sources. Now, it is a large and competitive field for which a specific technology is being developed, including special fibers (low birefringence, high birefringence, polarizing fibers) and components (fiber polarizers, modulators, and superluminescent diode sources).

The main advantages of using optical fibers in sensing applications can be summarized as follows:

1. large bandwidth and fast dynamic response;
2. immunity to, and no emission of, *electromagnetic interference* (EMI);
3. dielectric (electrically passive) construction;
4. chemical and environmental ruggedness;
5. small size and weight;
6. geometrical flexibility.

These features have led to the availability of fiber sensing elements that are competitive with traditional ones in a number of applications. A specific characteristic of these devices is a sensor head of an almost arbitrary shape that can be built in order to sense selectively a particular quantity or to accommodate the sensor in a given position; also, these devices can be installed in hostile environment with reduced shielding and protection.

Presently, fiber sensors are available in almost every field of technology [1–8]. Over 30 companies around the world offer sensors for measurands such as temperature, pressure, displacement, rotation, current and voltage, flow, and chemical parameters. Evidence of their importance is confirmed by the conspicuous number of papers that are published yearly on scientific reviews and congress proceedings [7, 8].

The principal disadvantage of this technology is the cross sensitivity of fibers to a number of environmental parameters, and most engineering work in this field is devoted to enhance selectivity, desensitizing a given sensor from unwanted external influences, or developing inherently immune configurations. Other problems arise from fiber fragility, as well as from the possible requirement of high-grade optical machining and of relatively expensive laser sources.

5.1.1 Classification

A fiber-optic sensor is a device that can transduce a physical quantity acting on an optical parameter into an electrical signal. Basically, it comprises a sensing head where the quantity under measurement acts on the lightwave propagating through the fiber, a readout setup where the modulated optical parameter is processed and transduced into a variation of the wave amplitude, and one or more detectors that supply the electrical output signal. Downlead trunks are often used for the connection to a remote measurement or detector location. The words *sensor* and *transducer* are usually employed interchangeably in literature, although the word *transducer* more correctly means the sensing head, whereas *sensor* should be the whole device. In a few cases, systems also embedding some electrical processing, such as the fiber gyroscope, are referred to as sensors, although the expression *electro-optical sensing system* could be more appropriate. The classification of fiber optic sensors can be developed from different points of view [9]. A straightforward approach, that we believe to be suitable for a survey and will be essentially followed in this chapter, is based on the physical quantity to be transduced. More accurate subdivisions should be focused on sensor topology, detection schemes, and radiation properties.

In this framework, it is customary to distinguish, first of all, *extrinsic* and *intrinsic* sensors. The first class comprises all sensors in which the lightwave is guided by the fiber but the interaction between the light and the quantity under measurement takes place outside the fiber. This kind of sensor takes advantage of the fiber as a suitable means to reach the measurement point with low attenuation and increased mechanical stability with respect to a bulk optics design. The sensor head is usually made by employing micro-optic components and its operation is not substantially different than that of its bulk counterpart. It can be pointed out that some of the best-developed sensors belong to this class: among others, we can quote voltage sensors based on the Pockels effect and current sensors based on the Faraday effect in suitable crystals. Their simple structure makes them reliable and cost effective.

The second class contains all sensors in which the fiber itself is used to transduce the measurand. If the measurand directly acts on the fiber, as in the case of Faraday sensors or of fiber gyroscopes, the sensor is called *direct.* On the contrary, if the measurand is locally transduced into another quantity acting on the fiber, so that sensing is a two-step process, the sensor is named *indirect.* For example, magnetostrictive current sensors that rely on the stress exerted on the fiber by a magnetostrictive coating that is sensitive to the measurand are indirect sensors.

Another common classification is based on technology. *All-fiber* sensors integrate the whole system, including the sensing head, the readout setup and the downlead trunks into a single piece of fiber, with the aim of achieving a high degree of stability. *Hybrid* sensors include bulk, micro-optic or integrated-optics elements of the type described in Sections 2.6.4, 2.6.5, 2.7.4 and 2.7.5.

A key point in the design of an optical system is the requirement of radiation coherence that determines the choice of the source of the propagation medium and of the optical components. From this point of view, a rough subdivision is based on the fiber sensor being *single mode* or *multi-mode.* A more fundamental classification, reflecting that adopted throughout this book, can be made in terms of the temporal degree of coherence, so that we can distinguish *high-coherence, low-coherence*, and *incoherent* sensors. This classification can be made according to the characteristics of the optical source, which can be a single mode laser, a multi-mode laser or an LED. In dealing with sensors, a division based on the operating principle is also suitable. In this context, it should be noted that there are devices which rely on radiation coherence though making use of a low-coherence source, as is the case for many fiber gyroscopes, which are in fact interferometers with a very small path unbalance. An effort will be made in the following sections to clarify to what extent a device operation relies on radiation coherence, as well as to point out the coherence requirements of the source.

Amplitude modulated and *phase modulated* sensors can be distinguished if we focus on the optical parameter carrying information. In amplitude modulated sensors the quantity to be transduced modifies the power output of the fiber through a loss mechanism. These sensors have rather low sensitivity but are rugged and inexpensive because they usually employ multi-mode fiber and have less critical requirements on setup stability. Phase modulated sensors are usually single-mode; according to the optical readout scheme, they can be further divided into *polarimetric* and *interferometric* devices. In the first class, a polarization analysis scheme reads the phase difference between two polarization eigenmodes; in the other case, a fiber interferometer is implemented to detect the phase difference between two modes traveling through different optical paths. Both schemes are less sensitive to fiber attenuation with respect to amplitude based transducers; they allow higher sensitivity and larger bandwidth at the expense of a more critical setup. In general, polarimetric sensors offer larger dynamic range, while interferometric sensors perform higher sensitivity especially in the low-frequency range. The choice between polarimetric and interferometric schemes depends on the specific application, as will be discussed in the following sections. Various subclasses can further be distinguished taking into account the demodulation principle, namely, direct detection, homodyne, or heterodyne. More sophisticated processing schemes have also been proposed in literature, and some of them will be considered in some detail in this chapter. For example, sinusoidal modulation of the optical path followed by synchronous phase demodulation is of widespread use in the fiber gyroscope and also serrodyne, where sawtooth phase modulation is employed to perform a piecewise linear frequency modulation of the source with a simplified setup. Recently, *rotating half-wave plate pseudoheterodyne* (RHWP) techniques have been proposed to enhance sensitivity in polarimetric fiber sensors without wasting the unambiguous operating range [10].

With reference to topology, the classes of *open loop* and *closed loop* sensors can be recognized. In the former, the output is directly obtained by detecting the optical parameter on which the measurand acts; in the latter, a feedback is introduced to null the output by a controlled effect, either homogeneous with the measurand or not, acting on the lightwave. This scheme is advantageous in terms of linearity and dynamic range and has been employed, e.g., in gyroscopes where the Sagnac phase delay due to rotation is compensated by a Faraday phase delay due to a magnetic field produced by a current flowing through a solenoid. This current is therefore the new output signal.

Also, it can be noted that most intrinsic fiber sensors have a *distributed* [11] nature because the effect of the quantity under measurement is cumulated over a rather long piece of fiber. This can be done in order to increase sensitivity or because one wishes to detect a mean value or even the distribution of a physical quantity on a certain region, such as a temperature profile in industrial equipment or the vibration of a large structure. In the latter case, suitable reading techniques (e.g., a variation of the OTDR method, described in Section 3.1 for telecommunication applications) must be employed to selectively interrogate specific portions of the sensor. On the other hand, *point* sensors can be obtained if a compact (though, possibly, inherently long, as in Faraday magnetic sensors) head is built with an accurate decoupling of the downlead fiber trunks.

Finally, the already stated cross sensitivity of fibers to a number of external measurands has been exploited in a few cases to make *multiple* sensors. Such devices can efficiently detect more than one variable, each modulating a different parameter of the traveling lightwave.

5.2 SINGLE-MODE SENSORS

In this chapter, we will consider single-mode fiber sensors and will focus almost exclusively on intrinsic sensors, both direct and indirect, because they are the most specific fiber-optic sensors. The main advantage of using single-mode fibers in sensing is that they are more suitable to implement configurations that achieve higher sensitivity and are not affected by amplitude fading. On the other hand, single-mode sensors usually require higher grade optical machining, better quality sources and components, and special design to minimize cross-sensitivity in order to wholly exploit their potential sensitivity and accuracy.

Single-mode sensors are generally based on differential modulation of the optical phase, either between a measuring path and a reference path, which are physically separated, or between two modes along the same geometrical path. The possible readout schemes are therefore interferometric or polarimetric. A general analysis and a comparison between the two approaches is given in [12].

It should be noted that, irrespective of the quite large number of transduced quantities, most single-mode sensors that are considered in this chapter essentially rely upon a reduced number of transduction mechanisms, namely:

1. reciprocal birefringence due to fiber elastic deformation, which can arise from a directly applied force, temperature variations, or an indirect mechanism, such as magnetostriction;
2. Sagnac effect (gyros), and
3. Faraday effect (current and magnetic field sensors).

Although many fiber optic single-mode sensors are barely at a laboratory or prototype level of development, some are well-assessed devices that can advantageously substitute for their traditional counterparts. Among commercially available devices, we should above all mention the fiber-optic gyroscope. Also, strain, pressure and temperature sensors are competitive for specific applications while current sensors constitute a viable alternative to traditional current transformers on very high voltage lines and in power plants.

It must be pointed out that all these sensors are presently used as stand-alone devices and that their commercial success depends on the specific benefits they can offer in terms of sensitivity, reliability, or cost in a particular application. In most cases, however, well-assessed standard technologies can do the job, and consequently fiber sensor deployment is not so common. The importance of the fiber-optic sensor market is expected to sensibly increase when the specific features of this medium (first of all, extended bandwidth and EMI immunity) will be exploited to build efficient integrated networks for multipoint sensing. This requires the study of suitable multiplexing techniques and network topologies and, above all, to constitute a complete set of devices for the measurement of all physical quantities that are related in a typical application (for instance, automotive or power line diagnostics).

Though at present fully optical multiplexing techniques have not yet been implemented in the field, many architectures have been proposed [5, 6, 17–23] and tested, on a reduced scale, in laboratory environment. For single-mode sensors, four addressing techniques have been proposed, namely: time, frequency, wavelength, and coherence multiplexing. In *time division multiplexing* (TDM) the information of every sensor is allocated in a specific time slot of a repetitive transmission period. In *frequency division multiplexing* (FDM) sensors, outputs are encoded on electrical subcarriers at different frequencies. *Wavelength division multiplexing* (WDM) consists of encoding information from different sensors on optical carriers of different wavelengths. Finally, coherence multiplexing involves the use of light with short coherence length so that signals returning from different sensors are mutually incoherent. Every technique has its own topological implementations and modulation/demodulation schemes [5, 6]. Though passive full-optic networks are more desirable, hybrid approaches, i.e., those that include electro-optics processing and signal conditioning, have also been proposed and have at present the advantage of being more viable and cost-effective.

5.2.1 The Fabry-Perot Fiber Interferometer

Different interferometric schemes have been extensively considered elsewhere in this book, including Mach-Zehnder, Michelson, Sagnac and Fabry-Perot types, and their fiber versions are currently used in several sensor implementations. Among them, the Fabry-Perot configuration deserves supplementary information, owing to the specific applications that will be discussed later in this chapter.

The basic Fabry-Perot interferometer has been analyzed in Section 2.1.5. Its fiber version [13] consists of a single-mode fiber with cleaved end faces to make an optical cavity as it is schematized in Figure 5.1. Alternatively, high-reflectivity end faces, either dielectrically coated [14, 15] or with butt-coupled mirrors, can be used.

The light travels back and forth and is partially transmitted at the end faces. Then, both the transmitted (T) and the reflected (R) output must be calculated as the superposition of a theoretically infinite number of contributions. The photogenerated current at output T can be found from Equations (2.70), (2.71), assuming normal incidence ($\vartheta_2 = 0$) and real r (arg[r] = 0) for simplicity. It has the well-known Airy dependence [16]:

$$I_t = \frac{I_{0t}}{1 + f \sin^2 \Theta/2} \tag{5.1}$$

where $\Theta = (4\pi/\lambda)nd$ is the roundtrip phase delay along the fiber ($\vartheta_2 = 0$), n is the fiber effective index and $f = 4R/(1 - R^2)$ is related to the finesse F by the expression

$$f = \frac{\pi\,(F)^{1/2}}{2} \tag{5.2}$$

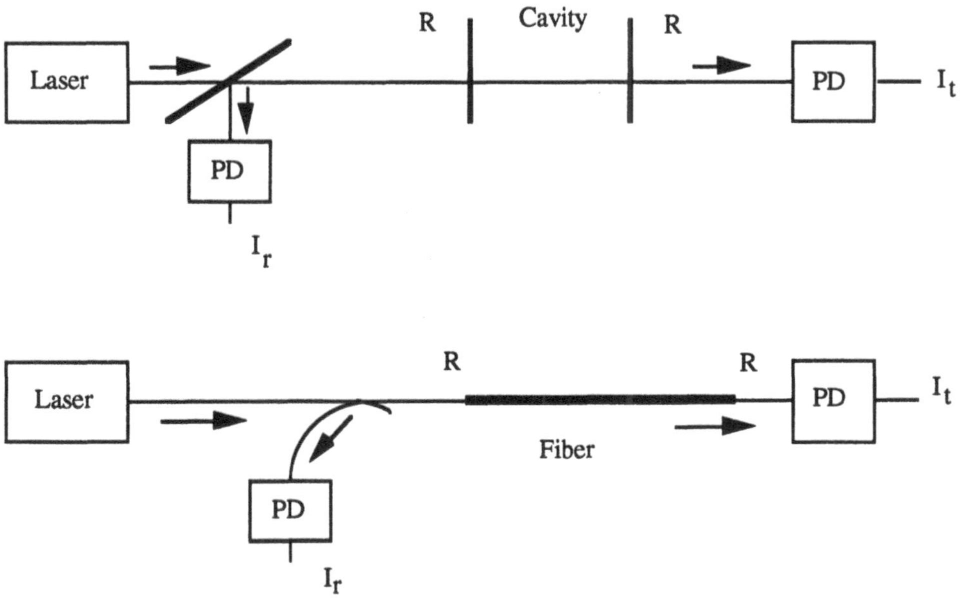

Figure 5.1 Bulk-optic and fiber-optic Fabry-Perot interferometer; PD = photodiode.

Similarly, it can be shown that for output R

$$I_r = I_{0r} \frac{f \sin^2\Theta/2}{1+f \sin^2\Theta/2} \tag{5.3}$$

A plot of the output signal I_t as a function of Θ has the same trend of Figure 2.6 and shows a periodic series of transmission peaks that are sharper as f increases and are centered around $\Theta = 2k\pi$ ($k = 0, 1, 2...$). It can be shown that the interferometer sensitivity, defined as $dI_t / d\Theta$, is maximum at $\Theta = \pm(4/3\ f)^{1/2}$ (modulo 2π), where it amounts to $\approx -0.32\ I_{0t}\ (f)^{1/2}$. If f is large, this figure can be many times greater than that of a standard two-beam interferometer, for which sensitivity is maximum at quadrature ($\Theta = \pi/2$), where it amounts to $dI / d\Theta = -I_0$.

On the other hand, in many fiber-optic sensor implementations of the Fabry-Perot, where the fiber is simply cleaved, f is low and the interferometer performance in transmission is not better than that of a two-beam interferometer. However, in this case the device can be advantageously used in reflection. Due to the lower mean output value of I_r with respect to I_t, this mode of operation results in an acceptable fringe visibility also for R

= 0.04 (uncoated glass/air interface). Unfortunately, an evident drawback is that the available signal is small.

5.2.2 Chapter Overview

Each of the following sections deals with a class of single-mode sensors that are grouped according to the nature of the transduced physical quantity. Whenever possible, sensitivity figures and remarkable experimental data are given, and an effort is made to make a comparison with standard technologies.

Section 5.3 deals with temperature sensors. It also gives general information about fiber optic interferometric and polarimetric schemes, also making a comparison in terms of dynamic range and sensitivity. The principle of compensation [24–26] in interferometric (phase compensation) and polarimetric (birefringence compensation) devices that is of wide use especially (but not exclusively) in this application is introduced. Also, two-variable sensing and white light [27] (i.e., low coherence) interferometry will be briefly considered.

Section 5.4 deals with mechanical sensors. Pressure and strain are among the first quantities to have been successfully measured with a fiber-optic device. The effect of coating on fiber sensitivity is a key topic in the design of such sensors and especially of hydrophones, which also represent an interesting example of distributed sensing. In the same paragraph, the state of the art of the fiber-optic gyroscope is reviewed; its development has represented one of the most challenging problems in interferometry and it is presently the best developed single-mode fiber-optic sensor. A comparison is made between different fiber configurations and the laser gyroscope, which is today the best electro-optic rotation sensor. Specific topics such as interferometer bias by phase modulation, the impact of reciprocity on interferometric operation, parasitic effects (such as scattering), and coherence requirements on the source are also addressed.

Section 5.5 is devoted to sensing of electromagnetic quantities, mainly current and magnetic field. We also consider in some detail the superposition of circular and linear birefringence and birefringence control, a topic of broad interest, especially in distributed sensors.

Finally, Section 5.6 reviews the emerging class of chemical sensors that is expected to experiment a conspicuous growth in the next years. Most of this section is devoted to an outline of the principles of operation of planar waveguide sensors, due to the potential impact of this approach in chemical optical measurements as well as in a number of measuring applications in different technology fields.

5.3 TEMPERATURE SENSORS

The use of single-mode optical fibers has been widely exploited for realizing temperature sensors. Interferometric, polarimetric, and combined interferometric-polarimetric schemes have been investigated.

5.3.1 Interferometric Schemes

The phase delay ϕ of the light travelling for a length L of fiber is given by $\phi = \beta L$, where $\beta = 2\pi n/\lambda$ is the propagation constant of the radiation in the fiber, λ the free-space radiation wavelength, and n is the effective refractive index, which can be approximated with the core refractive index if we consider the propagation of the fundamental mode for a normalized frequency $V > 2$. A temperature variation ΔT in the fiber induces a change in the optical phase delay of the light beam given by [28]

$$\Delta\phi = \frac{2\pi}{\lambda}\left[\frac{n}{L}\frac{dL}{dT} + \frac{dn}{dT}\right] L\,\Delta T \tag{5.4}$$

where $[(1/L)\,(dL/dT)]$ is the linear expansion of the fiber and dn/dT is the temperature dependence of the refractive index. More precisely, it can be demonstrated that [29]

$$\Delta\phi = \beta L\left\{\varepsilon_z + \frac{\Delta T}{n}\left[\frac{\partial n}{\partial T}\right]_\rho - \frac{n^2}{2}\,[(p_{11} + p_{12})\varepsilon_r + p_{12}\,\varepsilon_z]\right\} \tag{5.5}$$

where p_{11}, p_{12} are the elasto-optic coefficients of the core, ρ is the core density, while ε_z and ε_r are the temperature-induced axial and radial strains in the core. The first term comes from the fiber length change, while the other terms result from the temperature-induced refractive index changes related also to the photoelastic effect.

For uncoated silica fibers, the thermally induced phase shift is mainly due to the change in refractive index [28]. The change in fiber length due to thermal expansion may become dominant if the fiber is coated or bonded to a material with a large value of thermal expansion coefficient [28]. Considering the phase responsivity to temperature expressed as the ratio $R=\Delta\phi/L\Delta T$, it has been verified that a commercially available optical fiber shows a responsivity $R \approx 200$ rad/°C m, while the same fiber without the coating has a responsivity $R \approx 100$ rad/ °C m [29] at $\lambda = 633$ nm. In general, the temperature sensitivity can be increased or decreased with the proper choice of coating material [29, 30].

The phase change $\Delta\phi$ of the sensing radiation beam and, thus, the temperature variation ΔT, can be detected with highly sensitive interferometric techniques using a reference beam not affected by temperature changes.

Mach-Zehnder (MZ) [28], Michelson (M) [31, 32], and Fabry-Perot (FP) [15, 33–37] fiber-optic interferometers have been proposed as temperature sensors in conjunction with different signal processing systems. Mach-Zehnder and Michelson interferometers are two-arm schemes (Figures 5.2 and 5.3), which means that the sensing beam and the reference beam travel along two different fiber paths. The fiber arms must lie close to each other so that they are subjected to the same environmental disturbances. In the MZ scheme (Figure 5.2), a short length of optical fiber incorporated in the sensing arm of the interferometer is exposed to temperature variations, while the rest of the sensing arm as well as the reference arm must be thermally isolated. In the M scheme (Figure 5.3), the sensing arm (of length L_s) is usually longer than the reference arm (L_r) so that the sensing head incorporates only the extra path $L = L_s - L_r$ and is already physically separated by the connecting fiber leads. In two-arm configurations, a beam-splitter (either bulk or all-fiber) is used at the input to couple light from the source into both arms. A coupler is also necessary at the output to recombine sensing and reference radiation beams (or modes) so that they can beat on the photodetector, unless the measurement is performed merely by observing a displacement of the output fringe patterns [28]. In a MZ two separate splitters are required at the input and the output, while in a M scheme a single coupler is sufficient and serves also to separate the output from the input. For this feature the M interferometer is considered more suitable than the MZ for remote measurements.

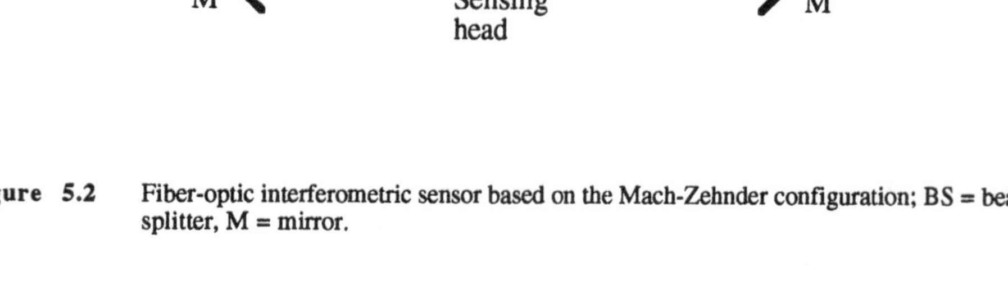

Figure 5.2 Fiber-optic interferometric sensor based on the Mach-Zehnder configuration; BS = beam-splitter, M = mirror.

If a laser source is used (high coherence-length light), for a given length L of exposed fiber, the output photodetected current is

$$I = I_o [1 + \cos(\phi_o + \Delta\phi)] \quad (5.6)$$

for the Mach-Zehnder configuration and

$$I = I_o [1 + \cos(\phi_o + 2\Delta\phi)] \quad (5.7)$$

for the Michelson scheme, where ϕ_o is the constant phase unbalance between the arms and $\Delta\phi$ is the phase change to be measured. Compared to the Mach-Zehnder scheme, the Michelson interferometer has thus double sensitivity. If both arms are affected in the same way by environmental disturbances, the phase difference $\Delta\phi$ in Equations (5.6) and (5.7) is given by Equations (5.4) or (5.5) and, thus, depends only on the temperature difference between the fiber arms. This also means that the device has been desensitized from spurious pressure and temperature variations in the surrounding environment. For $\lambda = 633$ nm and silica fibers, the responsivity expressed as the ratio $R = \Delta\phi/L\Delta T$ can be greater than 300 rad/°C m and the sensitivity, in terms of minimum detectable signal at S/N = 1, can be better than $2.5\ 10^{-4}$ °C at the quantum limit [38, 39].

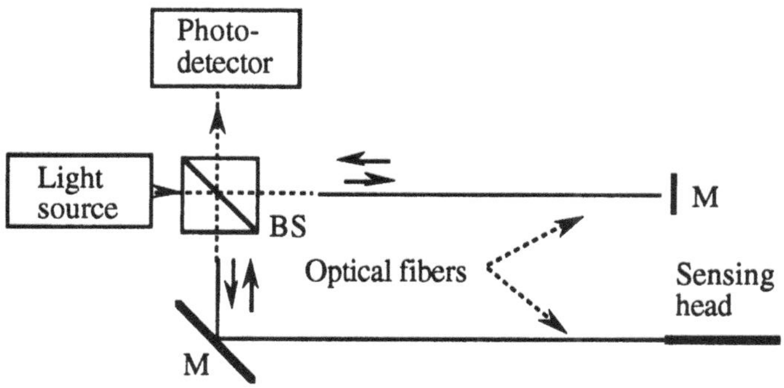

Figure 5.3 Fiber-optic interferometric sensor based on the Michelson configuration; BS = beam-splitter, M = mirror.

The fiber Fabry-Perot interferometer is, on the other hand, a single-arm scheme (Figure 5.4) and temperature is thus measured with a single fiber. The sensing head incorporates a Fabry-Perot resonator, made with a short piece of single-mode fiber, either by just cleaving and polishing the fiber ends or by also coating them with high-reflectance dielectric film or even by butt-coupling mirrors. In some cases, the Fabry-Perot is then attached or spliced to a longer single-mode fiber which guides light from the light source to the resonator and carries the reflected radiation to a photodetector. A single beam-splitter or coupler is required to separate the output from the input. The fiber FP interferometer has been used also in transmission [15, 34] (see Figure 5.1), but of course the configuration in reflection (Figure 5.4) is more suitable for remote measurements of temperature [33, 35, 36, 37].

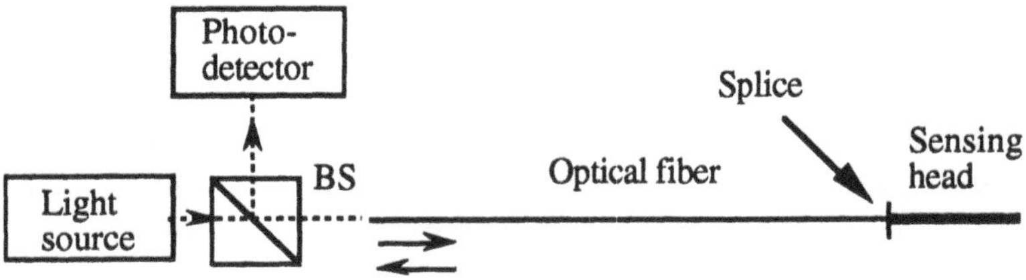

Figure 5.4 Fiber-optic interferometric sensor based on the Fabry-Perot configuration; BS = beam-splitter.

Although a single fiber is employed, a sensing beam and a reference beam can be still identified. The sensing beam travels through the Fabry-Perot resonator while the reference is the fraction of power reflected at the first interface limiting the resonator. The light undergoes a phase delay Θ per round trip through the Fabry-Perot resonator given by $\Theta = 4n\pi L/\lambda$, where L is the length of the resonator. A change in temperature ΔT induces a variation $\Delta\Theta$ in the phase delay which is $\Delta\Theta = 2\Delta\phi$, with $\Delta\phi$ given by Equation (5.4), causing the reflected power to sweep between maxima and minima of interference fringes.

The importance of interferometric temperature sensors relies on the possibility to perform temperature measurements with very high sensitivity and resolution using a device which is electrically and thermally noninvasive. However, interferometric sensing schemes have some major limitations that have to be carefully considered in the design of a practical instrument.

It must be pointed out that only the temperature difference between the sensing and reference path can be obtained; therefore, for absolute measurements it is necessary to have an external reference.

Moreover, the sensitivity to temperature is variable because of the nonlinear dependence of the photodetected current from relative phase changes. For MZ and M interferometers, the maximum sensitivity is reached when the interferometer is working in quadrature, which means that the optical phase unbalance between the arms is $\phi_o = \pi/2$. In this case, for small temperature changes, the photodetected current is linearly proportional to $\Delta\phi$.

Since the output of the interferometer repeats itself as the phase delay increases or decreases by 2π rad, the unambiguous range ΔT_u of the measurement is very limited. For example, for a responsivity R = 100 rad/°C m, the temperature range is $\Delta T_u \approx (0.0628/L)$ °C, with L expressed in meters. Thus, by increasing the length L of the sensing fiber, the device responsivity is enhanced but the unambiguous range is restricted. In the case of two-arm interferometers, when the instrument begins to operate the temperature difference between the arms must be smaller than ΔT_u in order to have an exact readout of the starting temperature. The value of the temperature difference can then be obtained from the overall cumulated phase delay. Otherwise, an initialization process is required. In addition, a temperature decrease can be distinguished from an increase only by tracking the overall phase changes.

A major disturbance in interferometric measurements based on a two-arm scheme comes from phase drift noise due to a different response of the sensing arm and of the reference arm to environmental perturbations. Similarly, phase fluctuations of the laser light can be a source of noise.

Some of the above mentioned problems can actually be solved. Different phase compensation techniques have been proposed for overcoming the problems of variable sensitivity and phase drift noise. A first method of phase compensation consists in winding a length of fiber on a *piezoelectric cylinder* (PZT): in this way, the optical path length and thus the phase delay can be properly adjusted to maintain a Mach-Zehnder and a Michelson interferometer in quadrature [32]. The same method is used to lock the output of a Fabry-Perot at the normal quadrature point on the cosine squared output fringes [13].

Fiber Fabry-Perot interferometers with low finesse sometimes include a PZT modulator in the system to modulate the light phase. In this pseudo-heterodyne detection scheme, the temperature variation affects the phase of the detected modulating signal, which can be measured with higher sensitivity [15, 34–37].

If it is desirable to avoid the presence of an electrical system (high voltage to drive the PZT) in the sensing region, it is more convenient to modulate the phase by controlling the source injection current. It is well known that the emission wavelength of a laser-diode can be modulated by changing the laser injection current. Since a change $\Delta\nu$ in the emission frequency corresponds to a phase change $\Delta\phi = (2\pi L/c)\Delta\nu$, the phase can be locked at the best operating value by using current feedback stabilization [32, 40]. However, with this technique, the frequency instability of the laser is read as a phase disturbance. To overcome this limitation, two separate control systems have to be used. For example, as reported for a thermometer based on a Michelson interferometer [32], the emission frequency of the laser can be locked to a reference (given by an external cavity) by means of an independent current-feedback stabilization. Then, a second loop controls the phase delay in order to maintain quadrature operation. This is performed by controlling the position of a mirror at the end of the reference arm [32]. The compensation based on frequency modulation of the laser diode source has been used for the read-out of a fiber Fabry-Perot temperature sensor, in which the change in the emission wavelength of the light source is used to keep the output intensity on the right order peak [33]. The change of bias current injected into the laser is thus related to the temperature variation to be measured.

By linearly ramping the laser injection current and thus the absolute frequency of the laser, the phase of the travelling light is modulated and a pseudo-heterodyne detection scheme can be realized [41]. Optical phase shifts induced by temperature in the interferometer have the effect to directly phase-modulate the carrier, and temperature detection may be obtained with conventional demodulation techniques [32].

5.3.2 Polarimetric Schemes

In many applications, the high temperature sensitivity of interferometric sensors is not required. On the other side, it is more desirable to have a less critical device operating on a wider range without ambiguity. Polarimetric sensors based on birefringence phenomena in single-mode optical fibers can actually satisfy these requirements [42]. Polarimetric temperature sensors have been developed by using *high linear-birefringent* (HB) fibers, which are characterized by two orthogonally polarized eigenmodes [32]. The differential propagation delay between the two eigenmodes of the fiber is used to sense external physical perturbations. The phase delay experienced by each eigenmode for a temperature change ΔT is given by [42]

$$\Delta\phi_i = \frac{2\pi}{\lambda}\left[\frac{n_i}{L}\frac{dL}{dT} + \frac{dn_i}{dT}\right] L\,\Delta T \qquad i = x, y \qquad (5.8)$$

where n_i is the effective refractive index of the i-th mode, and the subscripts x and y refer to the birefringence axes, often called the fast and slow axes of the fiber. In polarimetric sensors, also called birefringent interferometric sensors (Equation (5.8) has the same form of Equation (5.4)) the output phase response to temperature is the differential phase change $\Delta\phi = \Delta(\phi_x - \phi_y)$ between the modes due to a temperature variation ΔT, which is

$$\Delta\phi = \frac{2\pi}{\lambda}\left[\frac{\Delta n}{L}\frac{dL}{dT} + \frac{d(\Delta n)}{dT}\right] L\,\Delta T \qquad (5.9)$$

where $\Delta n = n_x - n_y$. The response is thus proportional to birefringence variations in the fiber due to temperature changes. If the fiber internal birefringence is due to high internal lateral stress (bow tie or Panda fiber), temperature variations directly affect the internal stress and, thus, the birefringence because of the intrinsic differential thermal expansion [42]. Although for this reason the sensitivity can be still quite high, the responsivity of a polarimetric sensor is 50–100 times lower than the responsivity of an interferometric sensor: however, this reduction corresponds to an equivalent increase of the unambiguous range. Moreover, it is easier to set a reference and have absolute measurements of temperature.

Polarimetric measuring schemes based on the use of highly birefringent fibers have another advantage over interferometric sensors. High-birefringence fibers, based on internal stress, are fairly insensitive to isotropic pressure variations in the low (i.e., ambient) pressure range: a fiber-optic temperature sensor using this kind of fiber can thus have reduced cross-sensitivity to the variation of the ambient pressure [42]. In addition, if only one of the HB fiber eigenmodes is excited, the state of polarization of the light travelling along the fiber is not affected by external perturbations. Thus, measurand-insensitive connecting leads can be realized with HB fibers, so that the polarimetric sensor becomes particularly suitable for remote measurements [43].

The typical scheme of a fiber-optic polarimetric thermometer is composed by an input fiber of arbitrary length attached to a short length of fiber, the sensing head, in such a way that the polarization axes of the two fibers are rotated by 45° with respect to each other (Figure 5.5).

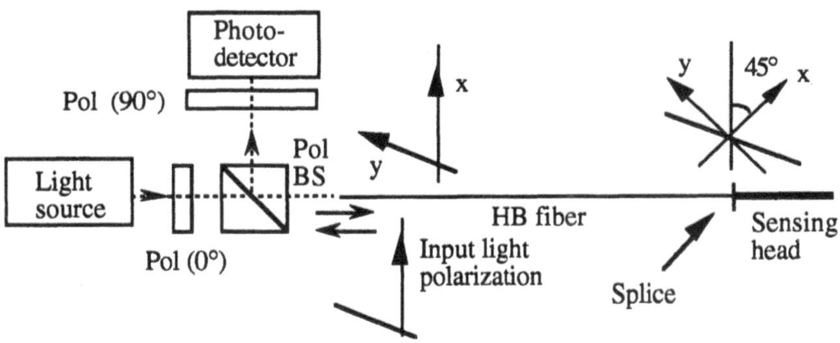

Figure 5.5 Typical scheme of a fiber-optic polarimetric thermometer; Pol BS = polarizing beam-splitter, Pol = polarizer; the inclination angle is between parentheses.

A linearly birefringent fiber is used for both parts and a polarizing beam-splitter is used to separate input and output. Temperature induced changes in the modal birefringence of the sensing head are then monitored in reflection. The reflected modes produce in general elliptically polarized light. Provided that down lead insensitivity is insured by exciting just one of the eigenmodes of the input fiber, perpendicularly polarized radiation (with respect to the state of the input light) is given only by birefringence changes in the sensing head. If a polarizer at 90° (with respect to the input) is placed in front of the photodetector, only this orthogonal contribution is detected and light eventually reflected at the splice does not reach the detector.

With regard to signal recovering, heterodyne detection schemes developed for interferometric schemes are also suitable for polarimetric configurations [32]. In order to further extend the unambiguous range and to distinguish between temperature increase and decrease, it is necessary to track the phase delay, due to changes in the birefringence of the sensing head, in a continuous way. A proposed dual-wavelength tracking system exploits the wavelength dependence of the linear birefringence β_l of the fiber [32]. By switching the wavelength of the light source, it is possible to sequentially obtain the quadrature components of the phase delay $\phi = \beta_l L$. Supposing that the output at λ_1 is of the form $1+\cos\phi$, then the output at λ_2 is of the form $[1 + \cos(\phi + \Delta\lambda\phi)$, where $\Delta\lambda\phi$ corresponds to $\Delta\lambda = \lambda_1 - \lambda_2$. If $\Delta\lambda\phi = (N+1/2)\pi$, the outputs at the two wavelengths are, respectively, $1 + \cos\phi$ and $1 + \sin\phi$.

Eventually, common-mode temperature and isotropic pressure changes in the area of the sensing head may affect the response even of a polarimetric sensor. To overcome this limitation a compensated polarimetric configuration has been proposed, in which the sensing head is composed by two equal lengths of HB fiber spliced together with the polarization axes rotated by 90° [44]. By operating the sensor in differential mode, that is half the sensing head is exposed to temperature while the other half is used as reference, a better desensitized device is obtained. Moreover, with this configuration a broadband source can be used [45], since the differences in the optical path length of the orthogonally polarized modes can be close to zero if the fiber sections are matched in length. The advantage of using low-coherence sources in interferometric sensors will be discussed in more detail in Section 5.3.5.

With the objective of cancelling the random fluctuations of the light phase due to external perturbations, which may occur along the fiber path connecting source and detector to the sensing head, new structures of fiber-optic sensors have been proposed, which are based on the method of polarization-rotated reflections (PRR) [46]. This idea is very similar to the principle supporting the compensated polarimetric scheme previously described. Both polarization eigenmodes of the HB fiber are excited, but the two modes exchange their polarization state with each other in the backward transmission with respect to the forward propagation. The reported temperature sensor based on this method is an extrinsic device, since HB fiber is used only for the connecting leads while the phase sensitive transducer incorporates bulk and micro-optic components. The sensing head consists of a Faraday rotator (because the temperature-induced birefringence is a reciprocal effect), a birefringent crystal (the actual temperature transducer), and a reflecting mirror. In principle, however, an intrinsic all-fiber device based on PRR could be envisioned.

5.3.3 Combined Polarimetric and Interferometric Schemes

Recently, new schemes have been proposed which combine the high sensitivity of the interferometric detection with the wide unambiguous measuring range of polarimetric sensors [47–49]. A two-channel Fabry-Perot interferometer is a simple scheme of combined sensor (Figure 5.6). Each polarization eigenmode of the high-birefringence fiber corresponds to an independent interferometer. Similarly to the single-beam Fabry-Perot interferometric sensor, an input HB fiber of arbitrary length is attached to a short length of fiber, the sensing head, in such a way that the polarization axes of the two fibers are perfectly aligned. Both eigenmodes of the input fiber are excited; therefore, interference occurs between parallel polarized beams that have been reflected at the interface between the fibers and at the distal end of the sensing head. A beam-splitter is required to separate the input from the output, and a second coupler is necessary to generate two output beams. Then, each photodetector analyzes radiation which is polarized along a different eigenaxis of the fiber.

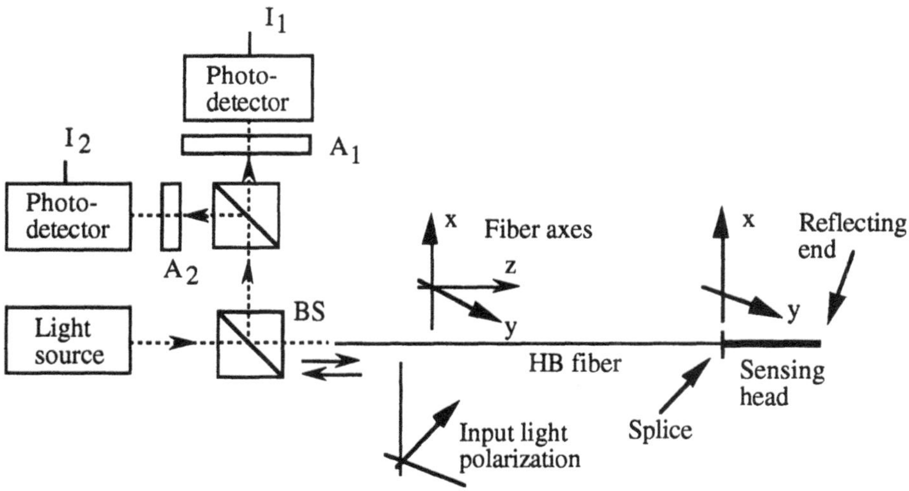

Figure 5.6 Scheme of a combined polarimetric and interferometric fiber sensor; BS = beam-splitter, A_1, A_2 = analyzers.

The photodetected currents are then

$$I_1 = I_o\,[1 + \cos(2\,\Delta\phi_x)] \tag{5.10}$$

$$I_2 = I_o\,[1 + \cos(2\,\Delta\phi_y)] \tag{5.11}$$

where $\Delta\phi_i$, i = x, y, are given by (5.8), with L being the length of the Fabry-Perot resonator, and do not depend on the phase delay accumulated along the input fiber. The phase difference between the outputs corresponds to the polarimetric signal given by (5.9) [47].

A more complex combined scheme exploits the temperature dependence of the linear birefringence which is induced in a single-mode optical fiber (with low residual birefringence) by coiling it under tension around a cylinder with high thermal expansion coefficient [38]. A temperature change modifies the radius of the cylinder and thus the birefringence. The state of polarization of light transmitted through the fiber coil depends on the differential phase delay between the two fiber-eigenmodes; thus, the polarimetric signal is obtained by detecting the transmitted light. At the same time, interference phenomena occur between the waves that are reflected by the ends of the fiber. The overall reflected light along one particular polarization axis experiences a round-trip phase delay which depends on the temperature and the reflected intensity is therefore the interferometric signal.

5.3.4 Two-Variable Sensing

As already discussed, the resolution of a fiber-optic interferometric thermometer may be compromised by environmental perturbations. Usually, the sensor is made more selective by shielding the device from unwanted physical effects or by using compensation techniques. The same problem occurs with pressure and strain sensors, which are negatively affected by temperature variations. Recently, cross-sensitivity of fiber sensors has been exploited for simultaneous determination of strain and temperature applied to HB fibers by means of combined interferometric-polarimetric measurements. A first reported example is based on the different effects of temperature and strain on the phase and the polarization state of the radiation [50]. By measuring both the interferometric phase and the polarization state of the output radiation, and then solving two independent equations in two variables, it is possible to determine both physical measurands: temperature and strain. As an alternative, two spatial modes can be excited in an optical fiber by using a wavelength that is lower than the fiber single-mode cut-off. The measurement of the phase difference between two spatial modes with the same polarization and of the phase difference between orthogonally polarized states of the same spatial mode enables us to determine two independent equations that depend on strain and temperature [51].

5.3.5 White-Light Interferometry and Polarimetry

For achieving good performances in interferometric or polarimetric sensors, it is preferable that the light coherence length is larger than the optical path imbalance of the interferometer. In general, laser sources operating above threshold are necessary to satisfy this requirement. For this reason, all the previously reported temperature sensors operate with laser sources. However, single-mode laser devices have some drawbacks. Semiconductor lasers undergo environmentally induced shifts in the emission wavelength, which negatively affect the sensor response. Laser sources show phase noise and the emission characteristics are very sensitive to optical feedback, which should be carefully avoided using an optical isolator. Novel interferometric-sensor structures have been recently reported, which can be operated with good performances using low-coherence emitters [45, 49, 52, 53]. The principle underlying the operation of these configurations, known as white light interferometry, consists essentially in the fact that if a low-coherence source is used, interference can be observed only if the optical path lengths within the interferometer are well-matched. Therefore, the position of zeroth order interference fringes is completely determined.

A practical system for temperature sensing applications should incorporate two interferometers in series. One is the sensing head and is exposed to the temperature variations to be measured; the other is the compensating section and is thermally isolated. Interference fringes can be detected with good visibility at the output if the path imbalance between the interferometers is much smaller than the coherence length of the light source. Zeroth-order fringes (corresponding also to maximum visibility) occur when the interferometers have exactly the same imbalance. Initialization problems of interferometric sensors are solved since it is possible to match the interferometers without ambiguity by modifying the optical phase delay of the reference interferometer until maximum visibility is obtained at the output. If the interferometers are matched at the beginning of the sensor operation, the output intensity depends sinusoidally on the variation in the phase delay of the sensing interferometer due to the temperature change. On the other hand, the increase of the path imbalance between the interferometers due to temperature changes reduces the fringe visibility. Therefore, by measuring both the phase and the visibility of the output intensity it is possible to increase the dynamic range. However, for the same reason, the maximum thermal range of operation is limited by the source coherence length. Broadband light sources have been successfully employed for temperature sensing in a compensated polarimetric scheme [45], in a combined interferometric-polarimetric configuration [49], and in a FP sensor [53].

5.4 MECHANICAL SENSORS

In this section, we present a review of single-mode fiber sensors for mechanical quantities. Pressure and strain sensors, and a few related devices such as acceleration and displacement transducers, are considered in Sections 5.4.1 and 5.4.2, while the fiber-optic gyroscope is the object of Section 5.4.3.

Fiber-optic strain gauges as well as fiber-optic transducers for pressure sensing and acoustic wave detection have been investigated since 1977 and they have reached an advanced stage of development. Thanks to the high sensitivity of optical interferometers, the minimum detectable strain and pressure for this type of fiber-optic sensors can be much lower that the threshold detectability of other transducers. As previously discussed for temperature sensors, sensitivity can be enhanced by increasing the length of the sensing fiber, and various phase modulation and demodulation techniques can be used for more convenient signal processing. While initially a great effort was dedicated to the study of interferometric schemes, more recently polarimetric schemes for strain and pressure sensing have also been largely investigated. It should be observed that while in the field of interferometric temperature sensors, as previously described, most of the research activity has been dedicated to the study of advanced optical configurations and signal processing techniques; the work on interferometric pressure sensors has mainly dealt with the investigation of proper coatings for sensitivity enhancement or reduction.

5.4.1 Strain Sensors

Interferometric Schemes

A fiber-optic, interferometric strain gauge is based on the change in optical path length caused by straining the fiber. Different strains imposed to the fiber arms of an interferometer cause a change $\Delta\phi$ in the phase delay which exists between the reference and the sensing radiation mode. The strain can then be determined from the motion of the fringes in an optical interference pattern or from the photodetected current following the expression given in (5.6) or (5.7), which depend on the phase delay $\Delta\phi$.

In order to derive the relationship between the applied strain ε and the induced phase change $\Delta\phi$, we recall that the phase delay ϕ of the light travelling for a length L of fiber is given by $\phi = \beta L$, and therefore the phase change $\Delta\phi$ can be expressed as [54]

$$\Delta\phi = \beta\,\Delta L + L\,\Delta\beta \qquad (5.12)$$

where the first term is due to the fiber length change ΔL and the second term comes out from a change in the propagation constant β. This second contribution can be specified as [54]

$$L\,\Delta\beta = L\frac{d\beta}{dn}\Delta n + L\frac{d\beta}{dD}\Delta D \qquad (5.13)$$

which means that it is due to two effects: the strain optic effect, which consists of a modulation of the fiber refractive index, and a waveguide mode dispersion effect due to a change in the fiber diameter D. Considering an axial strain ε_z on the fiber (thus, applied in the direction of light propagation), it can be demonstrated that [54]

$$\beta \, \Delta L = \beta \, L \, \varepsilon_z \tag{5.14}$$

and

$$\frac{d\beta}{dn} \Delta n = - \beta \, \varepsilon_z \frac{n^2}{2} [(1 - \mu) \, p_{12} - \mu \, p_{11}] \tag{5.15}$$

where μ is the Poisson's ratio and p_{12}, p_{11} are the significative components ($\neq 0$) of the strain optic tensor. Since usually the dispersion effect can be neglected [54], the phase responsivity $R = \Delta\phi/\varepsilon_z L$ to strain is then

$$R = \beta \, \{1 - \frac{n^2}{2} [(1 - \mu) \, p_{12} - \mu \, p_{11}]\} \tag{5.16}$$

For typical glasses, $n = 1.5$, $\mu = 0.25$, $p_{12} \approx p_{11} \approx 0.3$, which yields a responsivity of the order of $10^7 \ m^{-1}$. Experimentally measured values of responsivity for Mach-Zehnder [54] as well as for Fabry-Perot (FP) [55] optical fiber interferometers are in agreement with the theoretical result. Comparable sensitivity has been reported for an extrinsic fiber FP interferometer for measuring strain and microdisplacements [56]. Between a single-mode fiber, used as leading fiber, and a piece of multi-mode fiber an air gap is formed, which plays the role of FP cavity. Therefore a fiber movement, due to strain, is transformed in a modulation of the cavity length, which corresponds to an optical phase modulation. An interesting study of the dynamical response of a fiber-optic strain gauge based on the Mach-Zehnder interferometer is reported in [57]. It is demonstrated that, in addition to the high sensitivity, the sensor exhibits an extended range of operation for mechanical frequency.

Polarimetric Schemes

It has been reported that also the differential optical phase between the two orthogonally polarized modes of highly birefringent fiber can be substantially changed by stretching the fiber [58]. This phenomenon can thus be exploited to realize a fiber-optic strain gauge based on a polarimetric configuration. For this purpose, we recall the definition of modal birefringence B which corresponds in HB fiber to the refractive index difference between the two birefringence axes, that is, $B = n_x - n_y$. In bow-tie and panda highly birefringent fibers, the intrinsic modal birefringence B is due to the different thermal expansion coefficients of the stress-applying sections (α_1) and of the surrounding cladding (α_2), and it has been shown that [59]

$$B = (\alpha_1 - \alpha_2)\, T\, G \tag{5.17}$$

where T is difference between the ambient temperature and a characteristic temperature of the material inside the fiber, and G is a function of the mechanical and geometrical properties of the fiber. Similarly, if a uniform axial strain ε_z is imposed to the fiber, the stress-applying sections and the cladding are subjected to a different lateral strain because they have different Poisson's ratios, μ_1 and μ_2. Therefore, the strain-induced modal birefringence B_{st} can be expressed as [59]

$$B_{st} = -\,(\mu_1 - \mu_2)\, \varepsilon_z\, G \tag{5.18}$$

or in term of the intrinsic modal birefringence B, as

$$B_{st} = -\frac{(\mu_1 - \mu_2)\, \varepsilon_z}{(\alpha_1 - \alpha_2)\, T}\, B \tag{5.19}$$

which suggests that the ratio B_{st}/B is independent from the Young's modulus distribution in the fiber. This feature has been verified even experimentally and it is consistent with the observation that the existence of a different Young's modulus in the stress-applying sections and the cladding affects in the same way both the intrinsic birefringence and the strain-induced birefringence.

By placing $\phi = (2\pi/\lambda)\, B\, L$ and $\phi_{st} = (2\pi/\lambda)\, B_{st}\, L$, the differential optical phase between the two fiber eigenmodes is $\Delta\phi = \phi_{st} - \phi = [(2\pi/\lambda)\, (B - B_{st})\, L]$. The phase responsivity $R = \Delta\phi/\varepsilon_z L$ to strain can be expressed as

$$R = -\,B\,\frac{2\pi}{\lambda}\,\frac{(\mu_1 - \mu_2)}{(\alpha_1 - \alpha_2)\, T} = -\frac{2\pi}{L_p}\,\frac{(\mu_1 - \mu_2)}{(\alpha_1 - \alpha_2)\, T} \tag{5.20}$$

where $L_p = \lambda/B$ is the beat length of the fiber due to the intrinsic birefringence B. For a typical silica fiber, $[(\mu_1 - \mu_2)/(\alpha_1 - \alpha_2)\, T] \approx 25$ and for L_p =1.5 mm, $|R|$ is on the order of $10^5\ \mathrm{m}^{-1}$. As expected, the responsivity of the polarimetric scheme is lower than the responsivity of the interferometric schemes previously described.

5.4.2 Pressure Sensors

Some of the major advantages of using optical fibers for pressure sensing have been already discussed in the introduction. In addition, it is interesting to observe that thanks to fiber-optic flexibility, it is possible to fabricate sensors with different shapes, thus with the required properties of high sensitivity and compactness even for very specific applications [60].

Interferometric Schemes

When a pressure change ΔP is applied, the fiber is subjected to an axial strain ε_z and a radial strain ε_r [28]. Therefore, in order to derive the relationship between the applied pressure ΔP and the induced phase change $\Delta\phi$, we can use the method previously employed for obtaining strain responsivity. Again, the phase change is given by (5.12), in which the relation (5.14) still holds, but for the strain optic effect we obtain now [28]

$$\frac{d\beta}{dn}\Delta n = -\beta\frac{n^2}{2}[(p_{11}+p_{12})\varepsilon_r + p_{12}\varepsilon_z] \tag{5.21}$$

Neglecting the dispersion effect, the phase change per unit length of sensing fiber is then

$$\frac{\Delta\phi}{L} = \beta\left\{\varepsilon_z - \frac{n^2}{2}[(p_{11}+p_{12})\varepsilon_r + p_{12}\varepsilon_z]\right\} \tag{5.22}$$

For the determination of the phase responsivity $R = \Delta\phi/L\Delta P$ it would be necessary to express the relationships between strains and pressure, which depend on the pressure distribution, on the frequency of the acoustic wave, on the geometrical structure of the sensing head, and on the type of fiber. In particular, the pressure sensitivity depends strongly on the layers of different materials forming the fiber. We will summarize here the most important practical results, while a mathematical description can be found in [61–63].

Three different regimes must be considered in relation to the frequency of the acoustic wave [63, 64]. The low-frequency range corresponds to acoustic wavelengths Λ much greater than the largest dimension δ of the sensing head. In this range, the pressure applied on the fiber can be considered isotropic (hydrostatic pressure) and the dominant contribution to the phase change comes from the fiber length variation. The pressure sensitivity depends on the thickness and on the elastic moduli, either the Young's modulus and the bulk modulus, of the fiber outer jacket. High sensitivity can be attained by selecting low bulk modulus (i.e., soft) and high Young's modulus materials for the jacket, although for very thick coatings the sensor response depends only on the bulk modulus.

For some applications, it may be important to evaluate the effect of a certain material on the frequency response of the sensing fiber. As already mentioned about temperature sensors, for good performances it is important to have measurand-insensitive reference and connecting fibers. At low frequency, it is possible to achieve acoustic desensitization by coating the fiber with high bulk modulus materials, such as metals or glasses.

An intermediate-frequency regime must then be examined in which the acoustic wavelength Λ is comparable with the sensor largest dimension δ ($\Lambda \approx \delta$) but at the same time Λ is still much larger than the fiber diameter D [63, 64]. When working in this frequency range, we must take into account pressure gradients across the sensing head. Therefore, the geometrical structure of the sensor directly affects the characteristics of the sensor response. In the case of a fiber-optic coil, because the acoustic wave could excite some vibration modes of the coil, resonance peaks can be observed in the frequency response of the sensing head. Moreover, the response could depend on the relative orientation of the fiber coil with respect to the acoustic force. This effect could be exploited to realize a device which is able to detect also the traveling direction of the acoustic wave.

At high frequency, because of inertial effects, the fiber length cannot change and the fiber is axially constrained ($\varepsilon_z = 0$) [64]. In this regime, we must distinguish between two situations. As long as the acoustic wavelength Λ is larger than the fiber diameter D, the strains are isotropic within the fiber cross section. In this case, it has been demonstrated that the phase sensitivity of conventional fibers cannot be substantially increased by coating the sensing fiber; on the other hand, it seems possible to desensitize the fiber to pressure changes if a first layer made of low bulk modulus material is covered with a second coating of high bulk modulus (hard) material. If the acoustic wavelength becomes of the order or even smaller than the fiber diameter, the fiber is subjected to anisotropic strains and tends to become birefringent. In this ultrasonic frequency range, different phase changes occur along the parallel and perpendicular direction with respect to the acoustic wave; thus, the phase sensitivity depends on the light polarization [64].

We have assumed until now that only one of the interferometer arms played the role of the sensing head while the other arm was shielded from the variable under measurement (either pressure, strain, even temperature). In practice, very often it is difficult to successfully realize this assumption. An alternative sensor design consists in having both arms sensitive to the selected variable but with opposite responses to the same stimulus. Therefore, the sensor operates in a "push-pull" mode assuring a good common mode rejection of spurious, unwanted effects [65]. This approach has been exploited to develop mechanical sensors, such as seismometer [66], accelerometers [66], and hydrophones [67], utilizing fiber-optic strain gauges in a proper configuration. As an example of this concept, we report an interferometric fiber-optic sensor which incorporates in the sensor case a seismic mass supported by two soft rubber mandrels (springs), which are circumferentially wrapped with optical fiber [66]. Each fiber coil corresponds to one arm of a Michelson interferometer. The rubber mandrels transform a longitudinal displacement of the sensor case into strain on the optical fibers. At frequencies above the mass-spring resonance, the sensor operates as a seismometer while below resonance it works as an accelerometer with extremely high sensitivity. When the case moves, one of the mandrels is compressed, which results in an increase in mandrel diameter as well as in an increase in tension applied to the fiber. Therefore, one fiber coil is stretched while in the other coil the tension is partially released, which leads to the push-pull configuration.

Another example of this kind of design is the fiber-optic ellipsoidal flextensional hydrophone [67]. It has been reported that when a thin ellipsoidal shell of revolution (with the proper ratio between the axes' length) experiences hydrostatic pressure changes, the axes are subjected to strains with opposite sign. An optical fiber coil wound around the equatorial circumference of the spheroid and a second coil wrapped around the meridional circumference can be used to form the two arms of an interferometer, so that pressure changes (inducing a differential strain) are transformed into a differential optical phase shift.

Polarimetric Schemes

It is well known that fiber birefringence can be changed by applying a lateral compressive stress [68, 69]. For example, it has been proved that bending-induced birefringence is due to the lateral compressive stress produced in the radial direction of the curvature [69, 70]. Recently, bending-induced birefringence and bending-induced losses in a single-mode HB fiber have been combined in a fiber-optic sensing scheme for displacement detection to be used as aircraft altimeter [71]. The sensing head consists of two fiber coils working in a push-pull mode so that the phase delays due to compression are constructively added while the delay caused by thermal changes is cancelled.

However, other transduction effects can be exploited in polarimetric pressure sensors. It has been shown that high hydrostatic pressure has the capability to induce a large birefringence in a HB fiber [72]. This birefringence, and thus beat length change, is probably caused by the inhomogeneous elastic properties of the HB fiber, and in particular by the different compressibility of the doped stress-applying sections with respect to the surrounding cladding, which is usually pure silica. It has been demonstrated that if the pressure increases, the fiber birefringence decreases and the beat length increases. The beat length $L_p(p_o)$ at a pressure p_o and the beat length $L_p(p)$ at a pressure $p > p_o$ are related by the following equation [72]

$$\frac{1}{L_p(p)} = \frac{1}{L_p(p_o)} - \frac{\Delta p}{L\, p_{2\pi}} \tag{5.23}$$

where $\Delta p = p - p_o$, L is the fiber length exposed to high pressure, and $p_{2\pi}$ is the pressure required to induce a 2π phase differential delay between the fiber polarization eigenmodes. By placing $\phi_o = 2\pi L/L_p(p_o)$ and $\phi = 2\pi L/L_p(p)$, the differential optical phase between the two fiber eigenmodes $\Delta\phi = \phi_o - \phi$ is given by [73]

$$\Delta\phi = 2\,\pi\,L\,\frac{\Delta p}{L\,p_{2\pi}} \tag{5.24}$$

and the responsivity $R = \Delta\phi/L\Delta p$ to a hydrostatic pressure change Δp is then [73]

$$R = \frac{2\,\pi}{L\,p_{2\pi}} \tag{5.25}$$

For a HB-800 [73] bow-tie fiber (supplied, e.g., by York Fibers) at 632.8 nm the responsivity has been found equal to 9.6 rad/(MPa·m), while a much lower value has been determined for an elliptical-core fiber [73]. By exploiting this principle, polarimetric optical fiber sensors for electronic measurement of high pressure up to 200 MPa have been reported, and both transmission and reflection configurations have been investigated.

The effect of hydrostatic high pressure on the birefringence of highly birefringent optical fibers has been discussed also with a more general theory [74]. This theory points out that by imposing a uniform pressure on the fiber, the stress-applying sections and the cladding are subjected to a different lateral strain because of their different elastic properties. In analogy to the strain-induced birefringence, the pressure-induced birefringence B_p can be expressed as

$$B_p = (\varepsilon_1 - \varepsilon_2)\,G \tag{5.26}$$

where ε_1 and ε_2 are the lateral strains in the doped and undoped sections of the cladding. Therefore, B_p can be given as a function of various material properties of the cladding, of the stress-generating rods, and of the coating. This theory allows to optimize the design of a pressure sensor as well as the design of special fibers for pressure sensing.

In order to reduce the undesired sensitivity to ambient temperature of high-pressure sensors, a compensated polarimetric scheme has been employed [75]. For the transmission configuration, the thermal compensation is obtained by using two equal lengths of HB fiber, spliced together with their polarization axes rotated by 90° (see also pressure desensitization in temperature sensors in Section 5.3.3). In the reflective configuration, the method of *polarization rotated reflection* (PRR) has been proposed and implemented in all-fiber technology. Only one polarization eigenmode of the leading fiber is excited. A reference section is spliced at 45° with the leading fiber, and a sensing section is spliced at 90° with the reference section. The sensor tip is finally covered with a reflective coating. This configuration ensures desensitization from any physical perturbation which affects in the same fashion both sensing and reference sections [75].

As already discussed dealing with temperature sensors, one of the problems of phase-sensitive transducers is their periodic response, which corresponds to a limited unambiguous range of operation. Recalling that the higher the sensitivity, the smaller the unambiguous range, and also that the sensitivity can be increased by using a longer piece of sensing fiber, a proposed solution for having a polarimetric pressure sensor with high sensitivity and at the same time operating on a broad pressure range consists in fabricating a multisensor system, incorporating several sensing heads, each one characterized by different lengths of sensing fiber [76].

Pressure-induced birefringence has been measured in a HB *side-hole* fiber [77]. A side-hole fiber has the same structure of a panda HB fiber but it has two open channels in place of the stress-applying rods. With this fiber, at the 1300-nm wavelength, the experimental responsivity has been found equal to R = 77.8 rad/(MPa·m) with an accuracy better than ± 0.5 % in the pressure range 10–100 MPa. The side-hole fiber has shown good performances even when used for dynamic sensing. Precise dynamic measurements have been reported in the pressure range of 10^{-2} to 10^{3} Pa and in the frequency band of 25 to 150 MHz with the method of white-light interferometry [78].

5.4.3 Fiber-Optic Gyroscopes

The accurate measurement of angular rotation and of angular velocity is required in several military and industrial applications (such as in the inertial navigation systems used on aircrafts and spacecrafts [79]) and in industrial robotics, as well as in a number of scientific applications particularly in the geophysical and astronomical field and in relativity experiments.

The electromechanical gyroscope has been the most popular inertial rotation sensor for many decades and is still the most widely used. This device consists of a balanced spinning mass that is held in rotation by electromagnetic induction in a low-friction medium. The angular displacement gives rise to a fictitious force acting on the gyroscope axis and the resulting displacement is detected by an inductive or capacitive sensor. A typical accuracy requirement for aircraft navigation is of the order of 10^{-2}–10^{-3} °/h, which amounts to about 10^{-3}–10^{-4} times the angular speed of the earth. Mechanical gyroscopes achieving these performances are costly and delicate precision devices, requiring high-quality mechanical working facilities.

Rotation measurements by electro-optical means [80–85] are based on the Sagnac effect, which consists in a variation of the optical path that is measured between the two counterpropagating lightwaves in a ring cavity when the cavity itself is rotating with respect to an inertial reference frame. Though the Sagnac effect dates from 1903, only the advent of laser made it possible to make practical inertial rotation optical sensors. The most attractive characteristic of these devices is the absence of moving parts, which can result in improved reliability, reduced cost, reduced warm-up delay, and insensitivity to acceleration. In the classical view, the difference in the optical path as measured in the two propagation directions in a rotating cavity can be considered as a Doppler frequency shift. The time delay between the counterrotating optical waves has been derived in Section 2.2.5 and results in an optical phase difference Φ_s given by

$$\Phi_s = 8 \pi A \Omega / \lambda c \qquad (5.27)$$

where Ω is the component of the angular velocity perpendicular to the cavity plane, A is the area enclosed by the optical path and c is the speed of light. A more accurate analysis, based on general relativity, would show that this equation is correct if λ is assumed to be the wavelength in vacuum and not in the propagation medium, as one should classically expect. This result holds if the reference frame (where the detector is located) is consistent with the cavity, as one can usually assume. Equation (5.27) must be multiplied by the square of the refractive index n^2 of the propagation medium if the detector is still in a inertial reference frame or by $n^2 - 1$ if the detector is rotating and the cavity is still [86].

Three configurations of the electro-optic gyroscope have been proposed and are usually referred to as:

1. the *ring laser gyro* (RLG);
2. the *fiber-optic gyro* (FOG);
3. the *fiber-optic ring resonator gyro* (RFOG).

The Laser Gyro

The first configuration [87], which was proposed by Davies and coworkers in 1963, basically employs a HeNe triangular laser cavity to realize an active interferometer (Figure 5.7). The phase shift given by (5.27) results in a difference in the optical frequency for the two counterpropagating cavity modes that must satisfy the resonance condition. This frequency difference is proportional to the angular velocity, i.e.,

$$\Delta\nu = 4A\Omega/p\lambda \qquad (5.28)$$

where p is the perimeter of the cavity and $\Delta\nu$ can be detected by recombination of a fraction of the two modes on a photodetector, thus obtaining an output current signal

$$I = I_0 \, [1 + \cos\Phi_s(c/p)t] \qquad (5.29)$$

This scheme has been improved by the work of many research teams during the 1970s leading to a well-developed device that is being produced by several of firms in the U.S. and is currently employed on civilian airplanes (e.g., Boeing 757, 767). The most popular configurations [88] are now the dither RLG (DLAG) and the Zeeman four-frequency RLG (ZLAG) that overcomes some of the drawbacks of the basic scheme, first of all the impossibility to discriminate the rotation sign, and, secondly, the phase locking of the two modes which impairs the gyroscope performances for small angular rotation values.

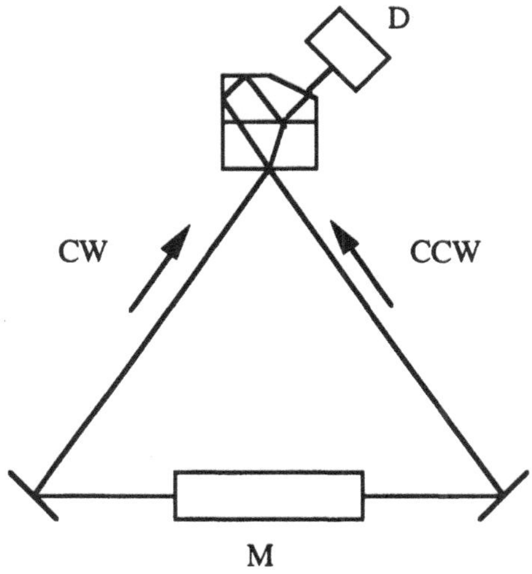

Figure 5.7 Basic scheme of the ring laser gyroscope (RLG); M = active medium, D = detector (adapted from [92]).

A fiber-optic version of this configuration can be envisaged [89] using the special Er- and Nd-doped fibers that have been developed for optical amplifiers and fiber lasers to be used in the telecommunication field. A fiber ring laser would be more compact and lightweight and could be used in robotics applications. The optical fiber could be wound on many turns, and optical pumping would be made by a laser diode connected to the fiber by a *wavelength selective coupler* (WDM). At present the main limitation of this approach lies in the too large linewidth of the fiber lasers and in the limited degree of development of the all-fiber devices.

The Fiber-Optic Gyroscope

The second configuration [90], which was first proposed by Vali and Shorthill in 1976 [91], is schematized in Figure 5.8. Among fiber gyroscopes it is now the best developed and is expected to rival the HeNe RLG for many applications in the next decade. It will be considered in some detail in the following.

In the FOG, the phase difference between the two counterpropagating waves is cumulated over a long (100 to 1000 m) fiber coil to increase responsivity and at the same time obtain a compact device. In the basic setup, the fiber coil is fed by a laser in both directions using a beam-splitter or a fiber-optic coupler. The phase shift, given by (5.27) by substituting $N \cdot A$ for A where N is the number of coil turns, is read at the photodetector R after recombination of the two traveling waves by the beam-splitter. Therefore the output detected current is of the type

$$I = I_0(1 - \cos\Phi_s) \tag{5.30}$$

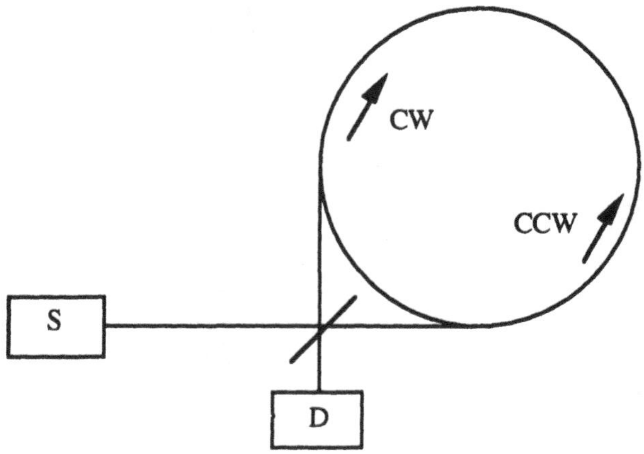

Figure 5.8 Basic scheme of the fiber-optic gyroscope (FOG); S = source, D = detector (adapted from [92]).

In Figure 5.9 [92] the Sagnac phase shift is shown for different values of the product N·A. From this diagram it can be concluded that one must appreciate about 30 μr in FOG (0.03 μr in RLG) to measure the earth angular rotation (Ω = 15 °/h) for typical values of geometrical dimensions. To reach the accuracy requirements for inertial guidance these figures are to be further scaled by a factor $10^2 - 10^3$. This amounts to read an optical path difference as small as a few fm. Since, in addition, this requirement must be satisfied in a dc coupled device, building an optical gyroscope requires an effort that has probably never been asked in any other interferometric application.

The phase noise at the quantum limit is equal to the inverse of the amplitude S/N evaluated at $\Phi_s = \pi/2$ [92]. Since $S = I_0$ and $N = \sqrt{2qI_0B}$, where q is the electron charge, it follows from (5.30):

$$\Phi_n = \sqrt{2qB/I_0} \tag{5.31}$$

or

$$\Phi_n = \sqrt{\frac{2h\nu B}{\eta P}} \tag{5.32}$$

where B is the measuring bandwidth, η is the photodetector quantum efficiency and P is the received power. A similar result can be found for the RLG [92–94]:

$$\Phi_n = \sqrt{\frac{2h\nu B(1-r)^2}{P}} \tag{5.33}$$

where r is the reflectivity of the output mirror.

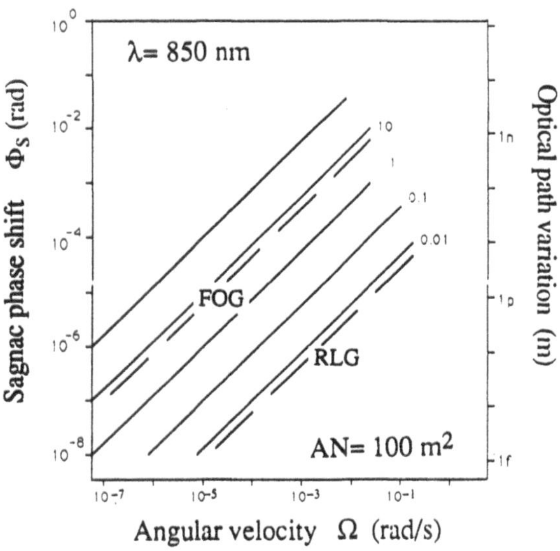

Figure 5.9 Sagnac phase shift Φ_s as a function of the angular rotation Ω for different values of the effective area N·A. The correspondent variation of the optical path is reported on the right scale. Dotted lines are pertinent to typical values of N·A for the FOG and for the RLG (adapted from [92]).

In Figure 5.10 [91] a comparison is made between the two configurations reporting Equations (5.32), (5.33) with P as a parameter. In practice, the standard (HeNe) RLG performs better (reaching the quantum limit) than the FOG in spite of its smaller effective area (N·A) because of its typical higher available power (e.g., 10 W vs. 0.1 mW) and because its cavity has an optical quality much better than that represented by a long fiber.

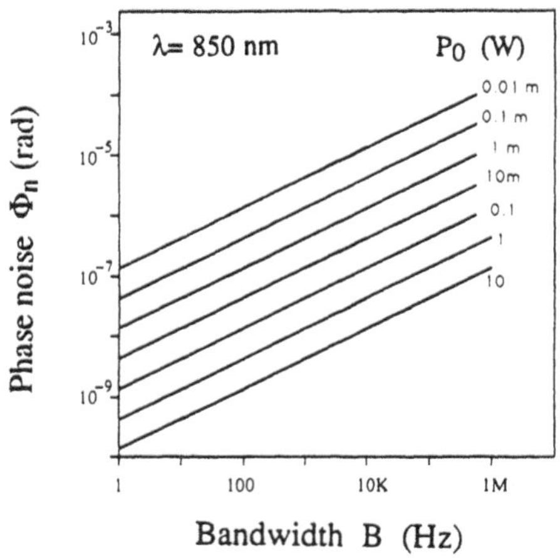

Figure 5.10 Phase-noise quantum limit for the Sagnac phase shift Φ_s as a function of the bandwidth B and with the equivalent power $P_0 = \eta P$ (FOG) or $P_0 = P/(1 - r)^2$ (RLG) as a parameter (adapted from [92]).

The availability of a frequency output is another specific advantage of the RLG configuration. It is suitable for digital processing and allows a dynamic range of 9–10 decades without significant linearity errors. As c/p is typically of about 1.5 GHz, it follows from Figure 5.9 that a resolution of about 1 Hz is required to measure a rotation with an accuracy of 0.5 °/h. Digital integration of the output signal allows to obtain the angular rotation with a precision that cannot be reached on an analog phase signal as that supplied by the standard FOG configuration. On the other side, the FOG configuration appeared soon attractive because of its modularity, scalability, reduced requirements on optical and mechanical machining, and potentially low cost. Therefore, much effort has been devoted to improve the basic FOG setup.

The first drawback to be addressed is the high output nonlinearity for small rotation angles ($\Phi_s \approx 0$). The output dependence on $\cos\Phi_s$ stated by Equation (5.30) can in principle be converted into a sine dependence in a number of ways [95]. A $\pi/2$ differential phase shift, inserted between the CW and CCW waves by using a Faraday rotator, would do the job, but it should be very stable and this approach is therefore unpractical. A better solution consists of frequency translating the output signal from dc to a carrier ω_m, so that the phase Φ_s can be measured with respect to a reference signal at the same frequency. Different methods have been proposed, including the use of a double frequency source, either a Zeeman splitted HeNe laser [96] or a frequency shifted source employing acousto-optic or electro-optic modulators. However, the constraint to use a laser diode source to implement an all-fiber device to have a compact, rugged and less ambient-sensitive sensor, have finally led to the development of a phase modulation scheme that is today generally employed [97]. In Figure 5.11 the typical setup of a practical FOG is reported. A PZT phase modulator is located near one end of the sensing coil, so as to impress a different phase to the counterpropagating waves. This device is made of a short hollow ceramic cylinder on which a few fiber loops are wound. Applying a voltage to the piezoceramic element, a radial elastic stress and an elasto-optic induced optical path variation arise in the fiber [98].

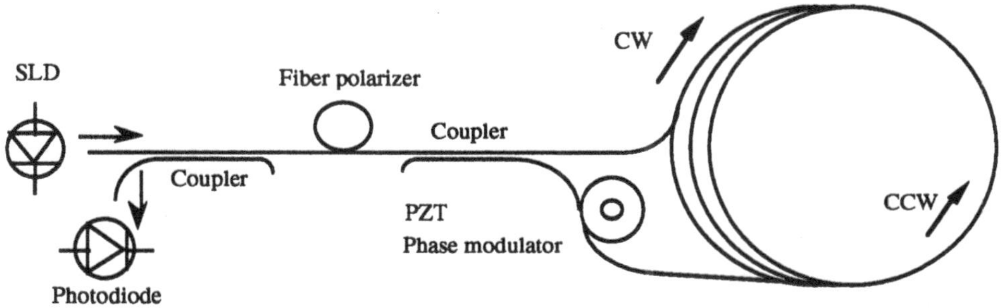

Figure 5.11 Practical configuration of the FOG with low-coherence source and all-fiber piezoceramic phase modulator (adapted from [92]).

Letting $\Phi(t)$ the instantaneous phase modulation:

$$\Phi(t) = \Phi_{m0} \cos \omega_m t \tag{5.34}$$

the *counterclockwise* (CCW) and the *clockwise* (CW) waves undergo a phase delay $\Phi(t)$ and $\Phi(t + L/c)$ respectively, where L is the fiber length. The relative phase difference on the detector is then

$$\Delta\Phi = \Phi_s + \Phi_m \sin \omega_m t \tag{5.35}$$

where the phase deviation Φ_m can be maximized by a proper choice of the modulation frequency.

The output current signal at the photodetector can then be expressed in the form

$$I = I_0 + I_0 \{J_0(\Phi_m) + 2\sum_k J_{2k}(\Phi_m) \cos 2K \omega_m t\} \cos \Phi_s +$$

$$+ I_0 \{2\sum_k J_{2k-1}(\Phi_m) \cos (2K-1) \omega_m t\} \sin \Phi_s \quad (5.36)$$

where the J's are the Bessel's functions.

It is evident from (5.36) that the output signal contains a dc term and, in addition, all the harmonics of ω_m. The phase Φ_s can then be measured by maximizing the term $J_1(\Phi_m)$ selecting $\Phi_m = 1.8$, and by detecting its amplitude by a lock-in amplifier that gives an output proportional to $\sin\Phi_s$. This technique allows to perform an accurate measurement of Ω with its sign, at least for a small rotation speed (0.1–1 °/h).

Different methods have been proposed to improve the linearity and the dynamic range of the FOG, including heterodyning [99–101] and special phase modulation techniques [102, 103]. An important class of FOGs exploits the closed loop approach. It usually consists in introducing a frequency shift between the counterrotating beams using acousto-optic modulators [104] or the Faraday effect. The feedback loop acts so as to null the output signal; this occurs when the phase shift due to the rotation is equal to that due to the relative frequency difference, which therefore represents a measure of the rotation rate. It can be shown that this approach leads to an output signal of the same form as that of the RLG, thus easily allowing digital processing. Practical implementations of the closed-loop FOG also include heterodyne or serrodyne [105, 106] schemes.

The second key topic to be considered is that, irrespective of the readout method, sensitivity at the quantum limit cannot be achieved without the elimination of all the sources of nonreciprocity, but the Sagnac effect, in the propagation of the CW and CCW waves. In fact, all the minute phase-difference contributions due to optical elements and their ambient-induced modulation give rise to zero-point fluctuations that can completely mask the Sagnac phase shift [107]. A considerable effort was devoted to address this point. To ensure reciprocity, three requirements have to be met:

1. the two counterpropagating waves must travel over the same optical path at rest;
2. they must have the same modal distribution;
3. they must have the same polarization state.

The last requirement is due to the polarization degeneration of the fundamental mode of a standard single-mode fiber. Actually, it supports two orthogonally polarized modes, which have equal propagation constants only in ideal conditions [108]. For what concerns point 1, it can be immediately recognized that the basic setup of Figure 5.8 is not fully reciprocal because the two beams do not travel symmetrically through the optical coupler, as one is cross-coupled twice while the other is directly coupled twice. It is therefore necessary to introduce a supplementary coupler and select the output port as in Figure 5.11 to ensure fully reciprocal propagation. The setup of Figure 5.11 is often referred to as the "minimum configuration" [96, 109] because it is the simpler setup offering stable (i.e., reciprocal) operation. For what concerns points 2 and 3, accurate modal and polarization filtering is required. Although a long, bent single-mode fiber is a good mode filter, the extinction ratio of a practical fiber polarizer can be a limiting factor [110]. Also, the polarizer does not prevent amplitude variations and signal fading; therefore, polarization controlling techniques, or, more recently, the use of polarization maintaining fibers [111, 112] are commonly used to improve the actual FOG performances. Incidentally, a completely different solution was proposed [113], which employs depolarized light as a mean to ensure a uniform modal distribution at the output irrespective of environmental perturbations. However, this approach requires a low-coherence source and a fiber depolarizer, on which very stringent requirements have to be made, and the use of unpolarized light is not likely to become a viable alternative.

Another source of error in the FOG is scattering [114, 115], both Rayleigh backscattering in the fiber and reflections from discontinuities in the coil, such as splices, as well as from the all-fiber components. Even larger effects are expected in hybrid implementations using, e.g., $LiNbO_3$ modulators [116] and couplers. As the scattered light is at least partially coherent, it can alter the phases of the original two waves that are coupled by random cross contributions. These effects have been reduced by using phase modulation, or by substituting the laser diode source with a relatively low-coherence *superluminescent diode* (SLD) [117] so that only a short section within a coherence length near the center of the fiber contributes to the phase error. Coherence reduction does not affect operation because of the very small path unbalance of the Sagnac interferometer.

A SLD source also alleviates for the optical Kerr effect. This third-order nonlinear source of error causes a difference in the propagation constant of the counterpropagating beams that is related to their power unbalance [118]. Theoretically, for a typical power of 100 μW in the fiber, one should control about 10 nW, requiring to design an optical coupler of a splitting ratio of 0.5 with a tolerance of 10^{-4}, a quite difficult task. However, a reduction of the source coherence can lower this requirement of about two orders of magnitude.

Several other sources of error have been considered in literature [119] such as the Faraday effect [120], that can arise from environmental magnetic fields or even from the earth magnetic field, and laser power and wavelength fluctuations due to injection phenomena [121]. Additional technological problems have to be considered for engineering prototypes, among them packaging that must protect the device from the environment without exerting undesired mechanical stress on the fiber coil.

As a final comparison between the two leading technologies of the electro-optic gyroscope, the performances of HeNe RLG and of FOG gyros, now commercially available, are shown in Figure 5.12 [122]. Also shown is the expected trend based on units now available as laboratory prototypes and that should be commercially available in a few years.

The Ring Resonator Gyroscope

The last gyroscope configuration (RFOG) [123] is based on a closed-ring optical cavity exactly as in the case of the RLG. However, in this approach the resonance-frequency change of the counterpropagating modes, given by one half the value of Equation (5.28), is measured not by pumping the optical medium but by launching the radiation of an external laser into the empty cavity through a coupler included in the optical path (Figure 5.13).

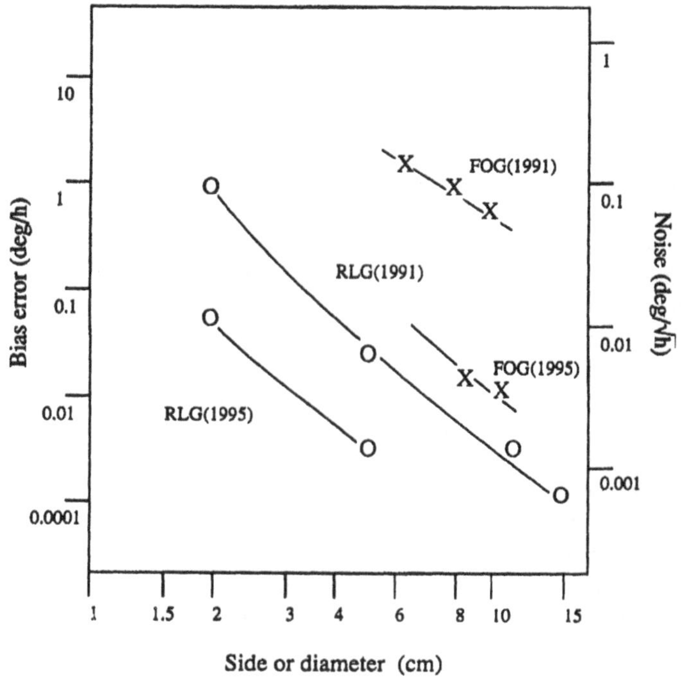

Figure 5.12 Zero-bias and noise performances of RLG and FOG as a function of side (RLG) or diameter (FOG) for commercially available devices (1991) and trend forecast (1995) based on present research work (adapted from [122]).

The mode spacing can be detected, as in a Fabry-Perot resonator, observing the variation of the output power at the free coupler port. If the cavity, or the injected wavelength, is tuned for maximum transmission at rest, a difference in the output power is related to a variation of the cavity length and thus to the angular velocity Ω. This approach has been proposed because removing the active medium from the cavity eliminates the problem of locking; in addition, a high-finesse resonator should perform better than a long open path as in the FOG. Different implementations have been proposed, starting from this basic setup. Closed loop is mandatory, in practice, to obtain an acceptable linearity range in spite of the nonlinear characteristic of the resonator. Acousto-optic modulators can be used for wavelength tuning, while piezoceramic actuators for cavity tuning. The key component in this configuration is the fiber coupler. It must have a very low cross-coupling ratio and very low loss in order to drain a very low power from the ring and allow a high finesse value. Among disturbing effects [124–126], one must consider Rayleigh backscattering, backreflections, and injection effects in the source. In practice, the actual performances of the RFOG are orders of magnitude lower than those of the FOG. Still far from the theoretically expected performances is another proposed approach which exploits Raman amplification in a fiber loop and shares some features with the active and with the passive ring gyro [127, 128].

Finally, it is worth noting that, being essentially interferometric devices, all-fiber-optic gyroscopes belong to the coherent class. However, while the RLG operation is based on highly coherent radiation, in the FOG and in the RFOG a high coherence source is neither required nor desirable. This relaxation of requirements on the source is a specific advantage of the passive configurations.

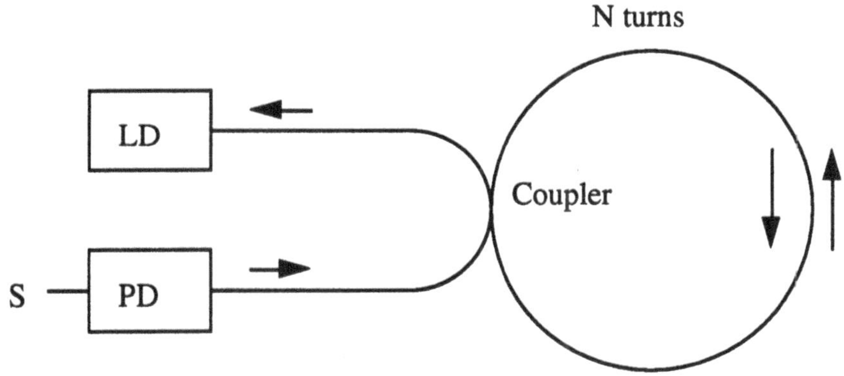

Figure 5.13 Basic scheme of the ring resonator fiber-optic gyroscope (RFOG); LD = laser diode, PD = photodiode.

Research Trends

Among laboratory research trends we can quote the application of phase conjugation [129] to restore the severely scrambled waves to their original state of polarization. This may be an interesting alternative to polarization control and polarization preserving components. Another attractive proposal consists in the use of squeezed radiation (see Section 2.3.5). This approach can result in a noise level below the shot-noise limit and is of potential interest as the FOG performances are approaching this limit. The use of squeezed light radiation has been investigated in interferometry showing that a significant improvement can be achieved if squeezed vacuum (i.e., a squeezed field with a very low photon mean value) is fed into the normally unused port of a Michelson or Mach-Zehnder interferometer [130]. The same technique can in principle be proposed for a fiber gyroscope. Recently, squeezing in a fiber ring has been performed using a pulsed pump. This radiation could be used to feed a Sagnac interferometer in an all-fiber system. However, many issues have to be clarified before it is attempted to use squeezed light in a practical device. First of all, it must be understood how the fiber gyroscope nonlinearity will affect operation, because it is expected [131] to further squeeze radiation, but in a nonfavourable direction, thus possibly wasting the potential benefit of this approach.

5.5 ELECTROMAGNETIC SENSORS

5.5.1 Introduction

In this section, we review the state of the art in the design of fiber-optic sensors for electric and magnetic quantities, namely, current, voltage, and magnetic and electric fields.

The main application area for these devices is in high-voltage power equipment, where one can take advantage of the high intrinsic electrical insulation of the fibers. Fiber-optic sensing allows a major reduction of the costly and bulky screening that is mandatory in standard current and voltage metering transformers. Moreover, it offers extended bandwidth and dynamic range. Finally, the potentially low cost and the reliability of fiber sensors are attractive in view of the increasing interconnection density in distribution lines, where real time control demands for a larger number of measuring points.

Common requirements in high-voltage applications are compactness, passive structure and noninvasiveness, so that conducting elements should be avoided in the sensing region. Both ac (single and three-phase) and dc sensors are of interest. In fixed metering applications in power plants or power lines, a typical expected accuracy is 1% with a bandwidth of 1 Hz and a linearity range of 4 decades ($1–10^4$ A) for current, somewhat reduced for voltage. For transient diagnostics and protection, a larger bandwidth is required (100 kHz) with reduced accuracy (5%). Lifetime should be in the range of 10 years. In mobile applications, accuracy and lifetime requirements are more relaxed; however, an important point is that the sensor should not be linked to the line under test.

Magnetic field sensors for stray flux measurement and early diagnostics in power transformers and generators must be very compact (about 1 cm) to map the field distribution and its distortions due to local short circuit. Specifications are less standard; however, a typical sensitivity figure is of the order of 10^{-2} gauss with a dynamic range of a few hundred gauss. Electric field sensors should have a typical sensitivity of 10^{-1} V/m with 1% accuracy.

Only a few devices are commercially available to date (e.g., Pirelli, Westinghouse). However, many devices are at a stage of development and field testing.

5.5.2 Current and Magnetic Field Sensors

Direct Sensing

Several magneto-optical effects are potentially available to implement magnetic field and current fiber sensors [132–136]. However, the Faraday effect is the most viable approach, and virtually the only one used in practical implementations, because of the quite high Verdet constant V that is available in several glasses (Figure 5.14) and even in ordinary fibers.

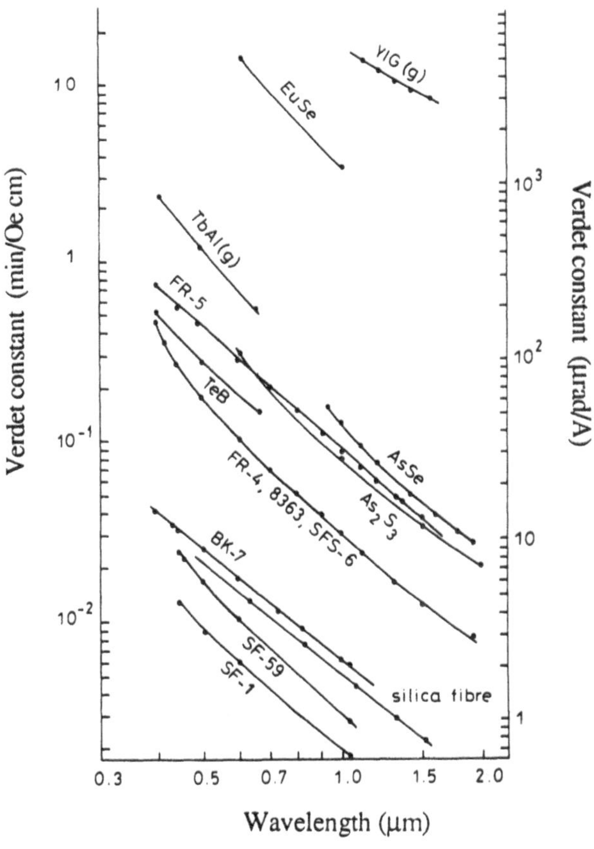

Figure 5.14 Verdet constant of some crystals and optical glasses as a function of wavelength (taken from [132]).

Discrete and micro-optic sensors employing YIG crystals have been developed in the last few years that have achieved a high degree of compactness (5×5×3 mm) [137] with a sensitivity of a fraction of μT (or, typically, mA) and a maximum bandwidth of hundreds of MHz. With regard to all-fiber devices, though in some cases special fibers doped with paramagnetic or ferromagnetic ions have been employed [138], in most cases silica fibers have been preferred because of their low cost and because they are available with well-controlled birefringence characteristics (low-birefringence fibers, polarization maintaining fibers, polarizing fibers). High responsivity can be obtained by cumulating the Faraday rotation over coils of many (e.g., 10 to 50) loops. As is well-known [133, 134], the Faraday effect consists of a nonreciprocal circular birefringence induced in the propagation medium by a magnetic field $\bar{H}$. It gives rise to a phase delay 2Φ between opposite circularly polarized fields (eigenstates of the optical system). The nonreciprocal rotation Φ of the polarization plane of a linearly polarized wave traveling along an optical path L is therefore given by

$$\Phi = \int_L V\, \bar{H} \cdot \bar{dl} \qquad (5.37)$$

It follows from (5.37) that when the optical path L is closed, F is proportional to the circulation of the vector H. In this case, if the magnetic field is generated by a current I linked to a fiber-optic coil (Figure 5.15a), we get

$$\Phi = VNI \qquad (5.38)$$

where N is the number of turns while

$$\Phi = 0 \qquad (5.39)$$

for an unlinked coil (Figure 5.15b).

Therefore, a closed path represents a true current sensor whose sensitivity is independent from geometry and vanishes for external contributions. On the other side, an open path (Figure 5.15c, d) represents a distributed magnetic field sensor; current is only indirectly measured through a scale factor which depends on the geometry. Until now, we have considered an ideal fiber, i.e., without birefringence effects other than Faraday rotation. This assumption will be removed later.

The measurement of H or I can be performed by reading Φ using an interferometric or a polarimetric scheme. The basic setup of the polarimetric analysis (low-coherence sensors) is shown in Figure 5.16 in the all-fiber implementation. The principal axes of the fiber polarizer are orientated at 45° with respect to the input linearly polarized radiation.

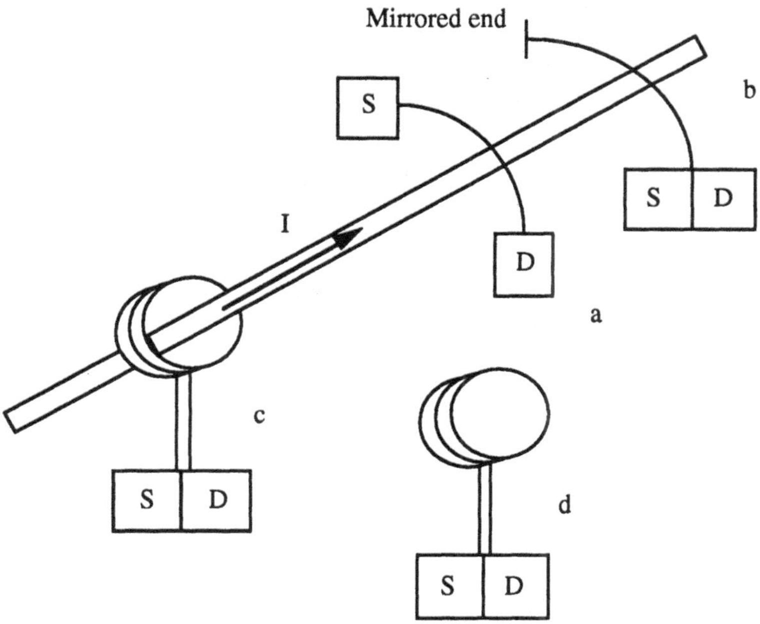

Figure 5.15 Fiber-optic current (a,b) and magnetic field (c,d) sensors based on the Faraday effect; S = source, D = detector (taken from [132]).

The photodetected signals I_1 and I_2 supplied by the photodetectors are

$$I_1 = I_0 \cos^2(45° + \Phi)$$
$$I_2 = I_0 \sin^2(45° + \Phi) \tag{5.40}$$

A signal S only dependent on Φ, regardless of the fluctuations of I_0 due to the optical power and launching, can then be computed as

$$S = \frac{I_1 - I_2}{I_1 + I_2} = \sin 2\Phi \approx 2\Phi \tag{5.41}$$

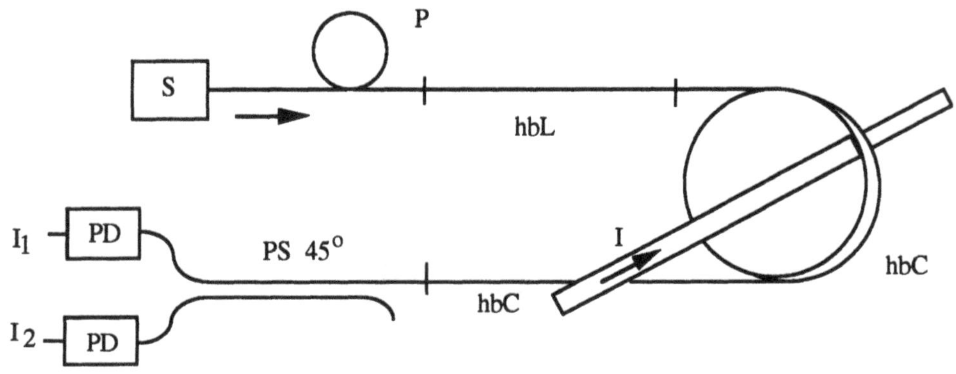

Figure 5.16 Polarimetric readout scheme; S = source, PD = photodiode, P = polarizer, PS = polarizing coupler, hbC = high circular birefringence fiber, hbL = high linear birefringence fiber (taken from [132]).

Excluding fiber related errors, noise contributions to S are the shot noise of the detected currents I_1 and I_2, Johnson noise of the photodetector amplifiers and excess noise of the source. The variance associated to signal S is [132]

$$\sigma^2(S) = \frac{2qB}{I_0}\left(F + \frac{4KT}{qRI_0}\right)(1 + S^2) \tag{5.42}$$

where q is the electron charge, KT/q is the voltage equivalent of temperature, R is the equivalent noise resistance of the preamplifier, and F is the excess noise factor, ranging from a fraction of a dB to more than 20 dB for a semiconductor laser biased near its threshold.

With a proper design, one can satisfy the condition $RI_0 \gg 4KT/q$ and reach the quantum limit, provided that $F \approx 1$. The noise equivalent current $\sigma(I)$ (S/N = 1) can then be obtained as

$$\sigma(I) = \frac{1}{2V\,N}\sqrt{\frac{2qB}{I_0}} \tag{5.43}$$

In Figure 5.17, $\sigma(I)$ is plotted vs. the bandwidth B for different values of I_0 and for N = 50, V = 2.5 µr/A (standard silica fiber).

The bandwidth B is intrinsically limited by the relaxation time of the material, which can be usually neglected with respect to the extrinsic limitation due to the transit time τ in the sensor. If r is the coil radius and n is the refractive index of the fiber, $\tau = 2\pi nNr/c$ resulting in a – 3 dB point

$$B(-3dB) = \frac{0.44}{\tau} = \frac{0.44c}{2\pi nNr} \tag{5.44}$$

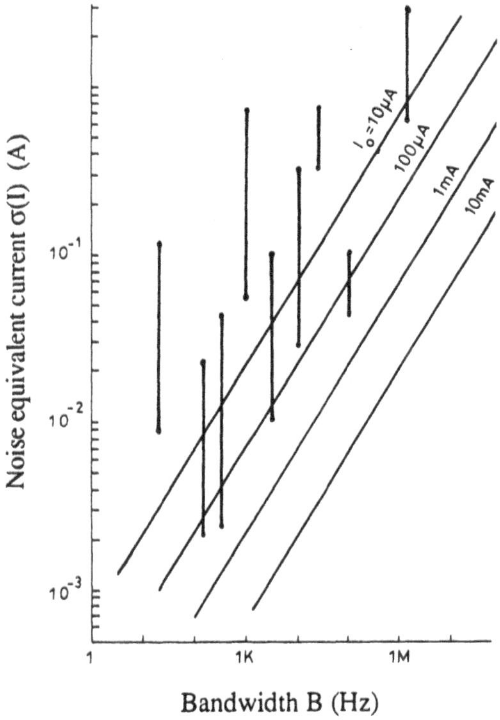

Figure 5.17 Noise equivalent current $\sigma(I)$ as a function of the bandwidth B and of the photodetected current I_0 for a silica fiber, N = 50.

A typical figure can be as high as 30 MHz, and in practice the bandwidth is often limited by the electronic processing circuitry, especially at low values of I_0. In many cases, filtering is introduced to increase sensitivity at the expense of response speed.

In the basic configuration, the dynamic range and linearity are limited by the sin 2Φ dependence of S. A simple correction can be obtained computing $S + S^3/6$ by analog or digital processing. A more efficient method, ensuring an unlimited dynamic range, is to double the optical detector section with a second beam-splitter oriented parallel to the input polarization obtaining two supplementary signals so that with the same kind of processing the signal S' = cos 2Φ is also available. Straightforward analog or digital processing [132] allows accurately recovering the angle Φ with its sign.

The basic scheme of the interferometric readout (coherent sensor) is shown in Figure 5.18. The nonreciprocal phase shift 2Φ can be read by a variety of configurations. However, the Sagnac interferometer is the first choice to cancel the minute reciprocal birefringence contributions, both residual and ambient induced.

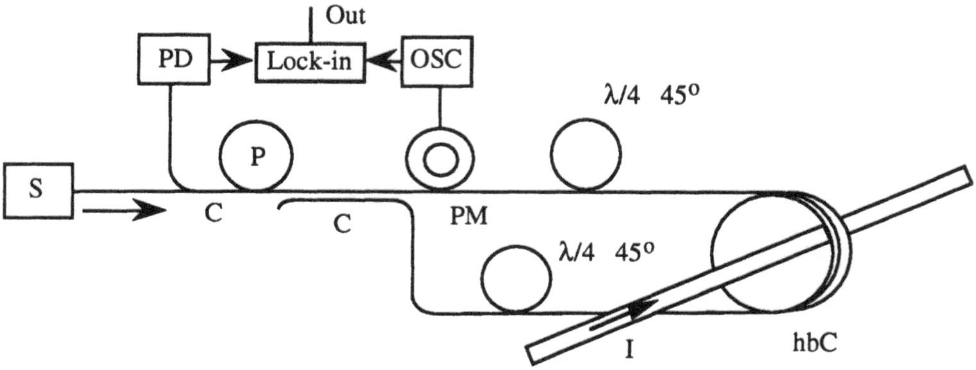

Figure 5.18 Interferometric readout scheme; OSC = oscillator, C = coupler (taken from [132]).

The measurement configuration is well established, being essentially the same developed for the fiber-optic gyroscope. It is modified for current sensing by inserting two quarter-wave retarders oriented at 45° with respect to the linear polarizer P. Therefore, two circularly polarized waves are launched into the coil and propagate in opposite directions. They are reconverted back into linear after propagation and recombine on the photodetector yielding a beating signal $I_s = I_0\,(1 + \cos 2\Phi)$. As in the gyroscope, the sensing scheme is reciprocal, and a phase modulator is employed to measure Φ and its sign. The component I_1 at the fundamental modulation frequency can be extracted from signal I_s, e. g., by using a lock-in amplifier, and we get for the output [132] S

$$S = J\, I_0 \sin 2\Phi \tag{5.45}$$

where J is maximized (≈ 0.581) by a proper choice of the modulation parameters. The measurement of the second harmonic is also useful to implement linearity improvement processing. The noise equivalent current $\sigma(I)$ is given at the quantum noise limit by

$$\sigma(I) = \frac{1}{2J\ V\ N}\sqrt{\frac{2qB}{I_0}} \qquad (5.46)$$

The diagram of Figure 5.17 can then be used, with the ordinates scaled by a factor J = 0.581, also for the interferometric readout. Although in principle polarimetric and interferometric readout schemes have nearly equivalent performances, it is well known that while the former works at baseband and is therefore sensitive to 1/f noise as well as to slow fluctuations of the polarization plane and of fiber birefringence, the interferometric readout scheme works at a carrier frequency and its performances at dc are considerably better. On the other hand, due to component limitations, particularly to the limited working frequency of the piezo phase modulator, it turns out that the polarimetric scheme is better suited for extended bandwidth and dynamic range operation.

It should be noted that the above configurations can be easily modified for closed-loop operation [139, 140]. In this case, a coil is wound around the fiber so as to give a Faraday rotation that is equal and opposite to that caused by the measurand. Thus, S is kept almost to zero by feedback, and the measurand is detected by the driving current of the coil. This approach allows to reduce nonlinearity errors and to improve dynamic range, at the expenses of a more involved setup and of a bandwidth reduction.

In the previous discussion, we have ignored birefringence effects in the sensing fiber other than the Faraday rotation. Parasitic effects [70, 141, 142, 143, 144] can sensibly reduce the maximum useful length L of the fiber and consequently the sensor responsivity. In a fiber with linear birefringence $\beta_l = \beta_y - \beta_x \neq 0$, the condition $L << 1/\beta_l$ must be satisfied in order to avoid a significant alteration of the polarization state. In ordinary fibers, $L \approx 1$ m because of residual and bending birefringence. Superposition of linear reciprocal birefringence β_l with circular reciprocal β_c and nonreciprocal $|\bar{H}|$ V can be evaluated by computing the Jones's matrix describing the evolution of the input vector during propagation [132]. After integration along the optical path, we get

$$M = \begin{pmatrix} \cos(\Delta z) + i\frac{\beta_l}{2\Delta}\sin(\Delta z) & \frac{1}{\Delta}\left(|\bar{H}|\ V + \frac{\beta_c}{2}\right)\sin(\Delta z) \\ -\frac{1}{\Delta}\left(|\bar{H}|\ V + \frac{\beta_c}{2}\right)\sin(\Delta z) & \cos(\Delta z) - i\frac{\beta_l}{2\Delta}\sin(\Delta z) \end{pmatrix} \qquad (5.47)$$

$$\Delta^2 = \frac{\beta_l^2}{4} + \left(|\bar{H}|\ V + \frac{\beta_c}{2}\right)^2 \qquad (5.48)$$

From the analysis of Equation (5.48) the following conclusions can be drawn:

1. Quenching of the Faraday effect: in a fiber with $\beta_c = 0$, the maximum rotation Φ is obtained for $\Delta z = \pi/2$. The useful length is therefore limited to $z = \pi/2\Delta$ or, if $\beta_l >> 2 |H| V$, to $z = \pi/\beta_l$. In these conditions, the Faraday rotation is reduced by $\Delta z = \pi/L$ with respect to the expected value $|H| V z$.
2. Quenching of the linear birefringence: if $\beta_c >> \beta_l$, the linear birefringence term $\beta_l(z/2)$ is reduced by $\sin \Delta z / \Delta z$. By developing Δ at the second order in $\beta_l / \beta_c = p << 1$ it is easy to show that the residual error with respect to a pure rotation is smaller than p in the field amplitude or p^2 in I_1 and I_2. In a practical sensor, it is therefore required to superpose a large circular birefringence to the linear residual birefringence. This can be obtained by fabricating a special high circular-birefringence fiber [145] or by using a twisted spun fiber.

In both cases, the rotation $\Phi_c = \beta_c z$ of the polarization plane adds to the Faraday rotation $|H| V z$, but represents only a constant bias and the useful signal can be easily extracted. This approach allows also to reduce mechanical disturbances acting on the fiber (vibrations, compression over the former, bending birefringence, thermal effects). Typically, torsion rates of about 30–100 turns/m are required, resulting in a stress that can be tolerated without damage by a standard fiber. The rotation Φ_c and its fluctuations can be cancelled from the output signal using balanced measuring schemes in which the nonreciprocity of the Faraday effect is exploited by propagating twice through the fiber in opposite directions [132].

Several modifications are possible to the above outlined schemes [146]. For instance, heterodyning has been proposed [147] to improve the noise rejection or to select the working bandwidth. Also, Michelson or Mach-Zehnder interferometers have been used.

A completely different approach consists in the annealing of the optical fiber to eliminate the elastic stress due to bending and allows to make sensors of remarkable compactness [148]. Another approach employs a fiber-optic Fabry-Perot interferometer. This solution allows, for a given fiber length, sensitivity to be improved by a factor of the order of the finesse of the resonator [149].

The noise performance reported by a few authors are summarized in Figure 5.17 where the bars connect the measured values (upper points) to the theoretical ones.

An important issue in the design of current sensors and magnetic field sensors is the dependence of the Verdet constant on temperature, particularly in power applications where thermal excursions can be wider than 100 °C. An analysis of this dependence is given in [150] and is of the order of 0.2%/°C for silica fibers.

Magnetic field and flux measurement in magnet gaps and electrical machinery require a high degree of compactness, with values for the coil radius of the order of 1 cm [132]. In addition, as the Faraday rotation vanishes on a closed path in any region where the magnetic field is irrotational, a different design approach is required that exploits the combined effect of the Faraday and of the linear (bending) birefringence [151–155]. It can be shown that if every single loop of a fiber coil immersed in a uniform magnetic field is exactly a λ-retarder, the overall effect consists of a pure nonreciprocal rotation. (More precisely, this rotation amounts to one half the maximum value that one would expect from a straight fiber of the same length.)

To have the downlead fibers on the same side with respect to the sensor, the geometrical arrangement of Figure 5.19 is preferred, making coils of N+1/2 turns of fiber. For an input linearly polarized field E_{x0} normal to the coil plane, the electric field components at the sensor output are then

$$E_x = -E_{x0} \cos \Phi_z, \; E_y = -E_{x0} \sin \Phi_z \tag{5.49}$$

(Figure 5.19a) or

$$E_x = -E_{x0} \cos\Phi_y \,, \; E_y = -iE_{x0} \sin \Phi_y \tag{5.50}$$

(Figure 5.19b), where

$$\Phi_{y,z} = V\, H_{y,z}\, \pi\, r\, (N+1/2) \tag{5.51}$$

Figure 5.19c shows a polarimetric readout scheme which allows one to obtain both components of the field H. In-phase and -quadrature analysis are performed in the upper and lower path, respectively, and the photodetected signals are $I_z = I_0 \cos^2(45° + \Phi_z)$ and $I_y = I_0 \cos^2(45° + \Phi_y)$. The minimum detectable magnetic field value can be shown to be

$$\sigma(\bar{H}) = \frac{1}{2V\left(N + \frac{1}{2}\right)\pi r} \sqrt{\frac{2qB}{I_0}} \tag{5.52}$$

and the graph of Figure 5.17 can be used with the scale factor $\sigma(\bar{H}) \approx 1.28\, \sigma(I)$, assuming that $\bar{H}$ is in Oe and r = 1 cm.

Indirect Sensing.

In the previous discussion we considered direct sensors. Let us now revert to indirect approaches. Among them, the most popular exploits magnetostriction in a suitable material (such as nickel) deposited onto the fiber or in a metal glass element around which the fiber is wound, or fixed, or in which it is embedded. This approach offers good responsivity and has been employed to make magnetometers [1]. However, it often leads to invasive devices.

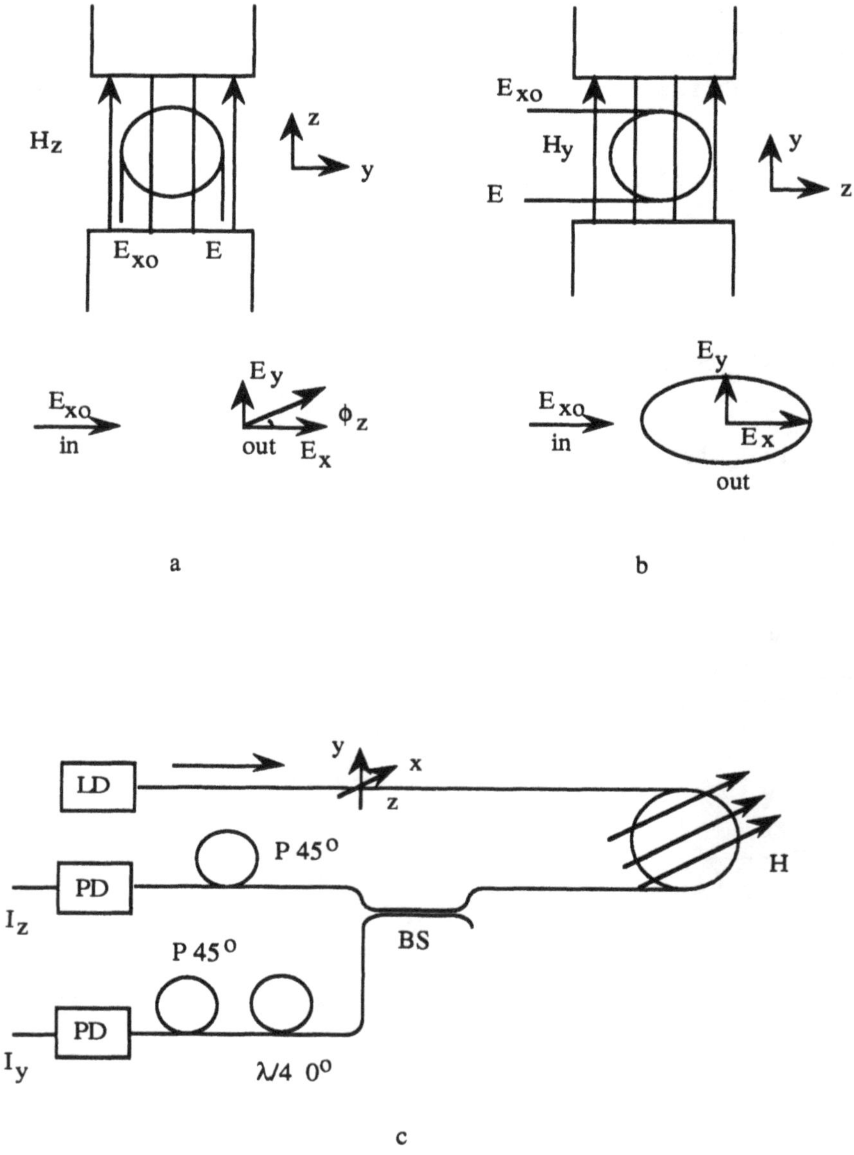

Figure 5.19 Magnetic field sensors based on (a,b) birefringence control and (c) vectorial readout scheme; LD = laser diode, PD = photodiode, BS = beam-splitter (taken from [132]).

The longitudinal strain ε_z in the coating is converted into a relative variation of the optical path

$$\Delta\Phi/\Phi = \{1 - (1/2)\, n^2[(1 - \mu)p_{12} - \mu p_{11}]\}\eta\, \varepsilon_z \tag{5.53}$$

where $\Phi = nkL$ and η is the transduction efficiency (strain in the fiber/strain in the coating) [54]. Since the magnetostrictive effect is quadratic in $|\bar{H}|$ ($\varepsilon_z \propto |\bar{H}|^2$), it is required to select a bias point in order to work in the linear regime. This can be done by applying a constant field H_b by means of an external solenoid. A more efficient solution consists in coating the fiber with ferromagnetic material of adequate coercive force $\bar{H}_c$(e.g., Al Ni Ci, ferrite, metal glass). The sensor can then measure variable magnetic fields $h = h_0 \sin \omega t$, since $\Delta\Phi$ becomes proportional to $2h|\bar{H}|_c \sin \omega t$. Alternatively, to measure static fields, one can apply to the sensor a reference field at a frequency ν so as to have a $\Delta\Phi$ component at the same frequency, and whose amplitude is proportional to the field under measurement.

Different readout approaches (both polarimetric [156] and interferometric) are still possible, usually Mach-Zehnder [157, 158] (Figure 5.20), open or closed loop, that employ different techniques to achieve dynamic range extension and improve ac and dc sensitivity.

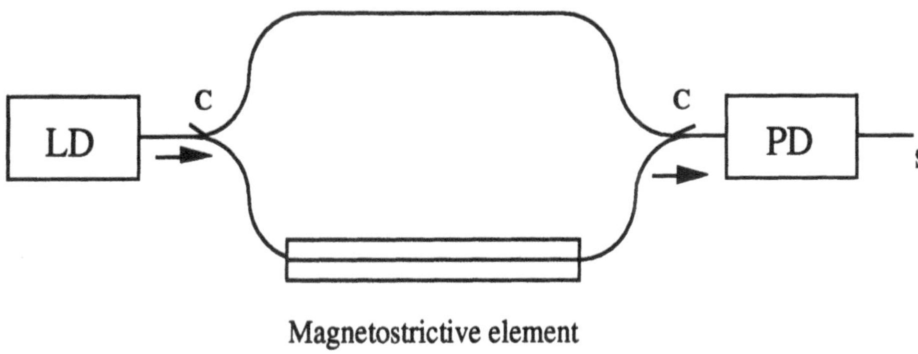

Figure 5.20 Magnetostrictive magnetic field sensor based on a Mach-Zehnder interferometer.

Sensitivity levels of 10 pT/(Hz)$^{1/2}$ have been reported with maximum bandwidth of 10 kHz for a typical coating length of about 10 cm. The characteristics and limitations of this class of sensors have been investigated in the literature [159]. They include demagnetization effects due to the high permeability of the coating material, responsivity reduction due to mechanical loading and nonflat frequency response due to mechanical resonances. An analysis of the intrinsic limit of sensitivity due to thermal effects and eddy currents in the metallic glass is reported in [160].

Other methods for measuring a magnetic field exploit the Lorentz force in a fiber with a conductive coating in which a current i is forced. The elastic strain or the amplitude of the induced vibration is interferometrically read. A linearity range of 3 decades for both static and quasistatic magnetic field has been reported in a 10-cm Al-coated sample [161]. A similar approach exploits a ferromagnetic coating to avoid current injection [162].

5.5.3 Voltage and Electric Field Sensors

While a number of current and magnetic field sensors are available that can satisfy the requirements for in-field deployment, electric field and voltage fiber sensors have not yet reached a comparable degree of development. In addition, it does not seem, at least for the moment, that the direct approach is viable, because the electro-optic effects are too small in glasses. Commercial fiber sensors and almost all published devices are indeed of the extrinsic class or even micro-optic. They are based on the transverse Pockels effect in crystals such as $LiTaO_3$, $LiNbO_3$, $Bi(GeO_4)_3$ and KDP.

The transverse Pockels [163] effect consists in a reciprocal linear birefringence induced by the electric field in a propagation medium. The refractive index difference between the principal axes is proportional to the applied electric field $\bar{E}$. The phase difference between two orthogonally linearly-polarized fields emerging from the crystal is measured with a polarimetric or interferometric configuration like those discussed in the next subsection, simply adding a quarter-wave retarder before the crystal [164, 165]. It must be noted, however, that the Pockels effect is intrinsically more complex than the Faraday effect and requires tensor analysis to be fully described. In fact, the length and the orientation of the crystal are critical and a major experimental problem consists in minimizing the sensor sensitivity to ambient-related physical quantities acting on it through the photoelastic, piezoelectric, and electrostrictive effects. Typical reported sensitivity values are of the order of 1 kV/m [165] with a band of about 10 MHz and are obtained in a crystal of about 1 cm in length. For voltage measurement a typical figure is 200 $\mu V/(Hz)^{1/2}$ for bismuth germanate crystals of about 5 cm in length.

More recently, the first examples of fiber-optic voltage and electric field indirect sensors have been proposed. They employ piezoelectric materials, either crystals (quartz, barium titanate, litium niobate) or ceramics, to make a former of suitable shape on which the fiber is wound. Alternatively, polymers such as PVF_2 and P(VDF-TFE) have been used as a coating or to embed the fiber [166, 167]. After suitable poling, an external electric field $\bar{E}$ gives rise to a stress field in the fiber (inverse piezoelectric effect) and the induced birefringence can be read by an interferometer. As an example, for a coaxial structure built around a fiber [167], the phase delay due to the elastic strain is $\Delta\Phi = knL \{\varepsilon_z - n^2/2[(p_{11} + p_{12})\varepsilon_r + p_{12}\varepsilon_z]\}$, where ε_z, ε_r are the axial and radial deformation, respectively, and p_{11}, p_{12} are elements of the elasto-optic tensor. Typical sensitivity performances reported in literature are of the order of 0.1 kV/cm for a fiber sample of about 2 cm with a bandwidth in excess of 100 kHz, while the theoretical limit (as given by the inteferometer equivalent noise, e.g., 10^{-7} r) is of a fraction of V/m. This gap is mainly attributable to ambient acoustic noise.

The analysis of the symmetry properties for the materials and the choice of suitable geometrical shapes allow to select the electric field component to be measured [167]. For instance, the use of a cylindrical quartz structure with the sensing fiber wound on its lateral surface allows to transduce selectively the component that is parallel to the cylinder axis (Figure 5.21).

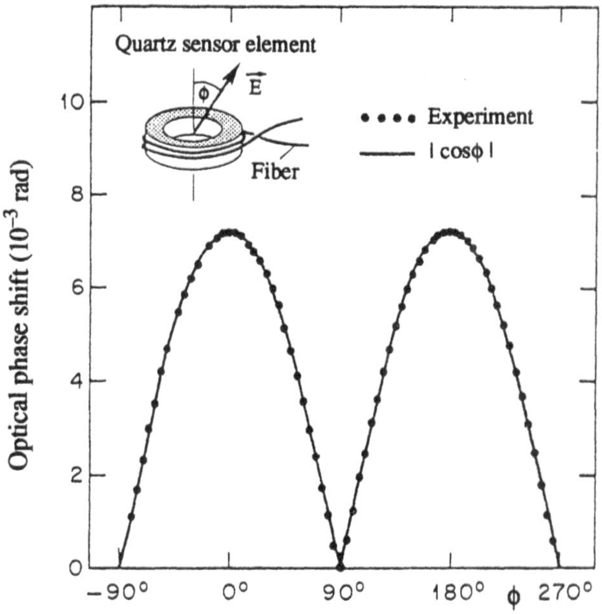

Figure 5.21 Electric field fiber sensor based on the inverse piezoelectrical effect: signal from a single element as a function of the field direction (taken from [41] with permission).

This feature has been exploited to make true voltage sensors. In fact, in voltage measurement a specific problem comes from the fact that this quantity is related to the electric field (the physical quantity that is directly sensed) by the line integral $V_{ab} = \int_L \underline{E} \cdot d\underline{l}$. It is therefore required to make a distributed measurement of the electric field or to select a geometry to infer V from a local measurement of $\underline{E}$. A solution that approximates the line integral by sampling the axial component of $\underline{E}$ has been proposed in [168]. The sensor is made of a number of small piezoelectric disks on which adjacent samples of fiber are wound to make a chain along path L; the symmetry properties for the element allows to measure just the axial component. The different phase delay contributions are added, and the total variation of the fiber length is given by [169]

$$\Delta L = \frac{1}{2}\, d_{11}\, 2\pi\, a\, N \sum_{i} E(r_i)\, D \tag{5.54}$$

where d_{11} is the suitable piezoelectric coefficient, N is the number of loops around the stress element, D the depolarization constant, a the fiber radius, $E(r_i)$ is the local field (averaged on element i), and the sum must be extended to all the stress elements. The total phase delay, detected at the output, allows obtaining the voltage V. The expected sensitivity is about 1V for a bandwidth of 100 kHz.

5.6 CHEMICAL SENSORS

5.6.1 Fiber Sensors

Recently, fiber-optic sensors for detection of chemical species have been extensively investigated. Although the majority of these chemical sensors employs multi-mode optical fibers [170], there are some interesting examples of chemical sensing using single-mode fibers in Mach-Zehnder [171] and in Michelson interferometers [172, 173]. In these applications, the sensing element is a piece of fiber "modified" in such a way that the concentration of the chemical analyte is transduced to a phase delay, which is then detected with interferometric techniques.

Hydrogen sensors have been reported based on the expansion of palladium metal exposed to hydrogen gas [171, 173]. The sensing device is realized either by coating the fiber (without jacket) with a thin film of palladium or by mechanically attaching a palladium wire to the fiber. Changes in dimensions of film or wire produce a strain in the fiber, which corresponds to a change in the optical path length, that is detected by placing the sensing fiber in one arm of the interferometer. These devices are basically strain transducers, transformed in chemical sensors by a proper choice of the coating.

Of course, similar considerations can be applied to temperature and pressure sensors [28]. As an example of a temperature-based sensor, a flammable gas (hydrocarbons) interferometric sensor has been demonstrated in which the sensing element is a piece of platinum-coated single-mode fiber [173]. The exothermic reaction between hydrocarbons and oxygen is catalyzed by platinum. The resultant heat is determined by measuring the phase delay of the optical beam traveling in the coated fiber. Considering pressure-based sensors, a spectrophone for monitoring the optical absorption of gas has been reported, which employs a fiber-optic coil placed in one arm of a Mach-Zehnder all-fiber interferometer [174] as the pressure transducer. The coil is closed in a container and senses the pressure modulation due to temperature changes which come from the optical absorption of the gas in the cell.

In principle, the concept of chemical sensing by sensitizing an optical fiber with an appropriate coating seems suitable for the detection of other chemical and biochemical analytes. An important step in the design of a chemical sensor is the choice of the right material for the given application, considering that the selectivity of the instrument is mainly determined by the coating. Promising materials might be, for example, polymers and enzymes.

Potentially, polarimetric detection schemes could be exploited also for chemical sensing, probably leading to easier setup but lower sensitivity compared to interferometric sensors. Moreover, any refractive index sensor could be applied to the measurement of analyte concentration. However, the refractometer as a chemical sensor is itself nonspecific and is unable to distinguish between different components of a mixture and interfering substances. Selectivity for an analyte thus depends on the use of a specific reagent (e.g., enzymes) or of another selective system, such as a semipermeable membrane or an adsorbing material, at the fiber-medium interface.

As a conclusion it is interesting to identify the major advantages and problems of chemical sensors based on single-mode fibers, even though some comments are valid also for multi-mode fiber sensors. First of all, interferometric sensors have a very high intrinsic sensitivity; thus very low concentrations can be potentially detected. Fiber-optic sensors can be envisioned for remote, *in situ*, and continuous measurements, even in hazardous environments, because the probe is electrically insulated and can be very small. Moreover, silica fibers are chemically inert and silica surface is adapt for receiving a wide range of coatings.

On the other side, cross sensitivity of interferometric sensors to other variables may limit the device resolution. In the interferometric chemical sensors previously described, if a spurious temperature difference between sensing and reference arm cannot be avoided, it is necessary to compensate for it. Moreover, even if both arms are at the same temperature, the coated fiber might have a thermal behavior different from the one of the bare fiber. This source of error could be compensated for by modifying the reference fiber so that both arms have the same mechanical characteristics but different chemical sensitivity.

In general, only partial chemical specificity can be impressed to these sensors and very likely the same coating will be affected by many analytes. In the case of complex samples, array of sensors in which each element is only partially selective for a different analyte may give better performances. The pattern of responses measured from the array may be used to identify the analyte concentrations of interest by means of chemometrics techniques (see, e.g., [175]).

5.6.2 Planar Waveguide Sensors

Integrated-optical chemical sensors have been demonstrated by Lukosz et al. since 1983 [176]. The transducer consists of a diffraction grating fabricated by an embossing technique on the surface of a planar SiO_2-TiO_2 optical waveguide [177, 178] and used as input [176, 179–185] or output [186, 187] light coupler. These sensors respond to changes in the refractive index of the liquid or gas sample, which covers the waveguide in the in- or out-coupling region, and respond also to adsorption on the grating of molecules out of the sample phase [179–182]. Chemical selectivity is obtained only if the waveguide is coated with a thin chemoresponsive layer [183–187]. This possibility has been exploited in particular for applying the integrated-optical sensors as immunosensors or immunoassays [184, 185].

Input and output configurations are reciprocal and the theoretical explanation underlying their operation is the same [186, 187]. We will discuss the working principle, first of the input coupler sensor and then, with reciprocity considerations, of the output setup. As it will be later described, the output configuration has some advantages which make it more suitable, with respect to the input setup for practical sensing applications.

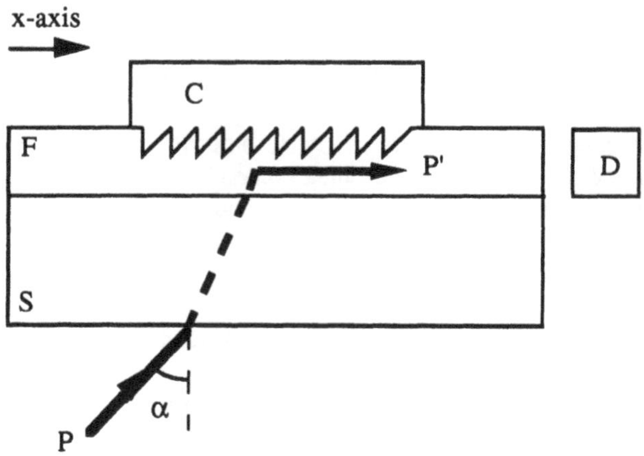

Figure 5.22 Grating coupler on planar waveguide as integrated-optical chemical sensor: input configuration; S: substrate; F: waveguiding film; C: covering sample; D: photodetector.

In the input configuration (Figure 5.22), the grating is employed to couple a linearly polarized laser beam (He-Ne, emission wavelength λ = 632.8 nm) into the optical waveguide. The angle of incidence α' of the laser beam on the grating can be changed since the optical transducer is mounted on a precision rotation stage. The propagation constant β_o of the incident wave along the x-axis is usually expressed in terms of the incidence angle α measured in ambient air as [176]

$$\beta_o = \frac{2\pi}{\lambda}\, n_{air} \sin\alpha \qquad (5.55)$$

where n_{air} is the air refractive index. The angles α, α' are obviously related by Snell's law.

As the beam interacts with the grating, different waves are diffracted and the wave corresponding to the diffraction order g has a propagation constant β_g given by

$$\beta_g = \beta_o + g\frac{2\pi}{\Lambda} = \frac{2\pi}{\lambda}\left[n_{air} \sin\alpha + g\frac{\lambda}{\Lambda}\right] \qquad (5.56)$$

where Λ is the grating period. The strongest coupling will occur between the g-order diffracted wave and the film guided wave with a propagation constant β (along the x-axis) matching the "in-coupling condition"

$$\beta_g = \beta \qquad (5.57)$$

This optimum incoupling condition is usually expressed for the effective refractive index N of the guided mode, $N = \beta / (2\pi / \lambda)$, as

$$N = n_{air} \sin\alpha_m + g\frac{\lambda}{\Lambda} \qquad (5.58)$$

where α_m is the optimum angle of incidence. In other words, for a given effective index N (of the single guided wave), which depends on refractive index and thickness of substrate, waveguiding film and covering medium, and for a selected diffraction order, the angle $\alpha = \alpha_m$ assuring optimum power coupling is given by Equation (5.58). It is worth mentioning that with the same polarized laser source and a proper optical setup, it is possible to excite either the TE_{00} or TM_{00} mode in the waveguide. These orthogonally polarized fields travelling in the waveguide have different effective index $N(TE_{00})$ and $N(TM_{00})$ and therefore they satisfy two different coupling conditions, which means $\alpha_m(TE_{00}) \neq \alpha_m(TM_{00})$.

For reciprocity, Equation (5.58) is also the optimum outcoupling condition for the output configuration and describes the relationship between the effective refractive index N of the guided mode and the angle α of the out-coupled light beam (Figure 5.23).

From the theory on diffraction gratings, it follows that the condition defined by Equation (5.58) is not infinitely sharp. If we consider, for example, the input configuration, we can define the in-coupling efficiency η as the power P' of the in-coupled guided mode detected by the photodetector D (Figure 5.22) divided by the power P of the incident laser beam. When all the waveguide parameters are defined, the efficiency η reaches a maximum η_m for an optimum angle of incidence $\alpha = \alpha_m$ satisfying the optimum in-coupling condition (5.58). By plotting the efficiency as a function of the parameter $\underline{N}$, defined as $\underline{N} = N - n_{air} \sin \alpha_m - g\, \lambda / \Lambda$, we describe the detuning from the optimum incoupling at $\underline{N} = 0$. The plot of η versus $\underline{N}$ has the form of a resonance curve with a small but finite width, and at resonance $\eta(\underline{N} = 0) = \eta_m$. Moreover, if the angle of incidence and the order g are defined, $\Delta\underline{N} = \Delta N$ and a change in the effective index ΔN corresponds to a change in the intensity of the guided mode.

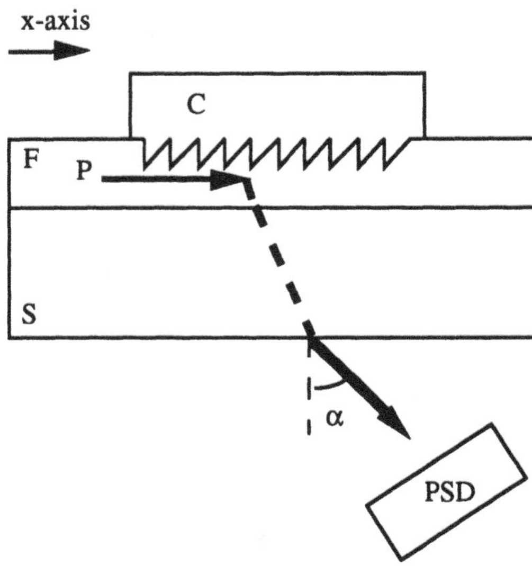

Figure 5.23 Grating coupler on planar waveguide as integrated-optical chemical sensor: output configuration; S: substrate; F: waveguiding film; C: covering sample; PSD: position sensitive detector.

In order to understand the transducing mechanism of this device, we have to consider that the effective index N of the guided mode in the coupling region depends strongly on the medium covering the grating. In fact, the effective index is given by the relation [188]

$$N = n_f \sin \theta \tag{5.59}$$

where n_f is the refractive index of the guide film and θ must satisfy the self-consistency condition (also known as the transverse resonance condition) [188]

$$\frac{4\,\pi}{\lambda}\, d_f\, n_f \sin \theta + \phi_{f\text{-}s} + \phi_{f\text{-}c} = 2\, m\, \pi \tag{5.60}$$

In this equation, d_f is guide film thickness, m is an integer (m = 0,1,2,..), while $\phi_{f\text{-}c}$ and $\phi_{f\text{-}s}$, depending on the angle θ and other waveguide parameters, are the phase shifts on total internal reflection at the film-cover interface and at the film-substrate interface, respectively [188]. By combining (5.59) and (5.60), it becomes clear that, in case of chemical sensing in the covering medium, three mechanisms can be responsible for a variation ΔN in the effective index:

1. a change Δn_c in the refractive index n_c of the covering medium for example due to a chemical reaction;
2. a refractive index change Δn_f in the waveguiding film, which can occur if micropores contained in the waveguide material are filled with the covering medium;
3. an adsorption of molecules, contained in the covering medium, on the waveguide surface forming an adlayer of thickness d'_f and refractive index n'_f. (This third mechanism can be distinguished from 1 as long as d'_f is smaller than the penetration depth of the evanescent wave, created at the interface film-external medium.)

As a consequence, if the change in N is small, the total effective index change can be expressed as [182]

$$\Delta N = \frac{\partial N}{\partial n_c}\Delta n_c + \frac{\partial N}{\partial n_f}\Delta n_f + \frac{\partial N}{\partial d'_f}\Delta d'_f \tag{5.61}$$

The analytical expressions of the sensitivity terms $\partial N/\partial n_c$, $\partial N/\partial n_f$, $\partial N/\partial d'_f$ have been reported in [176, 180, 182, 183]. In practice, using SiO_2-TiO_2 films on glass substrate (d_f = 100–200 nm), it has been demonstrated that it is possible to detect refractive index changes Δn_C as low as $2 \cdot 10^{-5}$ and changes in the adlayer thickness $\Delta d'_f$ larger than 0.4 nm [183]. As for the coupling condition, also the sensitivity depends on the mode polarization. Measurements of both $\Delta N(TE_{00})$ and $\Delta N(TM_{00})$ can be useful for example in adsorption studies for the determination of the adlayer refractive index n'_f as well as of the adlayer thickness d'_f [185–187].

The input configuration can be operated in two different ways [186]. A first straightforward way consists in measuring the incoupled power P'(t) as a function of the time t at a fixed angle of incidence. From P'(t) it is possible to derive the effective index change with time, and thus the chemical changes occurring in the covering sample. However, this method has a very limited dynamic range, which depends on the width of the in-coupling condition. To overcome this disadvantage, a *scanning mode* of operation can be selected. In this case, the in-coupled power P' is measured as a function of the angle of incidence α. Changes of effective index are determined by looking at the time evolution of the function $P'(\alpha)$. This method has wide dynamic range and high resolution for the minimum detectable index change. On the other side, temporal resolution can be affected by the scan time and the scan repetition rate, which are very likely limited by the mechanical rotation of moving parts.

All these problems can actually be solved by using the output configuration as follows [186, 187]. Light traveling in the waveguide is coupled out by the grating. The output beam is focused (ideally to a diffraction-limited spot-size) by means of a lens on a position-sensitive detector (PSD). A change ΔN in the effective index of the guided mode in the coupling region causes a change $\Delta\alpha$ in the out-coupling angle, $\Delta N = n_{air}\,\Delta(\sin\alpha) \approx \Delta\alpha$, and thus a displacement of the light spot on the PSD. As it is known, the PSD output depends on the position of the light spot and therefore the output voltage is a function of the effective index N. As a final observation, this configuration features high resolution and wide dynamic range as well as fast response time and no moving parts.

Integrated-optical grating couplers have been successfully demonstrated for humidity and gas detection [176, 179, 180] and for enzymatic reaction monitoring [181]. They have been operated as highly sensitive differential refractometer [183, 187] as well as immunosensor [184, 185].

REFERENCES

[1] Giallorenzi, T. G., J. A. Bucaro, A. Dandridge, G. H. Siegel, J. H. Cole, S. C. Rashleigh, and R. G. Priest, "Optical fiber sensor technology", *IEEE J. Quantum Electron.*, Vol. QE-18, 1982, pp. 626–664.

[2] Pitt, G. D., P. Extance, R. C. Neat, D. N. Batchelder, R. E. Jones, J. A. Barnett, and R. H. Pratt, "Optical-fibre sensors", *IEE Proc. J,* Vol. 132, 1985, pp. 214–248.

[3] Jackson, D. A., and J. D. C. Jones, "Fibre optic sensors", *Optica Acta*, Vol. 33, 1986, pp. 1469–1503.

[4] Dakin, J., and B. Culshaw, eds., *Optical fiber sensors: principles and components*, Vol. 1, Artech House, Norwood, MA, 1988.

[5] Culshaw, B., and J. Dakin, eds., *Optical fiber sensors: systems and applications*, Vol. 2, Artech House, Norwood, MA, 1988.
[6] Udd, E., ed., *Fiber optic sensors*, Wiley, New York, 1991.
[7] Proceedings of International Conferences on Optical Fiber Sensors: London, 1983; Stuttgart, 1984; San Diego, 1985; Tokio, 1986; New Orleans, 1988; Paris, 1989; Sidney, 1990; Monterey, 1992.
[8] Proceedings of SPIE on Fiber Optics, Fiber Optic Sensors, Fiber Optics and their Applications.
[9] Ning, Y. N., K. T. V. Grattan, W. M. Wang, and A. W. Palmer, "A systematic classification and identification of optical fibre sensors", *Sensors and Actuators* , Vol. 29, Part A, 1991, pp. 21–36.
[10] Martinelli, M., and P. Vavassori, "Novel method to pseudoheterodyne a polarimetric fiber optic sensor", *Appl. Opt.,* Vol. 28, 1989, pp. 2203–2204.
[11] Dakin, J. P., "Distributed Optical Fibre Sensors", *Proc. OFS 7*, 1990.
[12] Tambosso, T., and S. Donati, "Influence of the input state of polarization on interferometric and polarimetric measurements of birefringence", *Opt. Lett.*, Vol. 14, 1989, pp. 476–478.
[13] Kersey, A. D., D. A. Jackson, and M. Corke, "A simple fibre Fabry-Perot sensor", *Opt. Commun.*, Vol. 45, 1983, pp. 71–74.
[14] Petuchowski, S. J., T. G. Giallorenzi, and S. K. Sheem, "A sensitive fiber optic Fabry-Perot interferometer", *IEEE J. Quantum Electron.*, Vol. QE-17, 1981, pp. 2168–2170.
[15] Yoshino, T., K. Kurosawa, K. Itoh, and T. Ose, "Fiber-optic Fabry-Perot interferometer and its sensor applications", *IEEE J. Quantum Electron.*, Vol. QE-18, 1982, pp. 1624–1633.
[16] Born, M., and E. Wolf, *Principles of optics*, Pergamon Press, Oxford, 1975, pp. 323–333.
[17] Sakai, I., "Multiplexing systems for interferometric sensors with short imbalances", *Appl. Opt.*, Vol. 26,1987, pp. 4597–4602.
[18] Wentworth, R. H., "Theoretical noise performance of coherence-multiplexed inteferometric sensors", *IEEE J. Lightwave Technol.* Vol. LT-7, 1989, pp. 941–956.
[19] Kersey, A. D., A. Dandridge, and K. L. Dorsey, "Transmissive serial interferometric fiber sensor array", *IEEE J. Lightwave Technol.*, Vol. LT-7, 1989, pp. 846–854.
[20] Bucholtz, F., A. D. Kersey, and A. Dandridge, "Multiplexing of linear fiber-optic interferometric sensors", *IEEE J. Lightwave Technol.*, Vol. LT-7, 1989, pp. 514–519.
[21] Moslehi, B., M. R. Layton, and H. J. Shaw, "Efficient fiber-optic structure with applications to sensor arrays", *IEEE J. Lightwave Technol.*, Vol. LT-7, 1989, pp. 236–243.

[22] Dandridge, A., A. B. Tveten, A. D. Kersey, and A. M. Yurek, "Multiplexing of interferometric sensors using phase carrier techniques", *IEEE J. Lightwave Technol.*, Vol. LT-5, 1987, pp. 947–952.

[23] Sakai, I., "Multiplexing of optical fiber sensors using a frequency-modulated source and gated output", *IEEE J. Lightwave Technol.*, Vol. LT-5, 1987, pp. 932–939.

[24] Buckman, A. B., "Analysis of a novel optical fiber interferometer with common mode compensation", *IEEE J. Lightwave Technol.*, Vol. LT-7, 1989, pp. 151–157.

[25] Buckman, A. B., "General sensitivity enhancement and common-mode compensation principle for interferometric fiber-optic sensors", *IEEE J. Lightwave Technol.*, Vol. LT-8, 1991, pp. 1456–1460.

[26] Imai, Y., and Y. Ohtsuka, "Compensated sensing based on retardation characteristics in a bent birefringent fiber", *Appl. Opt.,* Vol. 25, 1986, pp. 4444–4447.

[27] Lefevre, H. C., "White light interferometry in optical fibre sensors", *Proc. OFS* 7, 1990.

[28] Hocker, G. B., "Fiber-optic sensing of pressure and temperature", *Appl. Opt.*, Vol. 18, 1979, pp. 1445–1448.

[29] Lagakos, N., J. A. Bucaro, and J. Jarzynski, "Temperature-induced optical phase shifts in fibers", *Appl. Opt.*, vol. 20, 1981, pp. 2305–2308.

[30] Lagakos, N., and J. A. Bucaro, "Minimizing temperature sensitivity of optical fibers", *Appl. Opt.*, Vol. 20, 1981, pp. 3276–3278.

[31] Corke, M., A. D. Kersey, D. A. Jackson, and J. D. C. Jones, "All-fibre Michelson thermometer", *Electron. Lett.*, Vol. 19, 1983, pp. 471–473.

[32] Corke, M., A. D. Kersey, and D. A. Jackson, "Temperature sensing with single-mode optical fibres", *J. Phys. E: Sci. Instrum.*, Vol. 17, 1984, pp. 988–993.

[33] Kist, R., S. Drope, and H. Wölfelschneider, "Fiber-Fabry-Perot (FFP) thermometer for medical applications", *Proc. OFS*, 1984, pp. 165–170.

[34] Seki, Y., and K. I. Noda, "Linearity stabilized fiber-optic thermometer using pseudo-heterodyne phase detection", *IEEE J. Lightwave Technol.*, Vol. LT-5, 1987, pp. 961–966.

[35] Lee, C. E., R. A. Atkins, and H. F. Taylor, "Performance of a fiber-optic temperature sensor from –200 to 1050°C", *Opt. Lett.*, Vol. 13, 1988, pp. 1038–1040.

[36] Lee, C. E., H. F. Taylor, A. M. Markus, and E. Udd, "Optical-fiber Fabry-Perot embedded sensor", *Opt. Lett.*, Vol. 14, 1989, pp. 1225–1227.

[37] Farahi, F., J. D. C. Jones, and D. A. Jackson, "High-speed fiber-optic temperature sensor", *Opt. Lett.*, Vol. 16, 1991, pp. 1800–1802.

[38] Chardon, D., and S. J. Huard, "A new interferometric and polarimetric temperature optical fiber sensor", *IEEE J. Lightwave Technol.*, Vol. LT-4, 1986, pp. 720–725.

[39] Tambosso, T., "Sensori a fibra ottica per misure di temperatura", *Alta Frequenza Riv. di Elettron.*, Vol. LVI, 1987, pp. 143–155.
[40] Dandridge, A, and A. B. Tveten, "Phase compensation in interferometric fiber optic sensors", *Opt. Lett.*, Vol. 7, pp. 279, 1982.
[41] Jackson, D. A., A. D. Kersey, M. Corke, and J .D. C. Jones, "Pseudoheterodyne detection scheme for optical interferometers", *Electron. Lett.*, Vol. 18, 1982, pp. 1081–1083.
[42] Eickhoff, W., "Temperature sensing by mode-mode interference in birefringence optical fibers", *Opt. Lett.*, Vol. 6, 1981, pp. 204–206.
[43] Corke, M., A. D. Kersey, K. Liu, and D. A. Jackson, "Remote temperature sensing using polarisation-preserving fibre", *Electron. Lett.*, Vol. 20, 1984, pp. 67–69.
[44] Dakin, J. P., and C. A. Wade, "Compensated polarimetric sensor using polarisation-maintaining fibre in a differential configuration", *Electron. Lett.*, Vol. 20, 1984, pp. 51–53.
[45] Kersey, A. D., M. A. Davis, and M. J. Marrone, "Differential polarimetric fiber-optic sensor configuration with dual wavelength operation", *Appl. Opt.*, Vol. 28, 1989, pp. 204–206.
[46] Enokihara, A., M. Izutsu, and T. Sueta, "Optical fiber sensors using the method of polarization-rotated reflection", *IEEE J. Lightwave Technol.*, Vol. LT-5, 1987, pp. 1584–1590.
[47] Akhavan Leilabady, P., J. D. C. Jones, and D. A., Jackson, "Combined interferometric-polarimetric fibre optic sensor capable of remote operation", *Opt. Commun.*, Vol. 57, 1986, pp. 77–80.
[48] Corke, M., J. D. C Jones, A. D. Kersey, and D. A. Jackson, "Combined Michelson and polarimetric fibre optic interferometric sensor", *Electron. Lett.*, Vol. 21, 1985, pp. 148–149.
[49] Newson, T. P., F. Farahi, J. D. C. Jones, and D. A. Jackson, "Combined interferometric and polarimetric fibre optic temperature sensor with a short coherence length source", *Opt. Commun.*, Vol. 68, 1988, pp. 161–165.
[50] Farahi, F., D. J. Webb, J. D. C. Jones, and D. A. Jackson, "Simultaneous measurement of temperature and strain: Cross-sensitivity considerations", *IEEE J. Lightwave Technol.*, Vol. LT-8, 1990, pp. 138–142.
[51] Lefevre, H. C., F. X. Desforges, P. Martin, and P. Simonpietri, "Two-mode fiber gauge for dual measurement of temperature and strain", *Proc. OFS* , 1990, pp. 387–390.
[52] Gerges, A. S., F. Farahi, T. P. Newson, J. D. C. Jones, and D. A. Jackson, "Interferometric fibre-optic sensor using a short-coherence-length source", *Electron. Lett.*, Vol. 23, 1987, pp. 1110–1111.
[53] Lee, C. E., and H. F. Taylor, "Fiber-optic Fabry-Perot temperature sensor using a low-coherence light source", *IEEE J. Lightwave Technol.*, Vol. LT-9, 1991, pp. 129–134.

[54] Butter, C. D., and G. B. Hocker, "Fiber optics strain gauge", *Appl. Opt.*, Vol. 17, 1978, pp. 2867–2869.

[55] Valis, T., D. Hogg, and R. M. Measures, "Fiber optic Fabry-Perot strain gauge", *IEEE Phot. Tech. Lett.*, Vol. 2, 1990, pp. 227–228.

[56] Murphy, K. A., M. F. Gunther, A. M. Vengsarkar, and R. O. Claus, "Quadrature phase-shifted, extrinsic Fabry-Perot optical fiber sensors", *Opt. Lett.*, Vol. 16, 1991, pp. 273–275.

[57] Martinelli, M., "The dynamical behavior of a single-mode optical fiber strain gage", *IEEE J. Quantum Electron.*, Vol. QE-18, 1982, pp. 666–670.

[58] Varnham, M. P., A. J. Barlow, D. N. Payne, and K. Okamoto, "Polarimetric strain gauges using high birefringence fibre", *Electron. Lett.*, Vol. 19, 1983, pp. 699–700.

[59] Chiang, K. S., D. Wong, and P. L. Chu, "Strain-induced birefringence in a highly birefringent optical fibre", *Electron. Lett.*, Vol. 26, 1990, pp. 1344–1346.

[60] Bucaro, J. A., and H. D. Dardy, "Fiber-optic hydrophone", *J. Acoust. Soc. Am.*, Vol. 62, 1977, pp. 1302–1304.

[61] Budiansky, B., D. C. Drucker, G. S., Kino, and J. R. Rice, "Pressure sensitivity of a clad optical fiber", *Appl. Opt.*, Vol. 18, 1979, pp. 4085–4088.

[62] Hughes, R., and J. Jarzynski, "Static pressure sensitivity amplification in interferometric fiber-optic hydrophones", *Appl. Opt.*, Vol. 19, 1980, pp. 98–107.

[63] Bucaro, J. A., "Optical fiber sensor coatings", *Optical Fiber Sensors, NATO ASI Series E, No. 132 Martinus Nijhoff Dordrecht*, 1987, pp. 321–338.

[64] Lagakos, N., A. Dandridge, J. H. Cole, A. B. Tveten, and J. A. Bucaro, "Ultrasonic acoustic sensing", *Fiber Optic Sensors II, A.M. Scheggi, Editor*, Proc. SPIE Vol. 798, 1987, pp. 94–101.

[65] Mcdearmon, G. F., "Theoretical analysis of a push-pull fiber-optic hydrophone", *IEEE J. Lightwave Technol.*, Vol. LT-5, 1987, pp. 647–652.

[66] Gardner, D. L., T. Hofler, S. R. Baker, R. K. Yarber, and S. L. Garrett, "A fiber-optic interferometric seismometer", *IEEE J. Lightwave Technol.*, Vol. LT-5, 1987, pp. 953–959.

[67] Danielson, D. A., and S. L. Garrett, "Fiber-optic ellipsoidal flextensional hydrophones", *IEEE J. Lightwave Technol.*, Vol. LT-7, 1989, pp. 1995–2002.

[68] Yoshizawa, N., T. Yabuta, N. Kojima, and Y. Negishi, "Jacketed optical fiber characteristics under lateral pressure", *Appl. Opt.*, Vol. 20, 1981, pp. 3146–3151.

[69] Annovazzi Lodi, V.,and S. Donati, "Stressed optical fibers and their use as polarizing components", *Alta frequenza*, Vol. LI, 1982, pp. 159–163.

[70] Ulrich, R., S. C. Rashleigh, and W. Eickhoff, "Bending-induced birefringence in single-mode fibers", *Opt. Lett.*, Vol. 5, 1980, pp. 273–275.

[71] Imai, Y., M. A. Rodorigues, and K. Iizuka, "Temperature insensitive fiber coil sensor for altimeters", *Appl. Opt.*, Vol. 29, 1990, pp. 975–978.

[72] Bock, W. J., and A. W. Domanski, "High hydrostatic pressure effects in highly birefringent optical fibers", *IEEE J. Lightwave Technol.*, Vol. LT-7, 1989, pp. 1279–1283.

[73] Bock, W. J., "High pressure polarimetric sensor using birefringent optical fibers", *IEEE Trans. Instrum. Meas.*, Vol. 39, 1990, pp. 233–237.

[74] Chiang, K. S., and D. Wong, "Hydrostatic pressure induced birefringence in a highly birefringent optical fibre", *Electron. Lett.*, Vol. 26, 1990, pp. 1952–1954.

[75] Bock, W. J., T. R. Wolinski, and A. Barwicz, "Development of a polarimetric optical fiber sensor for electronic measurement of high pressure", *IEEE Trans. Instrum. Meas.*, Vol. 39, 1990, pp. 715–721.

[76] Barwicz, A., and W. J. Bock, "An electronic high-pressure measuring system using a polarimetric fiber-optic sensor", *IEEE Trans. Instrum. Meas.*, Vol. 39, 1990, pp. 976-981.

[77] Jansen, K., and P. Dabkiewicz, "High pressure fiber-optic sensor with side-hole fiber", *Fiber Optic Sensors II, A.M. Scheggi, Editor*, Proc. SPIE, Vol. 798, 1987, pp. 56–60.

[78] Charasse, M. N., M. Turpin, and J. P. Le Pesant, "Dynamic pressure sensing with a side-hole birefringent optical fiber", *Opt. Lett.*, Vol. 16, 1991, pp. 1043–1045.

[79] J. D. Nuttal, "Requirements for gyroscopes for inertial navigation", *IEE Proc. J*, Vol. 132, 1985, pp. 250–254.

[80] H. C. Lefevre, "Fiber-optic gyroscope", in *Optical Fiber Sensors* , Vol. 2, edited by B. Culshaw, J. Dakin, Artech House, Norwood, MA, 1989, pp. 381–427.

[81] E. Udd, "Fiber optic gyros", in *Fiber Optic Sensors,* edited by E. Udd, Wiley, New York, 1990, pp. 245–254.

[82] R. B. Smith, editor, "Selected papers on fiber optic gyroscope", *SPIE Milestone Series,* Vol. MS8, Bellingham, WA, 1989.

[83] Special issue on gyroscopes and monomode optical components, *IEE Proc. J,* Vol. 132, pp. 249–308.

[84] Berg, R. A., H. C. Lefevre, and H. J. Shaw, "An overview of fiber optic gyroscopes", *IEEE J. Lightwave Technol.*, LT-2, 1984, pp. 91–107.

[85] Ezekiel, S., and H. J. Arditty, *Fiber-optic rotation sensors and related technologies,* Springer Verlag, Berlin, 1982.

[86] Vali, V., R. W. Shorthill, and M. F. Berg, "Fresnel-Fizeau effect in a rotating optical fiber ring interferometer", *Appl. Opt.* , Vol. 16, 1977, pp. 2605–2607.

[87] Aronowitz, F., "The laser gyro", in *Laser Applications*, edited by Monte Ross, Academic Press, New York, pp. 113–200.

[88] Chow, W. W., J. B. Hambene, T. J. Hutchings, V. E. Sanders, M. Sargent III, and M. O. Scully, "Multioscillator Laser Gyros", *IEEE J. Quantum Electron.*, Vol. QE-16, 1980, pp. 918–935.

[89] Yamasaki, S., F. Suzuki, T. Sakai, A. Wada, and R. Yamauchi, "All-fiber gyroscope using erbium-doped superluminescent fiber", *Proc. OFS 7*, 1990, pp. 143–147.

[90] Lin, S. C., and T. G. Giallorenzi, "Sensitivity analysis of the Sagnac-effect optical-fiber ring interferometer", *Appl. Opt.,* Vol. 18, pp. 915–930.

[91] Vali, V., and R. W. Shorthill, "Fiber Ring Interferometer", *Appl. Opt.,* Vol. 15, 1976, pp. 1099–1100.

[92] S. Donati, "Il giroscopio elettroottico: stato dell'arte e prospettive", *Alta Frequenza,* Vol. 2, 1990, pp. 143–154.

[93] Dorschener, T. A., H. A. Haus, M. Holz, I. W. Smith, and H. Statz, "Laser gyro at quantum limit", *IEEE J. Quantum Electron.*, Vol. QE-16, 1980, pp. 1376–1379.

[94] Manes, K. R., and A. E. Siegman, "Observation of quantum phase fluctuations in infrared gas laser", *Phys. Rev. A*, Vol. 4, 1971,pp. 376–386.

[95] Rashleigh, S. C., and W. K. Burns, "Dual-input fiber-optic gyroscope", *Opt. Lett.*, Vol. 5, 1980, pp. 482–484.

[96] Donati, S., and V. Annovazzi Lodi, "Fiber gyroscope with dual frequency laser", *Proc. ICALEO,* LIA, Vol. 34, 1982, pp. 85–89.

[97] Ezekiel, S., and H. J. Arditty, " Fiber optic rotation sensors review", in *Fiber Optic Rotation Sensors and Related Technologies,* edited by S. Ezekiel, H. J. Arditty, Springer Verlag, New York, 1982.

[98] G. Martini, "Analysis of a single mode optical fibre piezoceramic phase modulator", *J. Opt. Quantum Electron.*, Vol. 19, 1987, pp. 179–190.

[99] Hotate, K., N. Okuma, M. Higashiguchi, and N. Niwa, "Rotation detection by optical heterodyne fiber gyro with frequency output", *Opt. Lett.*, Vol. 7, 1982, pp. 331–333.

[100] Imai, Y., H. Koseki, and Y. Ohtsuka, "Heterodyne fiber-optic gyroscope using orthogonally polarized two-frequency beams", *Appl. Opt.,* Vol. 26, pp. 2962–2965.

[101] Culshaw, B., and I. P. Giles, "Frequency modulated heterodyne optical fibre Sagnac interferometer", *IEEE J. Quantum Electron.,* Vol. QE-18, 1982, pp. 690–693.

[102] Kim, B. Y., and H. J. Shaw, "Phase reading all-fiber-optic gyroscope", *Opt. Lett.*, Vol. 9, 1984, pp. 378–380.

[103] Kim, B. Y., and H. J. Shaw, "Gated phase-modulation approach to fiber-optic gyroscope with linearized scale factor", *Opt. Lett.*, Vol. 9, 1984, pp. 375–377.

[104] Davis, J. L., and S. Ezekiel, "Closed-loop, low-noise fiber-optic rotation sensor", *Opt. Lett.*, Vol. 6, 1981, pp. 505–507.

[105] Edberg, A., and G. Schiffner, "Closed-loop fiber-optic gyroscope with a sawtooth phase-modulated feedback", *Opt. Lett.,* Vol. 10, 1985, pp. 300–302.

[106] Kay, C. J., "Serrodyne modulator in a fibre-optic gyroscope", *IEE Proc. J*, Vol. 132, 1985, pp. 259–264.

[107] Ulrich, R., "Fiber-optic rotation sensing with low drift", *Opt. Lett.*, Vol. 5, 1980, pp. 173–175.

[108] Ulrich, R., and M. Johnson, "Fiber ring interferometer: polarization analysis", *Opt. Lett.*, Vol. 4, 1979, pp. 152–154.

[109] Berg, R. A., H. C. Lefevre, and H. J. Shaw, "All single mode fiber optic gyroscope", *Opt. Lett.*, Vol. 6, 1981, pp. 198–200.
[110] Burns, W. K., "Phase error bounds of fiber gyro with polarization-holding fiber", *IEEE J. Lightwave Technol.*, Vol. LT-4, 1986, pp. 8–14.
[111] Moeller, R. P., W. K. Burns, and N. J. Frigo, "Open-loop output and scale factor stability in a fiber-optic gyroscope", *IEEE J. Lightwave Technol.*, Vol. LT-7, 1989, pp. 262–269.
[112] Burns, W. K., R. P. Moeller, C. A. Villaruel, and M. Abebe, "All-fiber gyroscope with polarization-holding fiber", *Opt. Lett.*, Vol. 9, 1984, pp. 570–572.
[113] Bohm, K., P. Marten, K. Petermann, E. Weidel, and R. Ulrich, "Low-drift fibre gyro using a superluminescent diode", *Electron. Lett.*, Vol. 17, 1981, pp. 352–353.
[114] Cutler, G. C., S. A. Newton, and H. J. Shaw, "Limitation of rotation sensing by scattering", *Opt. Lett.*, Vol. 5, 1980, pp. 488–490.
[115] Chien, P., and C. Pan, "Fiber-optic gyroscopes based on polarization scrambling", *Opt. Lett.*, Vol. 16, 1991, pp. 189–190.
[116] Bulmer, C. H., and R. P. Moeller, "Fiber gyroscope with nonreciprocally operated, fiber-coupled $LiNbO_3$ phase shifter", *Opt. Lett.*, Vol. 6, 1981, pp. 572–574.
[117] Berg, R. A., H. C. Lefevre, and H. J. Shaw, "All single mode fiber optic gyroscope with long term stability", *Opt. Lett.*, Vol. 16, 1981, pp. 502–504.
[118] Frigo, N. J., H. F. Taylor, L. Goldberg, J. F. Weller, and S. C. Rashleigh, "Optical Kerr effect in fiber gyroscope: effects of nonmonochromatic sources", *Opt. Lett.*, Vol. 8, 1983, pp. 119–121.
[119] Hotate, K., "Noise sources and countermeasures in optical passive ring-resonator gyro", *Proc. OFS 7*, 1990, pp. 11–42.
[120] Hotate, K., and K. Tabe, "Drift of an optical fiber gyroscope caused by the Faraday effect: experiment", *IEEE J. Lightwave Technol.*, Vol. LT-5, 1987, pp. 997–1001.
[121] Annovazzi Lodi, V., and S. Donati, "Injection modulation in coupled laser oscillators", *IEEE J. Quantum Electron.*, Vol. QE-16, 1980, pp. 859–864.
[122] Donati, S., and G. Bendelli, "Giroscopi a fibra ottica e laser per applicazioni spaziali", *Atti di Elettroottica '92*, 1992, pp. 179–182.
[123] Meyer, R. E., S. Ezekiel, D. W. Stowe, and V. J. Tekippe, "Passive fiber-optic ring resonator for rotation sensing", *Opt. Lett.*, Vol. 8, 1983, pp. 644–646.
[124] Carrol, R., and J. E. Potter, "Backscatter and the resonant fiber-optic gyro scale factor", *IEEE J. Lightwave Technol.*, Vol. LT-7, 1989, pp. 1895–1900.
[125] Iwatsuki, K., K. Hotate, and M. Higashiguchi, "Effect of Rayleigh backscattering in an optical passive ring resonator gyro", *Appl. Opt.*, Vol. 23, pp. 3916–3924.
[126] Shupe, D. M., "Fiber resonator gyroscope: sensitivity and thermal nonreciprocity", *Appl. Opt.*, Vol. 20, pp. 286–289.

[127] Nakazawa, M., "Synchronously pumped fiber Raman gyroscope", *Opt. Lett.*, Vol. 10, 1985, pp. 193–195.
[128] Desurvivre, E., B. Y. Kim, K. Fesler, and H. J. Shaw, "Reentrant fiber Raman gyroscope", *IEEE J. Lightwave Technol.*, Vol. LT-6, 1988, pp. 481–491.
[129] Yeh, P., I. McMichael, and M. Khoshnevisan, "Phase-conjugate fiber-optic gyro", *Appl. Opt.*, Vol. 25, 1986, pp. 1029–1030.
[130] Annovazzi-Lodi, V., S. Donati, and S. Merlo, "Squeezed states in direct and coherent detection", *J. Opt. Quantum Electron.*, Vol. 24, 1992, pp. 285–301.
[131] Haus, H. A., K. Bergman, and Y. Lai, "Fiber gyro with squeezed radiation", *J. Optical Society of America B*, Vol. 8, 1991, pp. 1953–1957.
[132] Donati, S., V. Annovazzi Lodi, and T. Tambosso, "Magnetooptical fibre sensors for electrical industry: analysis of performances", *IEE Proc. J*, Vol. 135, 1988, pp. 372–382.
[133] Freiser, M. J., "Survey of magnetooptical effects", *IEEE Trans. Magnetics*, Vol. MAG-4, 1968, pp. 152–161.
[134] Pershan, P. S., "Magnetooptical effects", *J. Appl. Phys.*, Vol. 38, pp. 1482–1490.
[135] Tabor, W. J., "Magnetooptical Materials", in *Laser Handbook*, vol. 1, North Holland, 1972, pp. 1011–1127.
[136] Papp, A., and H. Harms, "Magnetooptical current transformers", *Appl. Opt.*, Vol. 19, 1980, pp. 3729–3745.
[137] Dieter, M. N., A. H. Rose, and G. W. Day, "Fast, sensitive magnetic-Field Sensors based on the Faraday effect in YIG", *IEEE J. Lightwave Technol.* Vol. LT-8, 1990, pp. 1838–1842.
[138] Edwards, H. O., K. P. Jedrzejewski, R. I. Laming, and D. N. Payne, "Optimal design of optical fibers for electric current measurement", *Appl. Opt.*, Vol. 28, pp. 1977–2094.
[139] Kersey, A. D., and A. Dandridge, "Optical fibre Faraday rotation current sensor with closed loop operation", *Electron. Lett.*, Vol. 21, 1985, pp. 464–465.
[140] Berwick, M., J. D. C. Jones, and D. A. Jackson, "Alternating current measurement and noninvasive data ring utilizing the Faraday effect in a closed loop fiber magnetometer", *Opt. Lett.*, Vol. 12, 1987, pp. 293–295.
[141] Smith, A. M., "Polarization and magneto-optic properties of single mode optical fiber", *Appl. Opt.*, Vol. 26, 1987, pp. 3153–3157.
[142] Rashleigh, S. C., and R. Ulrich, "Magneto-optic current sensing with birefringent fibers", *Appl. Phys. Lett.*, Vol. 34, 1979, pp. 768–770.
[143] Rashleigh, S. C., "Origins and control of polarization effects in single-mode fibers", *IEEE J. Lightwave Technol.* Vol. LT-1, 1983, pp. 312–331.
[144] Ulrich, R., and S. C. Rashleigh, "Polarization coupling in kinked single-mode fibers", *IEEE J. Quantum Electron.*, Vol. QE-18, 1982, pp. 2032–2039.
[145] Laming, R., and D. N. Payne, "Electric current sensors employing spun higly birefringent optical fibers", *IEEE J. Lightwave Technol.*, Vol. LT-7, 1989, pp. 2084–2093.

[146] see, e.g., *Proc. OFS7*, Sydney, 1990.
[147] Kersey, A. D., and D. A. Jackson, "Current sensing utilizing heterodyne detection of the Faraday effect in single mode optical fiber", *IEEE J. Lightwave Technol.*, Vol. LT-4, 1986, pp. 640–643.
[148] Tang, D., A. H. Rose, G. W. Day, and S. M. Etzel, "Annealing of linear birefringence in single mode fiber coils: application to optical fiber current sensors", *IEEE J. Lightwave Technol.*, Vol. LT-9, 1991, pp. 1031–1037.
[149] Maystre, F., and A. Bertholds, "Magneto-optic current sensor using a helical-fiber Fabry-Perot resonator", *Opt. Lett.*, Vol. 14, 1989, pp. 587–589.
[150] Ren, Z., Y. Wang, and P. A. Robert, "Faraday rotation and its temperature dependence measurements in low-birefringence fibers", *IEEE J. Lightwave Technol.*, Vol. LT-7, 1989, pp.1275–1278.
[151] Annovazzi-Lodi, V., and S. Donati, "Combined reciprocal and non-reciprocal birefringence in optical monomode fibres", *Opt. Quantum Electron.*, Vol. 15, 1983, pp. 381–388.
[152] Annovazzi-Lodi, V., and S. Donati, "Fiber current sensors for HV lines", *Proc. Symposium on Fber Optic Sensors II*, SPIE Vol. 798, 1987, pp. 258–265.
[153] Day, G. W., D. N. Payne, A. J. Barlow, and J. J. Ramskov-Hansen,"Faraday rotation in coiled monomode optical fibers: isolators, filters and magnetic sensors", *Opt. Lett.*, Vol. 7, 1982, pp. 238–241.
[154] Bergh, R. A., H. C. Lefevre, and H. J. Shaw, "Geometrical fiber configuration for isolators and magnetometers", in *Fiber-Optic Rotation Sensors*, Springer Verlag, 1982, pp. 400–405.
[155] Donati, S., V. Annovazzi-Lodi, G. Degli Esposti, and S. Merlo, "A novel magnetic field fiberoptic sensor", *Proc. VII Symposium ISH 1991*, pp. 123–126.
[156] Mermelstein, M. D., "Fiber-optic dc magnetometer utilizing a composite metallic glass resonator", *IEEE J. Lightwave Technol.*, Vol. LT-4, 1986, pp. 1376–1380.
[157] Dagenais, D. M., F. Bucholtz, K. P. Koo, and A. Dandridge, "Detection of low-frequency magnetic signals in a magnetostrictive fiber-optic sensor with suppressed residual signal", *IEEE J. Ligthwave Technol.,* Vol. LT-7, 1989, pp. 881–886.
[158] Kersey, A., D. A Jackson, and M. F. Corke, "Single mode fiberoptic magnetometer with dc bias stabilization", *IEEE J. Lightwave Technol.*, Vol. LT-3, 1985, pp. 836–840.
[159] Bucholtz, F., "Fiber optic magnetic sensors", in *Fiber optic sensors*, edited by E. Udd, Wiley, New York, pp. 389–394.
[160] Mermelstein, M. D., "Fundamental limit to the performance of fibre optic metallic glass dc magnetometers", *Electron. Lett.,* Vol. 21, pp. 1178–1179.
[161] Okamura, H., "Fiber-optic magnetic sensor utilizing the Lorentzian force", *IEEE J. Lightwave Technol.*, Vol. LT-8, 1990, pp. 1558–1564.
[162] Okamura, H., "Fiber optic magnetic gradiometer utilizing the magnetic translational force", *IEEE J. Lightwave Technol.*, Vol. LT-8, 1990, pp. 877–882.
[163] Nye, J. F., *Physical properties of crystals*, Oxford University Press, 1957.

[164] Rutgers, W., et al., *Kema Scient. & Tech. Rep.* 5-11 (1987), pp. 281–292.
[165] Robertson, S. R. M., and A. J. Rogers,"Measurement of dc electric fields using the electro-optic effect", *IEE Proc. J*, Vol. 132, 1985, pp. 195–198.
[166] Imai, M., S. Fujiwara, and Y. Ohtsuka, "Single mode electric-field-sensitive fiber jacketed with vinylidene fluoride-trifluoroethylene copolimer", *Opt. Lett.*, Vol. 13, 1988, pp. 838–840.
[167] Mermelstein, M. D., "Optical-fiber copolymer-film electric-field sensor", *Appl. Opt.,* Vol. 22, 1983, pp. 1006–1009.
[168] Bohnert, K., and J. Nehring, "Fiber optic sensing of electric field components", *Appl. Opt.,* Vol. 27, 1988, pp. 4814–4818.
[169] Bohnert, K., and J. Nehring, "Fiber-optic sensing of voltages by line integration of the electric field", *Opt. Lett.*, Vol. 14, 1989, pp. 290–292.
[170] Harmer, A., and A. Scheggi, "Chemical, biochemical, and medical sensors", in *Optical fiber sensors*, ed. by B. Culshaw, J. Dakin, Artech House, Norwood, MA, 1989, pp. 599–652.
[171] Butler, M. A., "Optical fiber hydrogen sensor", *Appl. Phys. Lett.*, Vol. 45, 1984, pp. 1007–1009.
[172] Farahi, F., P. Akhavan Leilabady, J. D. C. Jones, and D. A. Jackson, "Interferometric fibre-optic hydrogen sensor", *J. Phys. E: Sci. Instrum.*, Vol. 20, 1987, pp. 432–434.
[173] Farahi, F., P. Akhavan Leilabady, J. D. C. Jones, and D. A. Jackson, "Optical-fibre flammable gas sensor", *J. Phys. E: Sci. Instrum.*, Vol. 20, 1987, pp. 435–436.
[174] Leslie, D. H., G. L. Trusty, A. Dandrige, and T. G. Giallorenzi, "Fibre-optic spectrophone", *Electron. Lett.*, Vol. 17, 1981, pp. 581–582.
[175] Carey, W. P., K. R. Beebe, E. Sanchez, P. Geladi, and B. R. Kowalski, "Chemometric analysis of multisensor arrays", *Sensors and Actuators*, Vol. 9, 1986, pp. 223–234.
[176] Tiefenthaler, K., and W. Lukosz, "Integrated optical switches and gas sensors", *Opt. Lett.*, Vol. 10, 1984, pp. 137–139.
[177] Lukosz, W., and K. Tiefenthaler, "Embossing technique for fabricating integrated optical components in hard inorganic waveguiding materials", *Opt. Lett.*, Vol. 8, 1983, pp. 537–539.
[178] Heuberger, K., and W. Lukosz, "Embossing technique for fabricating surface relief gratings on hard oxide waveguides", *Appl. Opt.*, Vol. 25, 1986, pp. 1499–1504.
[179] Tiefenthaler, K., and W. Lukosz, "Integrated optical humidity and gas sensors", *Proc. SPIE*, Vol. 514, 1984, pp. 215–217.
[180] Tiefenthaler, K., and W. Lukosz, "Grating couplers as integrated optical humidity and gas sensors", *Thin Solid Films*, Vol. 126, 1985, pp. 205–211.
[181] Seifert, M., K. Tiefenthaler, K. Heuberger, W. Lukosz, and K. Mosbach, "An integrated optical biosensor (IOBS)", *Analyt. Lett.*, Vol. 19, 1986, pp. 205–216.

Index

[182] Spohn, P. K., and M. Seifert, "Interaction of aqueous solutions with grating couplers used as integrated optical sensors and their pH behavior", *Sensors and Actuators*, Vol. 15, 1988, pp. 309–324.
[183] Lukosz, W., and K. Tiefenthaler, "Sensitivity of integrated optical grating and prism couplers as (bio)chemical sensors", *Sensors and Actuators*, Vol. 15, 1988, pp. 273–284.
[184] Nellen, Ph. M., K. Tiefenthaler, and W. Lukosz, "Integrated optical input grating couplers as biochemical sensors", *Sensors and Actuators*, Vol. 15, 1988, pp. 285–295.
[185] Nellen, Ph. M., and W. Lukosz, "Integrated optical input grating couplers as chemo- and immunosensors", *Sensors and Actuators*, Vol. B1, 1990, pp. 592–596.
[186] Lukosz, W., Th. Brenner, V. Briguet, Ph. M. Nellen, and P. Zeller, "Output grating couplers on planar waveguides as integrated optical sensors", *Proc. SPIE*, Vol. 1141, 1989, pp. 192–200.
[187] Lukosz, W., Ph. M. Nellen, Ch. Stamm, and P. Weiss, "Output grating couplers on planar waveguides as integrated optical chemical sensors", *Sensors and Actuators*, Vol. B1, 1990, pp. 585–588.
[188] Kogelnik, H., "Theory of dielectric waveguides", in *Integrated optics*, edited by T. Tamir, Springer Verlag, Berlin, 1982, pp. 15–81.

www.ingramcontent.com/pod-product-compliance
Lightning Source LLC
Chambersburg PA
CBHW051254200926
51612CB00001B/46
* 9 7 8 0 8 9 0 0 6 6 0 2 7 *